Plant-microbiome Interactions for Climate-resilient Agriculture

Umesh Pankaj • Piyoosh Babele
Ashok Kumar Singh
Editors

Plant-microbiome Interactions for Climate-resilient Agriculture

Springer

Editors
Umesh Pankaj
College of Horticulture and Forestry
Rani Lakshmi Bai Central Agricultural University
Jhansi, Uttar Pradesh, India

Piyoosh Babele
Rani Lakshmi Bai Central Agricultural University
Jhansi, Uttar Pradesh, India

Ashok Kumar Singh
Rani Lakshmi Bai Central Agricultural University
Jhansi, Uttar Pradesh, India

ISBN 978-981-96-3533-7 ISBN 978-981-96-3534-4 (eBook)
https://doi.org/10.1007/978-981-96-3534-4

This Springer imprint is published by the registered company Springer Nature Singapore Pte Ltd.
The registered company address is: 152 Beach Road, #21-01/04 Gateway East, Singapore 189721, Singapore

This book is dedicated to our beloved parents and teachers

Preface

Climate change-driven uncertainties in weather patterns and incidences of persistent plant pathogens are already visible. Because of climate change, agriculture is facing the dual challenge of ensuring global food security and achieving it in a sustainable manner. The impact of climate change on agriculture is high, as crop plants are more frequently subjected to stresses of both natural (abiotic and biotic) and anthropogenic origin. In this context, the role of plant-microbe interactions in fostering climate-resilient agriculture is becoming increasingly recognized. This book on *Plant-Microbiome Interactions for Climate-Resilient Agriculture* offers a comprehensive overview of the negative effects of climate change-driven environmental stresses on crop productivity and describes various adaptation strategies evolved by plants to mitigate these stresses. The text is structured around key scientific advancements and emerging research that address the challenges of maintaining agricultural productivity and ecosystem health in a changing climate.

In the past few decades, tremendous progress has been made in understanding how plants respond to microbial colonization and how microbial pathogens and symbionts reprogram plant physiology and metabolism to improve plant growth and yield. In contrast, our knowledge of how extreme environmental conditions affect plant-microbe interactions is less understood at the mechanistic level, as most molecular studies are performed under simple and static laboratory conditions. This book explores the critical relationship between plants and microbes in the face of climate change, offering insights into how these interactions can be harnessed to bolster sustainable agricultural practices. The first chapter sets the stage by examining the profound impacts of climate change on agriculture and food security. From there, the book delves into the potential of climate-resilient sustainable agriculture as a solution to counteract food insecurity. Subsequent chapters explore a wide range of topics related to plant-microbe interactions, such as the influence of entomopathogenic fungi on soil microbial diversity, the beneficial effects of microbes in plant stress management, and the intricate biochemical processes that occur in the rhizosphere. We further examine how plant root exudates communicate with microbes under stress, as well as the critical role of secondary metabolites in combating environmental stresses.

With an emphasis on cutting-edge research, this book also covers the role of omics techniques in deciphering the complex relationships between plants and microbes. The inclusion of metabolomics and CRISPR/Cas systems opens up new possibilities for enhancing plant resilience and productivity in the face of climate change. Each chapter reflects the growing recognition of the need for integrated approaches to agricultural resilience, ecological sustainability, and food security. Through a multidisciplinary lens, we aim to highlight how the understanding and manipulation of plant-microbe interactions can pave the way for innovative agricultural strategies that mitigate the effects of climate change and promote long-term sustainability. As we move toward a future where climate change is an ever-present factor, it is imperative that we explore every avenue for fostering resilience in agriculture. This book provides a comprehensive resource for students, scientists, policymakers, and other stakeholders working to unlock the potential of plant-microbe interactions in creating a climate-resilient agricultural future.

We hope that the insights offered in this book inspire new approaches and strategies for addressing the interconnected challenges of climate change, food security, and sustainable agriculture. We argue that an increased understanding of the complex feedbacks between crop plants and microbial traits under field conditions will pave the way for harnessing the soil/rhizosphere microbiome to increase the resilience of future crop production. There is an urgent need to increase this important area of research to gain a system-level understanding of plant-microbe interaction to reflect the real picture occurring in nature. These require synergy in plant ecology, microbiome, modern omics technologies and systems, and synthetic biology approaches to achieve the goals of future food and nutrition security in the age of climate change.

Jhansi, Uttar Pradesh, India

Umesh Pankaj
Piyoosh Babele
Ashok Kumar Singh

Acknowledgments

First and foremost, we would like to extend our heartfelt gratitude to all the authors who have played a pivotal role in shaping the content of this book by their knowledge, research, and passion for the subject matter as reflected in every chapter. Their contributions have been critical in broadening the scope of this book in furthering the understanding of plant-microbe interactions and their potential for enhancing climate-resilient agricultural practices. We are truly honored to have collaborated with such a talented group of scholars.

We are deeply grateful to the Springer Nature publishing team, namely, Amelie von Zumbusch, Nandhini Viswanathan, Sanchi Bhimrajka, and Machi Sugimoto, for their unwavering support and commitment to bringing this work to the wider academic community. Your professionalism and dedication have been truly commendable.

We would also like to express our sincere appreciation to the Science and Engineering Research Board, Department of Science and Technology (SERB-DST) for proving the funding. Without your financial assistance, this book would not have been possible.

I would like to extend our gratitude to Rani Lakshmi Bai Central Agricultural University Jhansi, for providing the academic resources and environment that enabled us to complete this project. The support from our colleagues and the institution has been essential to the successful completion of this project. We thank the anonymous reviewers for their constructive comments and valuable feedbacks to improve the content of this book.

Finally, we offer our deepest thanks to our family members and friends for their constant motivation for bringing this book into reality.

Contents

Editors and Contributors

About the Editors

Umesh Pankaj is presently working as a Teaching cum Research Associate in Microbiology at Rani Lakhsmi Bai Central Agricultural University, Jhansi, Uttar Pradesh, India. His doctoral degree was awarded in the field of Agricultural Microbiology by Jawaharlal Nehru University, New Delhi, and research work carried out at CSIR-Central Institute of Medicinal and Aromatic Plants, Lucknow, India. His area of research is mainly focused on plant-microbe interaction with special emphasis on the restoration of salt-affected soil using halotolerant bio-inoculants like plant growth-promoting microbes and arbuscular mycorrhiza fungi. Currently, he is working on microbial bio-formulation development for drought-prone regions and their use for sustainable agriculture development. He is also engaged in teaching for the courses at post-graduate and undergraduate levels. Dr. Pankaj has published 12 research articles in peer-reviewed international journals, 2 edited books with CRC Press, Taylor and Francis Group, and 10 book chapters to his credit. He has received Young Microbiologist Award from Agro-Environmental Development Society, Uttar Pradesh. He is a life member of various scientific societies such as Association of Microbiologist of India (AMI), India Science Congress, etc.

Piyoosh Babele is currently working as Ramanujan Fellow at Rani Lakshmi Bai Central Agricultural University, Jhansi, UP. He was a former postdoctoral research scientist and visiting assistant professor at Vanderbilt University, USA. He earned his Ph.D. in Biochemistry from Banaras Hindu University Varanasi, India. His research fields are systems biology and metabolic engineering of photosynthetic organism cyanobacteria (for biofuels production), orphan millets (for improved water and nitrogen use efficiency), and nanobiotechnology.

He is a recipient of several fellowships such as Dr. D.S. Kothari Post-Doctoral Fellowship (UGC), Bridging Post-Doctoral Fellowship National Center for Biological Sciences-Tata Institute of Fundamental Research, Bengaluru, India, Merit award (American Chemical Society and SERB-DST), National Postdoctoral fellowship (SERB-DST), and BSR Research fellowship during Ph.D. in science for meritorious students (UGC).

Dr. Babele authored more than 40 research articles in reputed journals like *PNAS*, *Trends in Biotechnology*, *Trends in Microbiology*, *Metabolic Engineering*, etc. and contributed a number of book chapters in edited books. He is serving as associate and review editor in several journals like *BMC Biotechnology*, *Frontiers in Microbiology*, etc.

Ashok Kumar Singh is working as Vice-Chancellor of Rani Lakshmi Bai Central Agricultural University, Jhansi. He has more than 35 years of experience in the fields of agricultural research, education, and extension. In his research career, he served in several positions, such as Director of ICAR-ATARI, Kanpur, from 2005 to 2014; Assistant Director General (Agricultural Extension), ICAR, New Delhi, 2014; Deputy Director General (Agricultural Extension), ICAR, New Delhi, 2014–2022; Deputy Director General (Fisheries Science), (Additional Charge), ICAR, New Delhi, 2015–16; Deputy Director General (Horticultural Science), (Additional Charge), ICAR, New Delhi, 2016–17; Director & Vice-Chancellor (Additional Charge), ICAR-IARI, New Delhi, 2017–20; Agriculture Commissioner, Government of India, 2021–22.

Dr. Singh is a fellow of the National Academy of Agricultural Science (NAAS). He has been awarded a number of prestigious awards and recognitions, such as the Swami Sahajanand Best Extension Scientist Award of ICAR, is a member of the Uttar Pradesh Farmer Commission, and has published more than 150 research papers and articles in scientific journals.

His major research and development work has been in the areas of technology assessment, application, and outreach through a large network of NAREES. The conceptual framework and models like Farmer FIRST, ARYA, and work in the areas of climate-resilient agriculture, nutri-sensitive agriculture, pulse and oil seed promotion, digital agriculture, landscape diagnostics, etc.

Contributors

Niraj Agarwala Department of Botany, Gauhati University, Guwahati, Assam, India

Sayasta Ahmed Department of Molecular Biology & Bioinformatics, Tripura University (A Central University), Agartala, Tripura, India

Meenakshi Arya Department of Plant Pathology, Rani Lakshmi Bai Central Agriculture University, Jhansi, Uttar Pradesh, India

Piyoosh Babele Rani Lakshmi Bai Central Agricultural University, Jhansi, Uttar Pradesh, India

Ramen Barman Department of Botany, Gauhati University, Guwahati, Assam, India

Ravi Kant Bhatia Department of Biotechnology, Himachal Pradesh University, Shimla, Himachal Pradesh, India

Tanvi Bisht Department of Biotechnology, Himachal Pradesh University, Shimla, Himachal Pradesh, India

Debasish Chakraborty ICAR-Research Complex for NEH Region, Umiam, Meghalaya, India

Mahasweta Chakraborty ICAR-Research Complex for NEH Region, Umiam, Meghalaya, India

Anu Chaudhary Department of Biotechnology, Himachal Pradesh University, Shimla, Himachal Pradesh, India

Anuj Chaudhary Department of Biotechnology, Graphic Era Deemed to be University, Dehradun, Uttarakhand, India

Parul Chaudhary School of Agriculture, Graphic Era Hill University, Dehradun, Uttarakhand, India

Nishma Dahal Biotechnology Division, CSIR-Institute of Himalayan Bioresource Technology Palampur, Palampur, Himachal Pradesh, India

Pushpak Dahiya Department of Biotechnology, Himachal Pradesh University, Shimla, Himachal Pradesh, India

Abhishek Das Amity Institute of Microbial Technology, Amity University, Noida, Uttar Pradesh, India

Priyanka Das Department of Botany, Gauhati University, Guwahati, Assam, India

Barsha Datta Department of Molecular Biology & Bioinformatics, Tripura University (A Central University), Agartala, Tripura, India

Anand Dave Department of Microbiology, Seventh Day Adventist Arts and Science College, Ahmedabad, India

Susheel Dwivedi Kendriya Vidyalaya, Indian Embassy, Kathmandu, Nepal

Priyanshi Garg Department of Plant Pathology, Rani Lakshmi Bai Central Agriculture University, Jhansi, Uttar Pradesh, India

Samarendra Hazarika ICAR-Research Complex for NEH Region, Umiam, Meghalaya, India

Juveriya Israr Faculty of Biosciences, Institute of Biosciences and Technology, Shri Ramswaroop Memorial University, Barabanki, Uttar Pradesh, India

Durgesh Kumar Jaiswal Department of Biotechnology, Graphic Era Deemed to be University, Dehradun, Uttarakhand, India

Sapana Jha Laboratory of photobiology and Molecular Microbiology, Centre of Advanced Study in Botany, Institute of Science, Banaras Hindu University, Varanasi, India

Sampurna Kashyap Department of Botany, Gauhati University, Guwahati, Assam, India

Simran Kaur Amity Institute of Microbial Technology, Amity University, Noida, Uttar Pradesh, India

Neelam Kaushik Department of Biotechnology, Himachal Pradesh University, Shimla, Himachal Pradesh, India

Shruti Kaushik ICAR-Central Soil Salinity Research Institute, Karnal, Haryana, India

Ram Krishna ICAR-Indian Institute of Vegetable Research, Varanasi, Uttar Pradesh, India

Antresh Kumar Department of Biochemistry, School of Interdisciplinary and Applied Sciences, Central University of Haryana, Mahendergarh, Haryana, India

Hemlata Kumari Department of Biochemistry, School of Interdisciplinary and Applied Sciences, Central University of Haryana, Mahendergarh, Haryana, India

Neha Kumari Laboratory of Photobiology and Molecular Microbiology, Centre of Advanced Study in Botany, Institute of Science, Banaras Hindu University, Varanasi, India

Sonam Kumari Department of Biochemistry, School of Interdisciplinary and Applied Sciences, Central University of Haryana, Mahendergarh, Haryana, India

Navin Kumar Department of Biotechnology, Graphic Era Deemed to be University, Dehradun, Uttarakhand, India

Rakshak Kumar Department of Molecular Biology & Bioinformatics, Tripura University (A Central University), Agartala, Tripura, India

Jonnada Likhita Department of Plant Pathology, Rani Lakshmi Bai Central Agriculture University, Jhansi, Uttar Pradesh, India

Sneha Devi Marme Department of Botany, Gauhati University, Guwahati, Assam, India

Priyanka Mishra Department of Biotechnology, Graphic Era Deemed to be University, Dehradun, Uttarakhand, India

Vinay K. Mishra ICAR-Research Complex for NEH Region, Umiam, Meghalaya, India

Sankalp Misra Faculty of Biosciences, Institute of Biosciences and Technology, Shri Ramswaroop Memorial University, Barabanki, Uttar Pradesh, India

Umesh Pankaj College of Horticulture and Forestry, Rani Lakshmi Bai Central Agricultural University, Jhansi, Uttar Pradesh, India

Bhaskar Jyoti Parasar Department of Botany, Gauhati University, Guwahati, Assam, India

Gulafsha Parveen Department of Biotechnology, Graphic Era Deemed to be University, Dehradun, Uttarakhand, India

Suruchi Rai Department of Molecular Biology & Bioinformatics, Tripura University (A Central University), Agartala, Tripura, India

Palak Rana Laboratory of Photobiology and Molecular Microbiology, Centre of Advanced Study in Botany, Institute of Science, Banaras Hindu University, Varanasi, India

Alok Ranjan Department of Biotechnology, Patna Women's College, Patna, Bihar, India

Neha Saini Department of Biochemistry, School of Interdisciplinary and Applied Sciences, Central University of Haryana, Mahendergarh, Haryana, India

Christy B. K. Sangma ICAR-Research Complex for NEH Region, Umiam, Meghalaya, India

Neelam S. Sangwan Department of Biochemistry, School of Interdisciplinary and Applied Sciences, Central University of Haryana, Mahendergarh, Haryana, India

Aditya Sharma Department of Biotechnology, Himachal Pradesh University, Shimla, Himachal Pradesh, India

Indrani Sharma Department of Botany, Gauhati University, Guwahati, Assam, India

Neeraj Shrivastava Amity Institute of Microbial Technology, Amity University, Noida, Uttar Pradesh, India

Devendra Singh Faculty of Biotechnology, Institute of Biosciences and Technology, Shri Ramswaroop Memorial University, Barabanki, Uttar Pradesh, India

Garvita Singh Department of Botany, Gargi College, University of Delhi, Delhi, India

Rohit Kumar Singh Department of Biotechnology, Himachal Pradesh University, Shimla, Himachal Pradesh, India

Varsha K. Singh Laboratory of Photobiology and Molecular Microbiology, Centre of Advanced Study in Botany, Institute of Science, Banaras Hindu University, Varanasi, India

Yashowardhan Singh Department of Plant Pathology, Rani Lakshmi Bai Central Agriculture University, Jhansi, Uttar Pradesh, India

Rajeshwar P. Sinha Laboratory of Photobiology and Molecular Microbiology, Centre of Advanced Study in Botany, Institute of Science, Banaras Hindu University, Varanasi, India

Sandhya Sinha Department of Entomology, Chhattisgarh Agriculture College, Dhanora, Chhattisgarh, India

K. Sobha RVR & JCCE STP Foundation Chowdavaram, Guntur, Andhra Pradesh, India

Jyotsana Tilgam ICAR-National Bureau of Agriculturally Important Micro-organisms, Maunath Bhanjan, Uttar Pradesh, India

Harjeet Tomar Department of Biotechnology, Graphic Era Deemed to be University, Dehradun, Uttarakhand, India

Swati Tripathi Amity Institute of Microbial Technology, Amity University, Noida, Uttar Pradesh, India

Vishal Tripathi Department of Biotechnology, Graphic Era Deemed to be University, Dehradun, Uttarakhand, India

Khushboo Upadhayay ICAR-National Bureau of Agriculturally Important Micro-organisms, Maunath Bhanjan, Uttar Pradesh, India

Aayushi Verma Department of Biotechnology, Himachal Pradesh University, Shimla, Himachal Pradesh, India

Mahendra K. Verma RVR & JCCE STP Foundation Chowdavaram, Guntur, Andhra Pradesh, India

Shaloo Verma ICAR-National Bureau of Agriculturally Important Micro-organisms, Maunath Bhanjan, Uttar Pradesh, India

Sunil Kumar Verma Faculty of Biotechnology, Institute of Biosciences and Technology, Shri Ramswaroop Memorial University, Barabanki, Uttar Pradesh, India

Ritu Vishwakrma ICAR-National Bureau of Agriculturally Important Micro-organisms, Maunath Bhanjan, Uttar Pradesh, India

1 Climate Change and Agriculture: Impacts on Food and Nutrition Security with Special Reference to the North Eastern Himalayan Region

Mahasweta Chakraborty, Christy B. K. Sangma, Debasish Chakraborty, Samarendra Hazarika, and Vinay K. Mishra

Abstract

Climate change presents profound challenges to agricultural systems, significantly impacting food and nutrition security, particularly in vulnerable regions like the Northeastern Hill (NEH) Region of India, where agriculture is mainly rainfed with high dependence on monsoon precipitation. This chapter presents a detailed analysis of the climate change impact on agroecosystems affecting food and nutrition security, with a focus on the hilly and mountainous region of Northeast India. The NEH region is witnessing climate change with rising temperatures (0.2–1.2 °C per decade), erratic rainfall patterns (doubling of the coefficient of variation in recent times compared to the 1950–1980s), including fluctuations in the number of rainy days and periods, frequent occurrences of seasonal drought, and monsoon delays affecting the timely operation of agricultural practices in the region. These climatic variations have profound implications for agricultural productivity and sustainability, inducing shifts in crop cycles with changes in traditional cropping patterns, land use patterns, changes in crop suitability, and also the incidence of biotic and abiotic stresses in the region. Farmers in the NEH are increasingly facing challenges such as changes in the timing of sowing and harvesting of crops, leading to a mismatch of crop growth periods with the changing climate, increased input requirements for producing optimum yields, impacting food availability, and, moreover, the quality of food grains. Such changes are likely to seriously affect the socioeconomic condition of the farming community and pose risks to the overall nutritional security of the population, as staple crops and traditional diets may be compro-

M. Chakraborty · C. B. K. Sangma · D. Chakraborty · S. Hazarika · V. K. Mishra (✉)
ICAR-Research Complex for NEH Region, Umiam, Meghalaya, India
e-mail: VK.Mishra@icar.gov.in

U. Pankaj et al. (eds.), *Plant-microbiome Interactions for Climate-resilient Agriculture*, https://doi.org/10.1007/978-981-96-3534-4_1

mised. In response to these challenges, the chapter calls for concerted efforts from stakeholders to prioritize and implement climate-resilient agricultural practices by integrating traditional knowledge with modern agricultural practices in order to ensure sustainable food and nutrition security in the northeastern hill region of India under the ongoing climate crisis.

Keywords
Abiotic stress · Agroecosystems · Biotic stress · Climate variability · Climate-resilient agriculture · Northeastern hill region

1 Introduction

Climate change is defined by the United Nations Framework Convention on Climate Change (UNFCCC) as "change of climate which is attributed directly or indirectly to human activity that alters the composition of the global atmosphere and which is in addition to natural climate variability observed over comparable time periods." As the Indian agriculture production system faces the overwhelming charge of feeding its rapidly increasing population, climate change poses extra challenges to food security. The northeastern hill region of India, comprising states like Arunachal Pradesh, Assam, Manipur, Meghalaya, Mizoram, Nagaland, Tripura, and Sikkim, is characterized by its unique agroecosystems with mainly hilly and mountainous terrain experiencing diverse climate ranging from subtropical to temperate. The agroecosystems comprise of traditional farming practices, such as shifting cultivation (jhum), terrace farming, and mixed cropping systems. Rugged terrain and the predominance of rainfed agriculture make this region extremely vulnerable to the impacts of climate change. Climate change in this region manifests through a series of complex and interrelated phenomena, including altered precipitation patterns, increased temperatures, and more frequent extreme weather events such as floods and droughts. With the increase in the unpredictable nature of climatic patterns, agriculture in the northeastern hill region, which is mainly smallholder and subsistence-based, faces multiple challenges. Considering the significance of agriculture in the livelihoods of the population in the region, disruption in agricultural productivity leads to serious implications for food and nutrition security. Hence, there is an increasing need to assess these impacts to understand how climate change is affecting agricultural productivity, food security, and nutritional outcomes in the region.

2 Agricultural Practices and Their Role in Climate Change

Agriculture contributes significantly to CO_2, N_2O, and CH_4 emissions (Cole et al. 1997), of which land-use change, intensive tillage, increased use of fertilizers, and burning crop residues are the major practices that contribute to greenhouse gas (GHG) emissions. Due to increased anthropogenic activities such as urbanization, industrialization, deforestation, change in land use pattern, etc., the emission of

GHG is speeding up the climate change (Rao et al. 2020). Soils are reported to be capable of generating GHGs besides consuming them. Respiration fluxes from soil to atmosphere range from 83 to 108 Pg C $year^{-1}$ (Hursh et al. 2017), while at the same time, soil has a potential C sink capacity of 55–78 Gt C (Lal 2004). Forests absorb CO_2 from the atmosphere through photosynthesis, which is stored in the trees as well as in the soil. With the clearing/burning of forests, stored C is released into the atmosphere, mainly as CO_2. Gibbs et al. (2018) reported the emission of 4.8 Gt of CO_2 $year^{-1}$ (averaged over 2015–2017) brought about by the global loss of tropical forests. Among soil and crop management practices, conventional tillage also contributes significantly to GHG emissions (Lal 1991) through fossil fuel consumption and increased CO_2 emission from soil. It has been reported that tilled soils emit 21% more CO_2 than untilled soils (Abdalla et al. 2016), and the emission due to tillage is strongly correlated with the tillage intensity. Puddling and intermittent flooding in paddy fields are also considered to be major sources of CH_4 and N_2O emissions, considering the major share of this staple in food production, although the estimates may vary due to large spatial differences (Adhya et al. 2000). Crop residue burning and manure management also contribute to GHG emissions. A study in India showed that open field burning in almost a decade gap (i.e., in 1997–1998 and 2006–2007) emitted 3.73 and 4.06 MMT CO_2 equivalent of CH_4 and N_2O, out of which CH_4 constituted three-fourths, while one-fourth of GHG emissions was N_2O (Murali et al. 2010). Apart from these, modern intensive agriculture relying on heavy applications of nitrogen (N) fertilizer affects the global N cycle through N_2O emissions. Under the present scenario, emissions of greenhouse gases from agriculture are of serious concern, and it is expected to increase in the future decades with the intensification of agriculture involving tremendous pressure on resources, viz. irrigation water, fertilizer, and other inputs, and increased mechanization to feed the increasing population.

3 Climate Change Trends and Future Projections in the Northeastern Hill (NEH) Region

3.1 Climate Change Trends in the NEH Region

Climate change is a global issue with far-reaching consequences for ecosystems and human societies, but its pattern and intensity are highly location-specific (IPCC 2021; Li et al. 2018). The key manifestation of climate change is the alteration in rainfall and temperature patterns, which has significant implications (Sun et al. 2018; Gründemann et al. 2022). Mountain ecosystems, particularly those in the tropics, are highly vulnerable to climate change due to their unique topography, microclimates, and biodiversity (Korner and Spehn 2019; Schmeller et al. 2022). These ecosystems are crucial in maintaining ecological balance, providing vital resources, and supporting various life forms. Understanding the changes in weather variables in these ecosystems is imperative for effective conservation and sustainable development (Peters et al. 2019). Northeastern India (NEI) is a region of high ecological significance, known for its rich biodiversity and unique mountain ecosystems (Giri et al. 2020; Majumdar 2022). Climate change impacts in this region have already been observed, with several notable consequences (Deka et al. 2009; Jain et al. 2013; Chakraborty et al. 2014, 2017a; Saha et al. 2015, 2018). NEI experiences high rainfall variability due to its complex topography and the influence of monsoon systems (Sharma et al. 2023). Climate change has led to alterations in the timing, intensity, and spatial distribution of rainfall in this region. An increased frequency of extreme weather events, such as heavy rainfall and prolonged droughts, has been observed, posing significant challenges for the local population and ecosystem (Das et al. 2009; Dash et al. 2009; Sharma et al. 2023).

The region is situated in the high rainfall zone of the country, hosting the highest rainfall-receiving place, Mawsynram (near Cherrapunji) (Jain et al. 2013; Marak et al. 2020). But in recent years, the variation in the rainfall events has changed a lot with an increase in the heavy rainfall events and a decrease in the light and moderate ones (Dash et al. 2009; Prokop and Walanus 2015; Marak et al. 2020). In recent decades, even this region, which is blessed with bountiful rainfall, has experienced a few unique drought-like conditions apart from floods (Das et al. 2009). Evidence suggests a substantial rise in intense precipitation occurrences during the monsoon season, rendering it increasingly susceptible to flooding (Dash et al. 2009; Goswami et al. 2010; Varikoden and Revadekar 2020). The analysis of long-term (116 years) data reveals that during the last three decades (1991–2015), the annual rainfall has decreased, with an increase in the coefficient of variation (CV) (Saikia and Hazarika 2018). The CV has almost doubled in 1991–2015 (27.8%) as compared to 1901–1903 (14.7%). The monsoon rainfall has drastically decreased (8.1%) to 916.5 mm from 996.9 mm during the same time period, while the CV has increased to 34.8% from 24.8%. The rate of decrease in rainfall during the monsoon season since the 1980s is also statistically significant (−3.97 mm/year) (Kothawale and Rajeevan 2017).

Moreover, the number of rainy days has also decreased over NEI. The analysis of rainfall data over the Brahmaputra and Barak basins of Assam, part of NEI, unveiled a substantial downward trend in rainfall during the monsoon and post-monsoon seasons (Deka et al. 2013; Saha et al. 2018). Notably, both the annual and monsoon rainfall exhibited significant decreases during the last 30 years in the region. Recent studies report that even in one of the rainiest regions of the world, like NEI, droughts extend beyond the limits of the dry season inherent in the monsoon regime (Prokop and Walnus 2024). Further, there are also reports of a decrease in rainfall during the post-monsoon and winter seasons in different parts of the region, which poses a threat to continuing/expanding the rabi agriculture in the region (Chakraborty et al. 2017a; Saha et al. 2015, 2021). Therefore, during the crop seasons, there is a reduction in rainfall and rainy days, leading to drought-like situations, especially influenced by the orographic/physiography of the region, indicating an increase in water-related stress to agriculture in the region.

The NEI region is also experiencing an increase in the mean air temperatures (Deka et al. 2009; Chakraborty et al. 2017b) as well as its extremes (Chakraborty

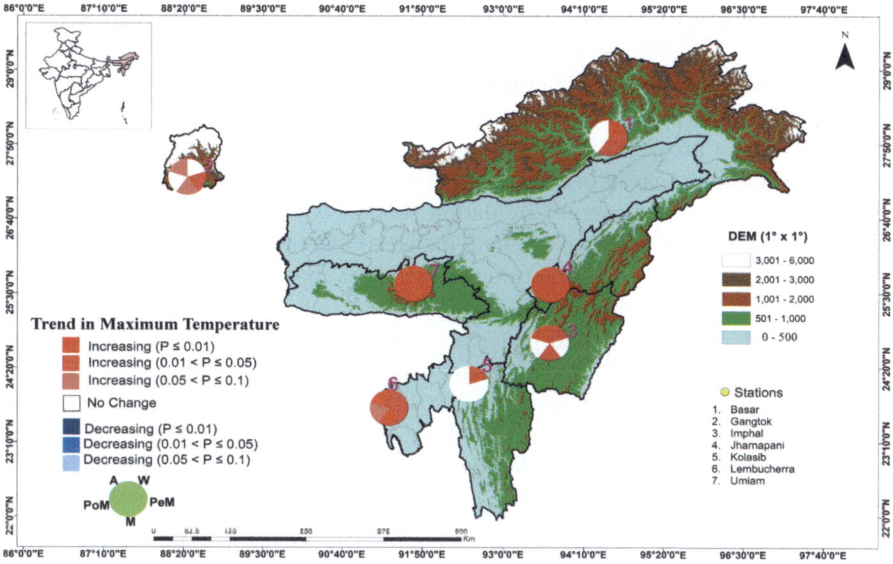

Fig. 1.1 The seasonal and annual trend in maximum temperature at different places of the NEI. (Source: Chakraborty et al. 2023)

et al. 2014; Saha et al. 2022). Deka et al. (2009) reported that, over NEI, maximum temperature (Tmax) is increasing at a greater rate (1.02 °C/decade) than minimum temperature (Tmin), whereas the Tmin is significantly increasing during the post-monsoon (0.64 °C/decade) and winter season (0.61 °C/decade). A significant increasing annual Tmin trend over various stations in NEI is also reported by Lairenjam et al. (2017) and Laskar et al. (2014). Though there exists spatial variability in the rate of change in the temperature, the increase in Tavg ranges from 0.2 °C to more than 1.0 °C per decade across the study region (Chakraborty et al. 2017b). These changes are mostly statistically significant for the maximum temperature in most of the locations and seasons (Fig. 1.1) (Chakraborty et al. 2023). Not only is the mean of the temperatures changing, but there have been lots of changes in the extreme values. Extreme warm days, warm nights, and heat-wave situations have been reported in the region (Chakraborty et al. 2018, 2019, 2020; Saha et al. 2022). The simultaneous changes in rainfall as well as temperature also affect the bioclimate of the region, as indicated by changes in indices like temperature-humidity index, heat index, and other indices like Compensated Summer Ombrothermic Index (CSOI), etc. (Chakraborty et al. 2014; Saha et al. 2016). The change-point analysis indicated that these changes took place mostly during the last half of the 1990s till 2010 (Chakraborty et al. 2017a, b). Overall, these changes are bringing a paradigm shift in the climatic regime of the region, which is having multifaceted impacts on agriculture in particular and the ecosystem in general.

3.2 Projections for Future Climate Change in the NEH Region

Future climate projections for the region remain uncertain; however, they consistently indicate a significant rise in temperatures by the end of the twenty-first century, alongside a projected 25% increase in the number of consecutive dry days during the monsoon season in Assam (Dobler et al. 2011). A comprehensive analysis of outputs from 22 General Circulation Models (GCMs) reveals that temperatures in the Brahmaputra basin are anticipated to increase by 1.3–2.4 °C by 2050 and by 2.0–4.5 °C by 2100. Furthermore, monthly evapotranspiration is projected to rise by 5–18% by 2050 and by 7–36% by 2100, indicating greater atmospheric water demand. Monthly rainfall patterns exhibit high variability, with projections ranging from a 14% decrease to a 15% increase by 2050 and from a 28% decrease to a 22% increase by 2100 (Mahanta et al. 2014). Recent studies indicate a projected temperature rise of approximately 3–8 °C under various Representative Concentration Pathways (RCPs) for the near future (2020–2049) and far future (2070–2099) (Kumar and Dimri 2018). Model projections also suggest significant changes in daily mean precipitation, with reductions of up to 60% during the winter season and increases of up to 40% during the post-monsoon period, depending on the RCP scenario. Notably, the warmer climate scenarios predict a more pronounced precipitation deficit across the region, especially when compared to the lower-emission RCP2.6 pathway. Further, under different Shared Socioeconomic Pathways (SSPs) scenarios, the climate extremes are also expected to increase in the region (Paul and Maity 2023). These projections suggest potential intensification of hydrological extremes, posing critical challenges for water resource management and agricultural sustainability in the region.

4 Impact of Climate Change on Soil Health in Agroecosystems

There exists a strong interdependence between soil and climate change. It is not only sensitive to climate change but is also a major driver of climate change. Climate change affects soil properties, as well as their resistance and resiliency, both directly and indirectly. Climate change significantly affects soil health through alterations in temperature, precipitation patterns, and the frequency of extreme weather events (Table 1.1). These changes impact the physical, chemical, and biological soil properties, influencing the capacity of soil to support plant growth and maintain ecosystem functions.

4.1 Impact on Soil Physical Health

Increased temperatures can lead to higher rates of soil organic matter decomposition, reducing soil structure, aggregate stability, and water-holding capacity. This makes the soil more prone to erosion. Higher soil temperatures can also increase the

Table 1.1 Few possible impacts of climate change on soil health of the Northeastern Hill (NEH) region

Soil health impact in NEH region	Description of the problem under climate change
Increase in soil loss due to erosion	The northeastern hilly region experiences substantial soil loss from water erosion (>40 t ha^{-1} $year^{-1}$), which is much higher than the national average (16 t ha^{-1} $year^{-1}$). This soil loss problem on the hilly slopes will be more intensified with an increase in the frequency of high-intensity precipitation.
Increased nutrient loss	Loss of fertile topsoil will lead to an increase in nutrient loss from the system through surface runoff and also leaching losses due to the increased frequency of high-intensity precipitation.
Loss of soil organic matter (SOM)	A rise in average temperature, decreased precipitation, and also increased frequency of high-intensity precipitation may cause loss of SOM through increased decomposition, reduction in primary net productivity, and loss through erosion in the long run, which may lead to gradual decline in soil health and crop productivity in the region.
Decline in soil water availability	Higher loss of organic matter in soil will affect the soil structure and water-holding capacity of soil. An increase in atmospheric temperature can also increase the rate of evapotranspiration. These will lead to a decline in soil water availability, which may emerge as a major challenge to sustain agricultural productivity amidst the changing rainfall patterns in the region.
Intensification of soil acidity	Majority of the soils (>80%) in the northeastern hill region are acidic in nature, causing nutrient imbalances such as aluminum toxicity, phosphorus deficiency, etc., which are serious crop production constraints in these soils. High carbonic acid concentration and intense leaching of basic cations from the soil system as a result of increased atmospheric CO_2 levels and rainfall events may intensify the soil acidity problem, challenging soil health and crop productivity in the region.

rate of evapotranspiration, leading to drier soil conditions and reduced soil moisture availability for crops. Altered precipitation patterns, including increased frequency and intensity of rainfall, can lead to soil erosion, especially in hilly regions. Heavy rainfall can wash away the topsoil, which is rich in nutrients and organic matter. Changes in rainfall distribution can affect the soil's ability to retain water, leading to either waterlogged conditions or drought. Both extremes negatively impact soil structure and plant growth. Extreme weather events like floods can cause significant soil erosion, loss of soil fertility, and deposition of sediments. They can also lead to waterlogging, which reduces soil aeration and negatively affects root growth. Landslides triggered by heavy rainfall or other climatic events can result in the loss of large amounts of soil and disruption of the soil profile.

4.2 Impact on Soil Chemical Health

Changes in precipitation and temperature can affect soil pH. Increased rainfall can lead to leaching of basic cations, resulting in soil acidification. Acid rain, resulting from increased atmospheric CO_2 and other pollutants, can intensify soil acidity, affecting nutrient availability and microbial activity. Excessive rainfall can lead to nutrient leaching, particularly of nitrogen, and also create anaerobic conditions in poorly drained soil, which may lead to Mn, Fe, Al toxicities, and B nutrient deficiency, while drought conditions can reduce nutrient uptake by plants due to limited soil moisture.

Higher temperatures can increase the decomposition rate of soil organic matter, releasing nutrients and influencing nutrient availability. However, if these nutrients are not taken up by plants, they can be lost through leaching or volatilization. An increase in temperature is also expected to deplete the soil organic carbon (SOC) pool, and the loss has been estimated to be around 28%, 20%, and 15% of SOC in the upper layers of hyper-humid, subhumid zone, and hyper-arid zones, respectively, in the Mediterranean region (Bottner et al. 1995). Follett et al. (2012) reported that a 1 °C increase in mean annual temperature causes a loss of 486 Tg of soil organic carbon and 180 kg of soil organic N ha^{-1} from the top 10 cm of soil over 30 years. Reduced organic matter also impacts the soil's ability to sequester carbon, contributing to higher CO_2 levels in the atmosphere, which further increases the climate change impacts. Soil CO_2 emission is also regulated by soil temperature and moisture. Soil organic matter contributes significantly to CEC, and changes in organic matter content due to climate change can alter the soil's ability to hold and exchange cations. Reduced organic matter can lead to lower CEC, affecting nutrient retention and availability. Soil temperature variations also directly affect nutrient movement and its acquisition by roots. Extreme weather events, in the form of high-intensity rainfall, increase the soil erosion and lead to the loss of topsoil, which is rich in nutrients and organic matter, thereby reducing soil fertility.

4.3 Impact on Soil Biological Health

Microorganisms in soil play a major role in carrying out the soil processes and services to maintain the soil fertility, support plant growth, and maintain the ecological balance. These processes occur at its peak with favorable climatic conditions. Climate change leads to alteration of the geographical range, diversity, distribution, and abundance of microbial species, influencing interactions among microbes and between microbes and plants. The structure and composition of microbial communities are also influenced directly or indirectly by changes in plant composition, root exudates, plant litter, etc. (Jain et al. 2020), eventually altering the soil processes and services. The temperature and moisture fluctuations will affect the rate of respiration of the microbes, altering the decomposition and CO_2 fixation (conversion of CO_2 to organic matter), and finally affecting nutrient cycling in the ecosystem. Prolonged high temperatures can affect microbial communities and reduce their diversity. Raised atmospheric CO_2 may lead the soil microorganisms to transmit harmful greenhouse gases or substances, viz. methane (CH_4), CO_2, and nitrous oxide, etc. Any changes in precipitation will greatly influence the composition of microbial communities, impacting their overall abundance (Castro et al. 2010). Drought conditions can reduce microbial activity and diversity, while waterlogging can create anaerobic conditions, inhibiting aerobic microorganisms and favoring anaerobic ones.

Earthworms and other soil fauna are also sensitive to changes in soil temperature and moisture. Flooding and waterlogging can reduce the populations of soil fauna by creating inhospitable conditions, leading to declines in their activity and diversity.

Climate change can influence the prevalence and activity of soil-borne pathogens. Higher temperatures and altered moisture conditions can create favorable environments for the proliferation of certain pathogens, leading to increased disease incidence in crops.

5 Climate Change and Crop Productivity

Projected Earth's climate shifts have raised concerns about the potential risks to crop production. The average surface temperature on Earth is increasing by 0.23 °C every 10 years (Huang et al. 2017), and the first to be impacted is agricultural production, as it is highly dependent on the climate and meteorological conditions, thereby increasing its vulnerability to climate change. Climate change has serious implications for agriculture and food security, and extreme weather conditions like heat waves, floods, droughts, etc., impact agricultural production and productivity in developing and low-income countries. The alteration of climatic conditions has a significant impact on weeds, diseases, and insect pests by increasing their geographical distribution, population, and number of generations, etc. It is reported that global grain production will decrease by about 10% with every 1 °C rise in temperature (Hossain et al. 2022). At the same time, there is a rising demand for food, which will continue to increase as the global population and consumption increase (Godfray et al. 2010). Despite the doubling of the world population and a decrease in global cropland per capita from 0.44 ha to less than 0.25 ha, global cropland area only increased by 12% over the last 50 years (1961–2009) (FAO 2012). So, food security will be a major issue in the years to come. One aspect of food security is crop production, which is the product of three factors: i.e., arable land, cropping intensity, and yield. One or more of these three factors must be increased to increase crop productivity to meet future needs (FAO 2002).

Cropping systems around the world will be impacted differently by the response of different crops to global climate change factors, due to regional differences in rates of daytime and nighttime temperatures, changes to the timing, frequency, and intensity of precipitation, and exposure to ozone (O_3) and air pollution. Other weather variables that affect agricultural productivity include temperature, rainfall, relative wind speed, humidity, sunshine duration, and evaporation rate. Warming is predicted to also slow down the global yield growth by about 1.5% per decade, while an increase in CO_2 concentration will boost yields by roughly the same amount (Easterling et al. 2007). This increase in temperatures and CO_2 concentrations in the long term can have a substantial impact on agricultural production (Varanasi et al. 2016), depending on the changes in climate conditions.

5.1 Impact of Elevated CO_2

The projected climate change and the increase in atmospheric CO_2 levels could have a significant impact on agricultural futures worldwide by altering the plant's cellular development, metabolic functions, and photosynthesis rate (Rezaei et al. 2015; Mall et al. 2017). The increase in atmospheric CO_2 levels could have some balancing effects against the negative effects of rising temperatures and reduced soil moisture. The C3 photosynthesis pathway will be less affected by the increased CO_2 fertilization effect in C3 species like wheat, rice, and most fruit and vegetable crops. The increase in CO_2 also benefits stomatal conductance, leading to an improvement in water-use efficiency of crops (Ainsworth and Long 2005). The increase in atmospheric CO_2 concentrations is estimated to increase yields by roughly 15% in C3 plants. Another reality concerning is that the nutritional quality of crops may be compromised by CO_2 fertilization with reduced levels of protein in crops harvested (Taub et al. 2008). The effect of rising atmospheric concentration is having different effects on different crops. For example, C4 grains are least responsive to CO_2, while the most responsive crops to CO2 concentration changes are root and tuber crops, with C3 grains being the most receptive. A rise in CO_2 levels can be beneficial for particular C3 crops such as rice, wheat, barley, soybean, etc. Nevertheless, this advantage from climate change can only be accomplished when other crop growth conditions are optimum. Doubling the CO_2 level may increase the yield and biomass of certain crops, such as wheat and soybean, by about 30%, whereas in crops like maize, there may be very low response (Rosenthal et al. 2012; Hatfield et al. 2011).

5.2 Impact of Elevated Temperature

The past few decades have seen an increase in crop yields due to better crop varieties, crop management, and fertilizer and irrigation use. However, a rise in temperatures would affect the growth period of crops, such as shortening the growth period of double-cropping rice, spring wheat, and soybean, and its per-unit yield may decline (Lobell et al. 2011). The restrictions in the crop growth phase will likely convert many two-cropping areas into three-cropping areas. The decrease in yield is attributed to the shortened grain-filling duration in cereal crops, which is a key factor contributing to reduced productivity under climate change (Farooq et al. 2015). However, the negative impact on production may vary depending on altitude, and in low-latitude areas, it may be the most pronounced (Gitz et al. 2016). An increase of 1 °C in global mean temperature is projected to decrease average global yields by 7.4% for maize, 3.2% for rice, 6% for wheat, and 3.1% for soybean (Zhao et al. 2017). The projected decline in wheat yield is estimated at $-6.0 \pm 4.2\%$ per 1 °C rise for France and $-5.5 \pm 4.4\%$ per 1 °C rise for the United States, which is similar to the global average. In contrast, the declines for Russia ($-7.8 \pm 6.3\%$ per 1 °C rise) and India ($-9.1 \pm 5.4\%$ per 1 °C rise) indicate greater vulnerability to rising temperatures (Zhao et al. 2017). However, for China, the world's largest wheat producer, projections indicate that wheat yields would decrease by only 2.6 ± 3.1% for each 1 °C rise in global average temperature (Zhao et al. 2017). Knox et al. (2012) estimated a 5% reduction in maize yields in African countries, while South Asian regions could see a 16% decline by 2050.

In addition, extreme heat and drought will increase soil water evaporation, accelerate organic matter decomposition, and lead to nutrient loss, all of which will diminish land productivity. Temperature will also affect the rates of photosynthesis, respiration, and grain filling. The optimum temperature for C4 crops such as maize and sugarcane is higher than for C3 crops such as rice and wheat. However, even C4 crops experience reductions in photosynthesis under elevated temperature conditions (Crafts-Brandner and Salvucci 2002). Daytime warming can either increase or decrease net photosynthesis (the balance of photosynthesis and respiration), depending on how current temperatures compare to the optimum. In contrast, nighttime warming leads to increased respiration rates without any benefit to photosynthesis. Atmospheric warming causes an exponential rise in the saturation vapor pressure of air. If relative humidity remains constant, warming increases the vapor pressure deficit (VPD) between the air and the leaf. A higher VPD reduces water-use efficiency, as plants lose more water for each unit of carbon gained (Ray et al. 2002). In response to very high VPD, plants may close their stomates, which reduces photosynthesis rates and raises canopy temperature (Willett et al. 2007). Extreme temperatures can directly damage plant cells and increase the risk of heat stress during the critical reproductive stage, potentially resulting in sterility, reduced yields, or even complete crop failure (Teixeira et al. 2012). Additionally, rising temperatures and increased atmospheric CO_2 may promote the growth and survival of various pests and diseases that affect agricultural crops (Ziska et al. 2010). Finally, increasing temperatures combined with higher atmospheric CO_2 levels may enhance the growth and survival of various pests and diseases that target agricultural crops (Ziska et al. 2010). Again, with changing climate, farmers may abandon certain crop varieties, leading to their permanent loss (Gitz et al. 2016). Climate change is also expected to increase the frequency of extreme weather events, including heat stress, droughts, and floods, which could severely threaten plant varieties cultivated in specific, localized regions.

6 Climate Change Impacts on Agriculture in the Northeastern Hill Region

Agriculture in northeast India is mostly dependent on monsoon precipitation, and therefore, changes in monsoon precipitation likely have significant implications for staple crops. This makes agriculture in the region highly vulnerable to climate change. Additionally, heavy rainfall events during the monsoon season result in landslides and floods, disrupting the food supply chain and market mechanisms in the region. Climatic variations have profound implications for agricultural productivity and sustainability, inducing shifts in crop cycles with changes in the traditional cropping patterns, land use patterns, changes in crop suitability, and also incidence of biotic and abiotic stresses in the region. Farmers in the region are increasingly facing challenges such as the changes in the timing of sowing and harvesting of crops leading to a mismatch of crop growth periods with the changing climate, increased input requirements for producing optimum yields, and impacts on food availability and moreover quality of food grains. Although there are limited extensive empirical observations regarding climate change and its effects on agricultural productivity in Northeast India, a few scattered reports offer insights into current trends and future projections for the region. In a study conducted by Schroder et al. (2024), simulations indicated an increase in mean annual soil erosion towards the end of the century in uphill shifting cultivation areas of Nagaland under all climate scenarios. Stronger influence was observed under medium-high (SSP370) and high-end (SSP585) climate change scenarios, with 2.2 and 3.1 times increases, respectively, compared to the historical baseline (1985–2014). Since soil erosion and degradation processes are very much interlinked, an increase in soil erosion indicates land degradation and loss of crop productivity in the region under climate change, particularly in areas having steep slopes. Although carbon fertilization effect with increasing atmospheric CO_2 is considered to contribute positively towards crop growth under climate change, soil nutrient losses and exposure of topsoils to oxidation losses of soil carbon through erosion and imbalanced nutrient availability in soil reduce the chance of possible benefits of carbon fertilization, ultimately lowering the crop yields. Simulation analysis conducted by Kumar et al. (2011) indicated that climate change in the region may impact the rice yields by about −10% to 5% and −35% to 5% in case of irrigated rice and rainfed rice, respectively. Decline in irrigated maize yields is expected to be about 40%. In the upper parts of Northeast India, potato yields might increase by up to 5% due to climate change, but in the central region, a yield loss of about 4% is expected, with even greater negative impacts in the southern parts. Irrigated mustard crops are also expected to experience yield reductions of up to 10% in the region. Climate change is also increasing the risk of pest and disease outbreaks in this hilly region (Patra et al. 2024; Roy et al. 2020; Sharma et al. 2016). Several factors, such as rising temperatures and precipitation changes, create conducive conditions for insects and pathogens. Temperature significantly influences the egg-laying and oviposition behavior of these pests, thereby affecting their reproductive processes and developmental rates (Régnière et al. 2012). While alterations in precipitation patterns can create favorable environments for insect pests, frequent and intense rainfall events may also wash away insect eggs, ultimately decreasing their populations (Kadam and Khaire 1995). Similarly, high temperatures, along with high rainfall and humidity, help create ideal conditions for the growth and proliferation of disease pathogens. Along with an increase in pest and disease outbreaks, climate change in the region may also lead to the emergence of new insects and pathogens. All these biotic and abiotic stresses, increased as a result of climate change, ultimately challenge the agricultural sustainability of the northeastern hill region.

7 Food and Nutrition Security in the Northeastern Hill Region

7.1 Current Status of Food and Nutrition Security in the Region

Food and nutrition security broadly encompasses an individual's ability to access nutritious, high-quality foods that are affordable and also readily available. It is a critical determinant of health and well-being. India is facing significant challenges regarding food and nutrition security. Although there has been some improvement in calorie intake status, malnutrition remains a serious issue. Food security challenges are often more severe in mountainous regions due to factors such as harsh biophysical conditions, environmental fragility, seasonal variations, poor market access, inaccessibility, limited economic opportunities, etc. As of recent assessments, food and nutrition security in Northeastern India presents a multifaceted challenge influenced by diverse factors including geographical isolation, socioeconomic disparities, and infrastructural limitations. According to the latest data from the National Family Health Survey (NFHS-5) and other regional reports, the Northeastern states continue to grapple with high levels of food insecurity and malnutrition. For instance, the prevalence of anemia in children is more than 50% in the states of Assam, Tripura, Arunachal Pradesh, and Sikkim; more than 50% of women are anemic in Tripura, Assam, and also in Meghalaya. Underweight children and prevalence of stunting among children under 5 years of age remain concerning in these states. Additionally, poverty in the region's population remains a significant barrier, affecting the ability of individuals to purchase nutritious food. Although the diverse agro-climatic conditions of the region benefit the cultivation of a variety of crops, allowing for crop diversification with traditional foods, including locally sourced fruits, vegetables, etc., rich in nutrients, the topographical variations and climate conditions in the region also contribute to inconsistent agricultural productivity, which challenges food and nutrition security.

7.2 Challenges in Achieving Nutrition Security Amidst Climate Change Impacts

Climate change is expected to have profound impacts on food and nutrition security at global, national, and regional scales. This complex phenomenon, which involves changes in weather patterns and the frequent occurrence of extreme weather events, is likely to impact food systems through various ways (Fig. 1.2).

Firstly, climate change can increase uncertainty in food production by limiting access to water, making weather patterns more unpredictable, and degrading the natural resources essential for crop growth, etc., thereby impacting food and nutrition availability. Climate-induced reductions in crop diversity and the availability of nutrient-rich foods contribute to persistent micronutrient deficiencies, including iron and vitamin A. Climate change may affect human health by increasing

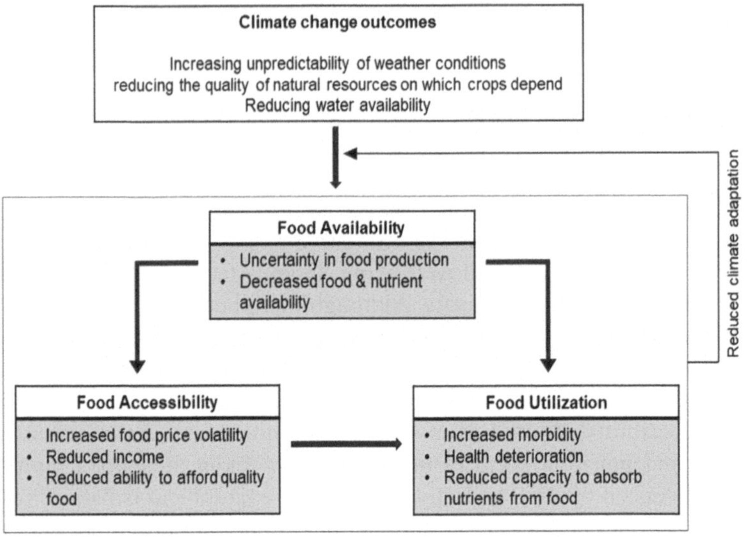

Fig. 1.2 Food and nutrition security challenges of Northeastern Hill Region under climate change

morbidity and the severity of diseases, which impacts an individual's capability to utilize food sources for nutrition. Uncertainty in food production increases food price volatility, which reduces accessibility of food by people, affecting the health, especially of tribal poor in the region. The deterioration in food availability and human health in the long run will result in reduced capacity to adapt to climate change, thereby increasing the risks of possible negative impacts of climate change on food and nutrition.

8 Conclusion and Way Forward

The intricate interplay between climate change and agriculture in the northeastern hill region presents a serious challenge with profound implications for food and nutrition security. This chapter presents how climate change, characterized by erratic rainfall, rising temperatures, and an increased frequency of extreme weather events, disrupts the agricultural systems in this region. The projected impacts include reduced crop diversity, diminished soil fertility, and heightened risks of pest and disease outbreaks, etc. These disruptions not only affect crop yields and agricultural productivity but also lead to food and nutrition insecurity among vulnerable populations. Addressing these challenges requires a comprehensive strategy. There is an urgent need for the integration of climate-smart agricultural practices involving the adoption of farming practices and technologies that enhance resilience to climate variability while simultaneously improving productivity. For example, the use of drought-resistant crop varieties, improved irrigation techniques, and soil conservation practices. Soil management that can potentially contribute to climate change mitigation and adaptation, while at the same time maximizing agricultural production and conserving soil and water quality, must be explored. The identification and development of location-specific, cost-effective, and eco-friendly mitigation and adaptation strategies are needed to sustain agricultural productivity in the region. In this regard, holistic strategies such as agroforestry and integrated farming systems play a major role in enhancing soil health and agricultural resilience. Efforts to strengthen food and nutrition security in the face of climate change require both agricultural and nonagricultural interventions, viz. promoting the cultivation of diverse traditional crops rich in nutrients that can address nutritional deficiencies and improve overall food quality, improving infrastructure and market access for smallholder farmers, and also promoting diversified livelihood sources that contribute to greater economic stability and food security. Increased investment in natural resources management, including soil, water, and energy, as well as the development of climate-resilient crop varieties and sustainable farming practices, is crucial for increasing agricultural production, diversifying local and traditional food systems, and ultimately for improving nutrition and health. This can be brought about by prioritizing climate change mitigation and adaptation in agricultural policies and programs. Collaborative research efforts that bring together scientists, policymakers, and farmers can generate actionable insights for mitigating the impacts of climate change. Finally, raising awareness and promoting education on climate change and its implications for agriculture can help in the collective action. Empowering farmers with information and resources can enhance their adaptive capacity, enabling them to better manage the risks associated with climate variability. By fostering a shared understanding of the challenges and potential solutions, stakeholders need to work together more effectively to address the complex issue of climate change and its impact on food and nutrition security, thus protecting the resources and livelihoods of the northeastern hill region.

References

Abdalla K, Chivenge P, Ciais P, Chaplot V (2016) No-tillage lessens soil CO_2 emissions the most under arid and sandy soil conditions: results from a meta-analysis. Biogeosciences 13(12):3619–3633

Adhya TK, Bharati K, Mohanty SR, Ramakrishnan B, Rao VR, Sethunathan N, Wassmann R (2000) Methane emission from rice fields at Cuttack, India. Nutr Cycl Agroecosyst 58:95–105

Ainsworth EA, Long SP (2005) What have we learned from 15 years of free-air CO_2 enrichment (FACE)? A meta-analytic review of the responses of photosynthesis, canopy properties and plant production to rising CO_2. New Phytol 165:351–371

Bottner P, Coûteaux MM, Vallejo VR (1995) Soil organic matter in Mediterranean-type ecosystems and global climatic changes: a case study—the soils of the Mediterranean Basin. In: Global change and Mediterranean-type ecosystems. Springer New York, New York, pp 306–325

Castro HF, Classen AT, Austin EE, Norby RJ, Schadt CW (2010) Soil microbial community responses to multiple experimental climate change drivers. Appl Environ Microbiol 76(4): 999–1007.

Chakraborty D, Singh RK, Saha S, Roy A, Sethy BK, Kumar A, Ngachan SV (2014) Increase in extreme day temperature in hills of Meghalaya: its possible ecological and bio-meteorological effect. J Agrometeorol 16(Special issue I):147–152

Chakraborty D, Saha S, Singh RK, Sethy BK, Kumar A, Saikia US, Das SK, Makdoh B, Borah TR, Nomita Chanu A, Walling I, Rolling Anal PS, Chowdhury S, Daschaudhuri D (2017a) Spatio-temporal trends and change point detection in rainfall in different parts of north-eastern Indian states. J Agrometeorol 19(2):160–163

Chakraborty D, Saha S, Singh RK, Sethy BK, Kumar A, Saikia US, Das SK, Makdoh B, Borah TR, Chanu AN, Walling I, Rolling Anal PS, Chowdhury S, Daschaudhuri D (2017b) Trend analysis and change point detection of mean air temperature: a spatio-temporal perspective of North-Eastern India. Environ Process 4(4):937–957

Chakraborty D, Sehgal VK, Dhakar R, Varghese E, Das DK, Ray M (2018) Changes in daily maximum temperature extremes across India over 1951–2014 and their relation with cereal crop productivity. Stoch Env Res Risk A 32(11):3067–3081

Chakraborty D, Sehgal VK, Dhakar R, Ray M, Das DK (2019) Spatio-temporal trend in heat waves over India and its impact assessment on wheat crop. Theor Appl Climatol 138(3–4):1925–1937

Chakraborty D, Sehgal VK, Dhakar R, Varghese E, Das DK (2020) Increase in spatial spread of extreme warm day temperature events: regional to national perspective for India (1951–2014). Curr Sci 118(12):1930–1938

Chakraborty D, Saha S, Singh NU, Bag K, Das SK, Sangma C, Sethy BK, Singh IM, Das B, Kalita H, Ajaykumar K, Singh N, Singh HJ, Singh HD, Hazarika S, Mishra VK (2023) Trend and variability in weather variables over the NER: regional to local perspective, PME Publication No. ICARNEH-ML-TB-2023-28. ICAR Research Complex for NEH Region, Umiam, p 38

Cole CV, Duxbury J, Freney J, Heinemeyer O, Minami K, Mosier A, Paustian K, Rosenberg N, Sampson N, Sauerbeck D, Zhao Q (1997) Global estimates of potential mitigation of greenhouse gas emissions by agriculture. Nutr Cycl Agroecosyst 49:221–228

Crafts-Brandner SJ, Salvucci ME (2002) Sensitivity of the C4 plant, maize, to heat stress. Plant Physiol 129:1773–1780

Das A, Ghosh PK, Choudhury BU, Patel DP, Munda GC, Ngachan SV, Chowdhury P (2009) Climate change in northeast India: recent facts and events–worry for agricultural management. In: ISPRS archives XXXVIII-8/W3 workshop proceedings: impact of climate change on agriculture, pp 32–37

Dash SK, Kulkarni MA, Mohanty UC, Prasad K (2009) Changes in the characteristics of rain events in India. J Geophys Res Atmos 114(D10). https://doi.org/10.1029/2008JD010572

Deka RL, Mahanta C, Nath KK (2009) Trends and fluctuations of temperature regime of north east India. In: ISPRS archives XXXVIII-8/W3 workshop proceedings: impact of climate change on agriculture, pp 376–380

Deka RL, Mahanta C, Pathak H, Nath KK, Das S (2013) Trends and fluctuations of rainfall regime in the Brahmaputra and Barak basins of Assam, India. Theor Appl Climatol 114:61–71

Dobler A, Yaoming M, Sharma N, Kienberger S, Ahrens B (2011) Regional climate projections in two alpine river basins: Upper Danube and Upper Brahmaputra. Adv Sci Res 7(1):11–20

Easterling WE, Aggarwal PK, Batima P, Brander KM, Erda L, Howden SM, Kirilenko A, Morton J, Soussana J-F, Schmidhuber J, Tubiello FN (2007) Food, fibre and forest products. In: Parry ML, Canziani OF, Palutikof JP, Linden PJVD, Hanson CE (eds) Climate change 2007: impacts, adaptation and vulnerability. Contribution of Working Group II to the Fourth Assessment Report of the Intergovernmental Panel on Climate Change. Cambridge University Press, Cambridge, pp 273–313

FAO (2002) World agriculture: towards 2015/2030, summary report Food and Agriculture Organization of the United Nations

FAO (2012) Statistical yearbook 2012 World Food and Agriculture Organization of the United Nations

Farooq M, Hussain M, Wakeel A, Siddique KHM (2015) Salt stress in maize: effects, resistance mechanisms, and management a review. Agron Sustain Dev 35:461–481

Follett RF, Stewart CE, Pruessner EG, Kimble JM (2012) Effects of climate change on soil carbon and nitrogen storage in the US Great Plains. J Soil Water Conserv 67(5):331–342

Gibbs D, Harris N, Seymour F (2018) By the numbers: the value of tropical forests in the climate change equation. World Resources Institute, p 4

Giri K, Mishra G, Rawat M, Pandey S, Bhattacharyya R, Bora N, Rai JPN (2020) Traditional farming systems and agro-biodiversity in eastern Himalayan region of India. In: Microbiological advancements for higher altitude agro-ecosystems & sustainability. Springer Nature Singapore, pp 71–89

Gitz V, Meybeck A, Lipper L, Young C, Braatz S (2016) Climate change and food security: risks and responses. Food and Agriculture Organization of the United Nations. https://doi.org/10.1080/14767058.2017.1347921

Godfray C, Beddington J, Crute I, Haddad L, Lawrence D, Muir J, Pretty J, Robinson S, Thomas S, Toulmin C (2010) Food security: the challenge of feeding 9 billion people. Science 327:812–818

Goswami BB, Mukhopadhyay P, Mahanta R, Goswami BN (2010) Multiscale interaction with topography and extreme rainfall events in the northeast Indian region. J Geophys Res Atmos 115(D12). https://doi.org/10.1029/2009JD012275

Gründemann GJ, van de Giesen N, Brunner L, van der Ent R (2022) Rarest rainfall events will see the greatest relative increase in magnitude under future climate change. Commun Earth Environ 3(1):235

Hatfield JL, Boote KJ, Kimball BA, Ziska LH, Izaurralde RC, Ort D, Thomson AM, Wolfe D (2011) Climate impacts on agriculture: implications for crop production. Agron J 103:351–370

Hossain A, Knorr G, Jokat W, Lohmann G, Hochmuth K, Gierz P, Gohl K (2022) Climate model results of different Miocene and Preindustrial atmospheric CO_2 concentrations experiments of AWI-ESM 2 in NetCDF format [dataset]. PANGAEA. https://doi.org/10.1594/PANGAEA.943430

Huang B, Thorne PW, Banzon VF, Boyer T, Chepurin G, Lawrimore JH, Menne MJ, Smith TM, Vose RS, Zhang H (2017) Extended Reconstructed Sea Surface Temperature, version 5 (ERSSTv5): upgrades, validations, and intercomparisons. J Clim 30(20):8179–8205

Hursh A, Ballantyne A, Cooper L, Maneta M, Kimball J, Watts J (2017) The sensitivity of soil respiration to soil temperature, moisture, and carbon supply at the global scale. Glob Chang Biol 23(5):2090–2103

IPCC (2021) Climate change 2021: the physical science basis. Contribution of Working Group I to the Sixth Assessment Report of the Intergovernmental Panel on Climate Change (Masson-Delmotte V, Zhai P, Pirani A, Connors SL, Péan C, Berger S, Caud N, Chen Y, Goldfarb L, Gomis MI, Huang M, Leitzell K, Lonnoy E, Matthews JBR, Maycock TK, Waterfield T, Yelekçi O, Yu R, Zhou B (eds)). Cambridge University Press

Jain SK, Kumar V, Saharia M (2013) Analysis of rainfall and temperature trends in Northeast India. Int J Climatol 33(4):968–978

Jain S, Jain J, Singh J (2020) The Rhizosphere Microbiome: Microbial Communities and Plant Health. In: Varma A, Tripathi S, Prasad R (eds) Plant Microbiome Paradigm. Springer, Cham. https://doi.org/10.1007/978-3-030-50395-6_10

Kadam JR, Khaire VM (1995) Raining and relative humidity: key factors to suppress *Earias vittella* (Fabricius) infestation on okra crop. J Entomol Res 19(3):201–205

Knox J, Hess T, Daccache A, Wheeler T (2012) Climate change impacts on crop productivity in Africa and South Asia. Environ Res Lett 7(3):034032

Korner C, Spehn EM (eds) (2019) Mountain biodiversity: a global assessment, vol 7. Routledge

Kothawale DR, Rajeevan M (2017) Monthly, seasonal and annual rainfall time series for all-India, homogeneous regions and meteorological sub-divisions: 1871–2016, Research report no. RR-138. Indian Institute of Tropical Meteorology (IITM), Pune

Kumar D, Dimri AP (2018) Regional climate projections for Northeast India: an appraisal from CORDEX South Asia experiment. Theor Appl Climatol 134:1065–1081

Kumar SN, Aggarwal PK, Rani S, Jain S, Saxena R, Chauhan N (2011) Impact of climate change on crop productivity in Western Ghats, coastal and northeastern regions of India. Curr Sci 101(3):332–341

Lairenjam C, Hodam S, Bandyopadhyay A, Bhadra A (2017) Historical and temporal trends of climatic parameters in North East India. Global NEST J 19(4):547–561

Lal R (1991) Tillage and agricultural sustainability. Soil Tillage Res 20(2–4):133–146

Lal R (2004) Soil carbon sequestration impacts on global climate change and food security. Science 304(5677):1623–1627

Laskar SI, Kotal SD, Bhowmik SR (2014) Analysis of rainfall and temperature trends of selected stations over North East India during last century. Mausam 65(4):497–508

Li D, Wu S, Liu L, Zhang Y, Li S (2018) Vulnerability of the global terrestrial ecosystems to climate change. Glob Chang Biol 24(9):4095–4106

Lobell DB, Bänziger M, Magorokosho C, Vivek B (2011) Nonlinear heat effects on African maize as evidenced by historical yield trials. Nat Clim Chang 1:42–45

Mahanta C, Zaman AM, Shah-Newaz SM, Rahman SM, Mazumdar TK, Choudhury R, Borah PJ, Saikia L (2014) Physical assessment of the Brahmaputra river. IUCN, The International Union for Conservation of Nature, Dhaka

Majumdar RK (2022) State of biocultural diversity, climate change and ethnic meat and fish products of Northeast India: a review. In: Impact of climate change on hydrological cycle, ecosystem, fisheries and food security. CRC Press, pp 425–437

Mall RK, Gupta A, Sonkar G (2017) Effect of climate change on agricultural crops. In: Current developments in biotechnology and bioengineering crop modification, nutrition, and food production. Elsevier, Amsterdam, pp 23–46

Marak JDK, Sarma AK, Bhattacharjya RK (2020) Innovative trend analysis of spatial and temporal rainfall variations in Umiam and Umtru watersheds in Meghalaya, India. Theor Appl Climatol 142:1397–1412

Murali S, Shrivastava R, Saxena M (2010) Green house gas emissions from open field burning of agricultural residues in India. J Environ Sci Eng 52(4):277–284

Patra S, Chakraborty D, Verma VK, Pande R, Sangma RHC, Chakraborty M, Layek J, Hazarika S (2024) Influence of shifting thermal regimes on tomato fruit borer, *Helicoverpa armigera* (Hubner) in the Eastern Himalaya: implications for pest management strategies. Int J Biometeorol 68:1–11

Paul AR, Maity R (2023) Future projection of climate extremes across contiguous Northeast India and Bangladesh. Sci Rep 13(1):15616

Peters MK, Hemp A, Appelhans T, Becker JN, Behler C, Classen A, Detsch F, Ensslin A, Ferger SW, Frederiksen SB, Gebert F (2019) Climate–land-use interactions shape tropical mountain biodiversity and ecosystem functions. Nature 568(7750):88–92

Prokop P, Walanus A (2015) Variation in the orographic extreme rain events over the Meghalaya Hills in Northeast India in the two halves of the twentieth century. Theor Appl Climatol 121:389–399

Prokop P, Walanus A (2024) Orographic effects on droughts in a monsoon climate with the world's highest rainfall. Int J Climatol 44(7):2438–2461

Rao CS, Srinivas T, Rao RVS, Rao NS, Vinayagam SS, Krishnan P (2020) Climate change and Indian agriculture: challenges and adaptation strategies. ICAR-National Academy of Agricultural Research Management, Hyderabad, p 584

Ray JD, Gesch RW, Sinclair TR, Hartwell Allen L (2002) The effect of vapor pressure deficit on maize transpiration response to a drying soil. Plant Soil 239(1):113–121

Régnière J, Powell J, Bentz B, Nealis V (2012) Effects of temperature on development, survival and reproduction of insects: experimental design, data analysis and modeling. J Insect Physiol 58(5):634–647

Rezaei EE, Webber H, Gaiser T, Naab J, Ewert F (2015) Heat stress in cereals: mechanisms and modelling. Eur J Agron 64:98–113

Rosenthal DM, Slattery RA, Miller RE, Grennan AK, Cavagnaro TR, Fauquet CM, Gleadow RM, Ort DR (2012) Cassava about—FACE: greater than expected yield stimulation of cassava (*Manihot esculenta*) by future CO_2 levels. Glob Chang Biol. https://doi.org/10.1111/j.1365-2486.2012.02726.x

Roy S, Barooah AK, Ahmed KZ, Baruah RD, Prasad AK, Mukhopadhyay A (2020) Impact of climate change on tea pest status in Northeast India and effective plans for mitigation. Acta Ecol Sin 40(6):432–442

Saha S, Chakraborty D, Choudhury BU, Singh SB, Chinza N, Lalzarliana C, Dutta SK, Chowdhury S, Boopathi T, Lungmuana AR, Ngachan SV (2015) Spatial variability in temporal trends of precipitation and its impact on the agricultural scenario of Mizoram. Curr Sci 109(12):2278–2282

Saha S, Chakraborty D, Singh SB, Chowdhury S, Syiem EK, Dutta SK, Choudhury BU, Boopathi T, Singh AR, Ramakrishna Y, Roy A (2016) Analyzing the trend in thermal discomfort and other bioclimatic indices at Kolasib, Mizoram. J Agrometeorol 18(1):57–61

Saha S, Chakraborty D, Paul RK, Samanta S, Singh SB (2018) Disparity in rainfall trend and patterns among different regions: analysis of 158 years' time series of rainfall dataset across India. Theor Appl Climatol 134(1–2):381–395

Saha S, Chakraborty D, Yadav GS, Das A, Das B, Chowdhury S, Lalhmachhuana P, Chauduri DD, Choudhury BU, Datta M, Shakuntala I (2021) Rainfall analysis across the north east Indian state of Tripura. J Agrometeorol 23(4):471–475

Saha S, Chakraborty D, Hazarika S, Shakuntala I, Das B, Chhabra A, Sadhu S, Chakraborty D, Mukherjee J, Singson L, Mishra VK (2022) Spatiotemporal variability of weather extremes over eastern India: evidences of ascertained long-term trend persistence and effective global climate controls. Theor Appl Climatol 148:643–659

Saikia US, Hazarika S (2018) Climate change: a North East India perspective, Research folder. National Innovations in Climate Resilient Agriculture (NICRA), Division of Natural Resource Management, ICAR Research Complex for NEH Region, Umiam

Schmeller DS, Urbach D, Bates K, Catalan J, Cogălniceanu D, Fisher MC, Friesen J, Füreder L, Gaube V, Haver M, Jacobsen D (2022) Scientists' warning of threats to mountains. Sci Total Environ 853:158611

Schröder LS, Rasche L, Jantke K, Mishra G, Lange S, Eschenbach A, Schneider UA (2024) Combined effects of climate change and agricultural intensification on soil erosion in uphill shifting cultivation in Northeast India. Land Degrad Dev 35(2):670–686

Sharma G, Partap U, Dahal DR, Sharma DP, Sharma E (2016) Declining large-cardamom production systems in the Sikkim Himalayas: climate change impacts, agroeconomic potential, and revival strategies. Mt Res Dev 36(3):286–298

Sharma D, Das S, Goswami BN (2023) Variability and predictability of the Northeast India summer monsoon rainfall. Int J Climatol 43(11): 5248–5268.

Sun F, Roderick ML, Farquhar GD (2018) Rainfall statistics, stationarity, and climate change. Proc Natl Acad Sci 115(10):2305–2310

Taub D, Miller B et al (2008) Effects of elevated CO_2 on the protein concentration of food crops: a meta-analysis. Glob Chang Biol 14:565–575

Teixeira KJA, Snyder PK, Twine TE, Cuadra SV, Costa MH, DeLucia EH (2012) Climate-regulation services of natural and agricultural ecoregions of the Americas. Nat Clim Change Lett. https://doi.org/10.1038/NCLIMATE1346

Varanasi A, Prasad PV, Jugulam M (2016) Impact of climate change factors on weeds and herbicide efficacy. Adv Agron 135:107–146

Varikoden H, Revadekar JV (2020) On the extreme rainfall events during the southwest monsoon season in northeast regions of the Indian subcontinent. Meteorol Appl 27(1):e1822

Willett KM, Gillett NP, Jones PD, Thorne PW (2007) Attribution of observed surface humidity changes to human influence. Nature 449:710–712.

Zhao C, Liu B, Piao S, Wang X, Lobell DB, Huang Y, Huang M, Yao Y, Bassu S, Ciais P, Durand JL (2017) Temperature increase reduces global yields of major crops in four independent estimates. Proc Natl Acad Sci 114(35):9326–9331.

Ziska LH, Tomecek MB, Gealy DR (2010) Competitive interactions between cultivated and red rice as a function of recent and projected increases in atmospheric carbon dioxide. Agron J 102:118–123

Climate-Resilient Sustainable Agriculture to Evade Food and Nutrition Insecurity

2

Piyoosh Babele, Umesh Pankaj, K. Sobha, and Mahendra K. Verma

Abstract

Climate change imposes various extremes, i.e., elevated CO_2 and ozone levels, drought, heat waves, erratic and intense rainfall patterns, storms, floods in low-lying lands, and the emergence of pathogens and pests. Higher temperatures and water scarcity result in soil salinity and sodicity. The changing climate, somehow, affects all forms of life, but its impact on crop plants is most obvious, forfeits as crop plants are the main source of food and nutrition. Hence, to respond against the adverse effects of climate change, climate-smart agricultural practices and the development of climate-resilient crops are of pivotal importance for the sustainability of all life forms. Conventional crop breeding has had considerable success, but it is slow and largely limited to exploiting the existing genetic variation in crop plants and their very close relatives. Biotechnology and genetic engineering give us the prospect of creating dramatic alterations in crops to withstand a number of abiotic and biotic stresses, which is difficult to attain using conventional breeding approaches. Modern biotechnological and genetic engineering approaches, such as genome editing, RNA-mediated gene silencing, armored with next-generation sequencing and genome mapping, have paved the way for precise and faster genetic modifications of plants. In the present review, we discussed the impact of climate change on agricultural systems. We summarize the applications and promises of modern biotechnological tools and technologies in generating

P. Babele
Rani Lakshmi Bai Central Agricultural University, Jhansi, Uttar Pradesh, India

U. Pankaj
College of Horticulture and Forestry, Rani Lakshmi Bai Central Agricultural University, Jhansi, Uttar Pradesh, India

K. Sobha · M. K. Verma (✉)
RVR & JCCE STP Foundation Chowdavaram, Guntur, Andhra Pradesh, India

U. Pankaj et al. (eds.), *Plant-microbiome Interactions for Climate-resilient Agriculture*, https://doi.org/10.1007/978-981-96-3534-4_2

climate-resilient crops to achieve food and nutrition security in a sustainable manner. We encourage researchers and other stakeholders to engage in activities beyond the laboratory if they hope to see what is technologically possible translated into practice at this critical point in agricultural transformation.

Keywords
Climate change · Resilient crops · Food and nutrition security · Genetic engineering · Genome editing

1 Introduction

Climate change poses a severe threat to the environment, biodiversity, and agriculture, and it already affects human society and the livelihoods of farmers globally. Emissions of greenhouse gases through anthropogenic activities are the main reason for climate change. The addition of greenhouse gases (GHG) to the atmosphere warms up the Earth's surface. Asia is facing alarming challenges due to climate change and variability, as illustrated by various climatic models predicting that the global mean temperature will increase by 1.5 °C between 2030 and 2050 if it continues to increase at the current rate (IPCC 2019). In the arid areas of the western part of China, Pakistan, and India, it is also projected that there will be a significant increase in temperature (IPCC 2019). During the monsoon season, there would be an increase in erratic rainfall of high intensity across the region. The impacts of climate change on agriculture include rise in temperatures and an increased incidence or severity of extreme weather events. Since agricultural production mostly depends on the selection of crops suitable for the climatic conditions of specific regions, agriculture is always considered a climate-dependent bio-industry with some regional characteristics. In India, variability in temperature and rainfall and climate change-driven extreme weather events such as flood, drought, heat stress, cold waves, and storms have several negative impacts on crop productivity. The recent report on climate forecasts for agriculture shows that if current GHG emissions and climate change trends persist, there will be an increasing frequency of extreme weather conditions leading to increased pest and pathogen incidence, abiotic stress, and a reduction in the crop yield of many major crops (Aryal et al. 2019). Climate-related disruptions in the food system, and thus the loss of crop productivity and nutritional quality, result in food and nutrition security issues. Food and nutrition security are major factors in determining human health and well-being (Ahmad et al. 2019).

The above-mentioned facts indicate that agriculture is facing the threats of climate change, which can drastically affect food and nutrition security. Therefore, to address these challenges, innovative solutions using concerted research efforts must be implemented to improve yield, nutritional quality, and stress tolerance in staple crops (FAO 2015; Ahmad et al. 2019). Attempts are being taken to increase the adaptive capacity of major food crops by developing tolerant cultivars utilizing various approaches. Traditional plant breeding techniques are practiced using the capacity of plants to acquire resilient traits from wild relatives (Myers et al. 2017).

Comparatively, as a faster and more precise method, modern biotechnology and genetic engineering tools have revolutionized plant breeding and crop improvement by creating genetically modified climate-resilient crops (Yu et al. 2018). Genomics-assisted breeding, next-generation sequencing methods, and genome editing are some of the tools utilized to create high-yielding, better-adapted crop varieties that are resilient to climatic changes. This review mainly aims to summarize the impacts of climate change and variability on the agriculture and food system. We catalog the biotechnological advancements that have been made towards producing climate-resilient crops that are better adapted to climate change. We will summarize the efforts undertaken thus far and describe the limitations and opportunities that exist with modern crop biotechnology technologies.

2 Climate Change and Impact on Agriculture Productivity

Agriculture is an important sector of developing economy where climate change is a challenge to agricultural production in areas with food insecurity, particularly in Asian nations. The farmers' way of life has been negatively impacted by a variety of climate-driven extremes, including drought, heat waves, irregular and strong rainfall patterns, storms, floods, and emerging insect pests. While there is unpredictability in climatic patterns, future climate estimates indicated a large rise in temperature and irregular rainfall with increasing intensity (Fig. 2.1) (Ahmad et al. 2019). In the Asian continent, a 2.8 °C increase in the maximum temperature and a 2.2 °C decrease in the minimum temperature are predicted for the middle of the century (2040–2069). The requirement to maximize climate-smart and resilient agricultural practices and technology for sustainable productivity is a response to the worst-case scenarios of climate change (Rasul et al. 2019). Under climate

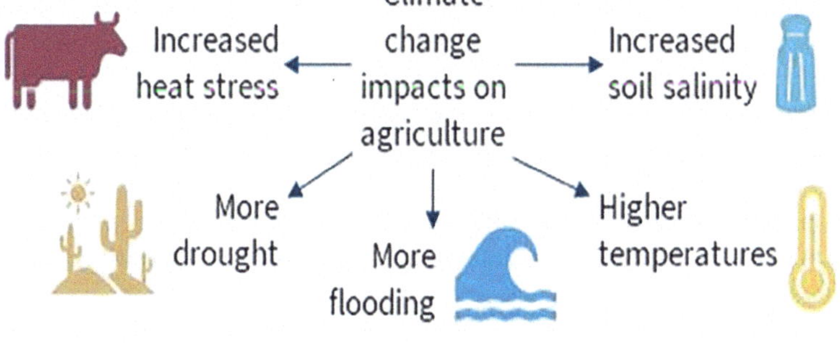

Fig. 2.1 Impact of climate change on environmental process and agriculture production

change scenarios, adaptation technology has the potential to increase the total productivity and profitability of the rice-wheat cropping system by modifying crop management practices such as sowing time and density, nitrogen application, and irrigation application. In addition, this study summarizes recent research on the negative effects of climate change on agricultural productivity as well as related key problems, difficulties, and opportunities for sustainable agricultural productivity to guarantee food security in Asia (Rezaei et al. 2018). Agroforestry, mixed livestock systems, climate-resilient plants, livestock and fish breeds, farming of monogastric livestock, early warning systems, decision support systems, carbon sequestration, climate, water, energy, and soil-smart technologies, and the promotion of biodiversity are a few examples of the following opportunities that have the potential to lessen the negative effects of climate change (Ahmad et al. 2019). Climate change has significant implications for agriculture, affecting various aspects of crop production, livestock management, and overall food security (Arunrat et al. 2018). Here are some key points about the relationship between climate change and agriculture:

(a) *Changing weather patterns:* Climate change can lead to altered rainfall patterns, increased frequency and intensity of extreme weather events (such as droughts, floods, and storms), and changes in temperature. These shifts can disrupt agricultural activities, affecting planting seasons, crop growth, and harvests.
(b) *Crop productivity:* Rising temperatures and changing rainfall patterns can impact crop yields. While some crops may benefit from a longer growing season in certain regions, others may experience reduced yields due to heat stress, water scarcity, increased pests, and diseases. Changes in temperature can also affect pollination, leading to reduced fruit set and crop failures.
(c) *Water availability:* Climate change can alter the availability and distribution of water resources, which are crucial for agricultural irrigation. Changes in rainfall patterns can lead to water scarcity in some regions, affecting crop irrigation and livestock production. In areas where glaciers and snowpack serve as water sources, reduced snowfall and earlier melting can disrupt water availability for irrigation.
(d) *Pests and diseases:* Climate change can influence the distribution and prevalence of pests, weeds, and diseases that affect crops and livestock. Warmer temperatures may allow pests to expand their range, leading to increased infestations and crop damage. Changes in rainfall patterns can also create favorable conditions for the spread of diseases and fungi.
(e) *Shifts in suitable agricultural regions:* As the climate changes, the suitability of certain regions for specific crops may change. Some areas may become less suitable for traditional crops due to increased heat or water scarcity, while other regions may see new opportunities for cultivation. This can result in shifts in agricultural practices, such as changing crop choices or adopting new farming techniques.

(f) *Food security:* Climate change poses risks to global food security. The combination of reduced crop productivity, increased frequency of extreme weather events, and changing agricultural conditions can disrupt food production and supply chains. Vulnerable populations, particularly in developing countries, may face increased food insecurity and malnutrition.

Climate change and variability's effects on rice and wheat crops, due to less timely water and rainfall availability, as well as irregular and strong rainfall patterns over the past few decades, have significantly decreased crop productivity in many parts of Asia (Hussain et al. 2018; Aryal et al. 2019). Despite the green revolution's enhanced crop output, it will be difficult to maintain that production and raise food security for Asia's poor rural inhabitants in the face of climate change (FAO 2015; Ahmad et al. 2019). Damage brought on by climatic changes may endanger food security and national economic production in the least developed countries (Myers et al. 2017). Due to changes in climate patterns, yield decreases in various crops (such as rice and wheat) have differed across regions (Yu et al. 2018). According to Obermeier et al. (2017), CO_2 fertilization can boost crop yield and counteract the severe consequences of increased temperatures on C3 plants, but it is unable to lessen those effects (Arunrat et al. 2018). Rising temperatures and unpredictable rainfall have a negative impact on crop growth and development (Rezaei et al. 2018; Asseng et al. 2019). Major producers of food security in Asia are rice and wheat. To fulfill the constantly rising food demand, it will be difficult to boost wheat production by 60% by 2050 (Rezaei et al. 2018). An increase in temperature at lower latitudes is thought to be the cause of decreased crop productivity in arid to semiarid environments.

Drought and flooding have decreased yields of rice, wheat, and maize, and it is anticipated that these problems will have a greater impact on crop output in the future (Chen et al. 2018). According to Yang et al. 2017, rice is sensitive to nighttime temperature changes, causing yield and biomass to drop by 16–52% if the temperature rises by 2 °C above the critical temperature of 24 °C. Semiarid to dry regions in Asia are in danger and already struggle with low productivity and severe stress. Because of the negative effects of rising temperatures and sporadic, intense rains, the quality of wheat produce—including its protein, sugar, and starch content—and grain production have decreased (Yang et al. 2017). Climate variability has significantly reduced wheat yield in India, China, and the North Nile Delta of Egypt (up to 17.6%), which is attributed to an increase in temperature, erratic rainfall, and an increase in insect pest infestation (Arunrat et al. 2018; Shah et al. 2018; Aryal et al. 2019; Kheir et al. 2019). According to Chun et al. (2016), the yield of rice in rain-fed areas of South Asia has already fallen and may do so by 14% under the RCP 4.5 scenario and 10% under the RCP 8.5 scenario by 2080. Due to their negative effects on the booting and anthesis stage across Asia, particularly in Pakistan and China (Zafar et al. 2018; Ahmad et al. 2019), high temperatures and dryness have decreased rice production. Similar to drought, heat stress poses a serious threat to rice because it reduces the number of productive tillers, causes grains to shrink, and ultimately lowers grain output (Wang et al. 2019). Upland rice

(10 million ha) and rain-fed lowland rice (>13 million ha) would both be impacted by climate change in Asia.

3 Need for Climate-Resilient Agriculture

Several climate models forecast that the global mean temperature will rise by 1.5 °C between 2030 and 2050 if it continues to increase at the current rate; Asia is facing severe difficulties as a result of climate change and variability (IPCC 2019). It is also anticipated that there will be a large rise in temperature in dry regions of western China, Pakistan, and India (IPCC 2019). There would be an increase in irregular, heavy rainfall throughout the area throughout monsoon season. A decrease in winter rainfall would result in an increase in aridity throughout South and Southeast Asia. Globally, sea level will rise by 2100 ft (0.1 m) as a result of climate anomalies (IPCC 2019). In Asia, irregular and uncertain rainfall patterns, an increase in heat waves, hot and dry days, and a worsening of dust storms and tropical cyclones are all forecasted (Gouldson et al. 2016). Natural disasters, such as extreme temperatures, storms, and wildfires (23%), floods (37%), drought (19%), and pest and animal disease infestations (9%) that cost 10 USD billion, are the main causes of agricultural productivity (crops and livestock) losses in Asia (FAO 2015). Tropical cyclones in the Pacific have been more frequent and intense during the past few decades. South Asia was the world's most food-insecure region, with 262 million people suffering from malnutrition (FAO 2015; Rasul et al. 2019). Rural populations are especially susceptible to climate change in isolated dry plains and deserts since there are fewer natural resources available.

Climate variability (in terms of temperature and rainfall) and climate-driven extremes (such as floods, droughts, heat stress, cold waves, and storms) have several negative effects on the agriculture sector in Asia (FAO 2016), particularly on the cropping system, which plays a significant role in ensuring food security (Cai et al. 2016; Aryal et al. 2019). As a result, there are problems with and challenges with regard to food security in Asia. Climate change is threatening the rice-wheat cropping system, a significant cropping system that supplies half of Asia's food requirements (Ghaffar et al. 2022). Wheat and rice harvests are negatively impacted by climate change in terms of both quantity and quality (Din et al. 2022; Wasaya et al. 2022). For instance, the detrimental effects of rising temperatures have resulted in decreased wheat grain production and protein content (Asseng et al. 2019). The crop's growing season has been shortened by the temperature increase, and crop evapotranspiration has consequently decreased wheat output (Azad et al. 2018). The rising average temperature during the growing season is blamed for the negative effects of climate change and variability on China's winter wheat output (Geng et al. 2019). The quality characteristics of wheat grains, particularly their protein content, sugar, and starch percentages, are being negatively impacted by climate change. High temperatures and increased carbon dioxide levels improve wheat grain growth features while lowering their protein content (Asseng et al. 2019).

The amount of protein and soluble carbohydrates in the wheat crop is also decreased by drought stress (Rakszegi et al. 2019; Hussein et al. 2022). Under drought stress, a decrease in the starch content of wheat grains has also been noted (Noori and Taliman 2022). Wheat grains' protein, soluble sugar, and starch contents all decrease as a result of heat stress (Zahra et al. 2021; Iqbal et al. 2022; Zhao et al. 2022). Because of the increase in temperature and the resulting decrease in protein content, sugars, and starch, climate change has a negative impact on the quality of wheat products as well. According to estimates from Kheir et al. (2019), a 1–4 °C increase in temperature might result in a 17.6% reduction in wheat yield in the Egyptian North Nile Delta. Crop management practices and climate variability have both had an impact on crop phenology in China (Liu et al. 2018). Wheat phenology in India and China has been negatively impacted by both potential climate change scenarios and human management techniques (Lv et al. 2013; Ren et al. 2019). The increased aphid population on wheat crops as a result of the rising warmth has decreased yield (Tian et al. 2019). Diseases linked to climate change have a clear and strong linkage. For instance, the Fusarium species that causes Fusarium head blight of wheat crops is more likely to attack in a hot, humid climate (Shah et al. 2018).

4 Biotechnological Tools for Climate Resilient

One of the most crucial agricultural enhancement methods used today is plant breeding. The idea of plant breeding predates agriculture and is said to have originated more than 10,000 years ago with the domestication of crops. It deals with the science of purposefully establishing inheritable alterations in plants. Selecting plants with better qualities, employing these superior plants for ongoing multiplication, or crossbreeding the superior plants to create genotypes with traits from both parents are all examples of plant breeding. Advanced genetics, genomics, and molecular biology are used in modern plant breeding techniques to produce genotypes with the desired phenotype (Aryal et al. 2019). Prior to any breeding program, a number of aspects must be considered, including the mode of reproduction, the inheritability of genes, the mode of action of genes, the amplitude of gene effects, the relationship between the gene(s) and agronomic qualities, the ability to combine traits, and screening tools available.

(a) *Classical Plant Breeding:* In traditional plant breeding, plants with superior qualities are found in a population of diverse individuals, and they are then employed for propagation for a number of generations until those traits become fixed in the population. It is the science of selecting breeding plants with desirable traits and eradicating those with bad traits, to put it simply (Pingali 2012). Characteristics of abiotic stress tolerance might be discovered in landraces or wild relatives. The breeder's initial goal is to find these features, which they will subsequently use to benefit future generations. Finding the phenotypes that closely correlate with a specific sort of stress is a significant issue for this type of strategy. Thus, selecting for the greatest amount of genetic variety is necessary when breeding for any such features (Stevenson et al. 2013). The deliberate interbreeding of closely or distantly related species, known as "crossing," is another technique for creating crop varieties with desired properties. The crossing may occur intraspecifically (between members of the same species), interspecifically (between members of different species), or intragenetically (between members of different genera). It is not required for genes from one species to meld so seamlessly into the genome of another species, though (Ray et al. 2013). The majority of the time, the resulting progeny is sterile or never develops. Plants are crossed to introduce desired characteristics into the genetic make-up of the other.

(b) *Mutation Breeding:* By exposing seeds to certain mutagenic agents, such as ethyl methane sulfonate (EMS) or ionizing radiation (like X Ray), mutation breeding, also known as variant breeding, produces mutagenic plants. These mutagens cause genomic mutations at random, some of which are occasionally advantageous. After screening mutant populations, mutant plants with desirable features are chosen. To create desired mutant lines, selfing or backcrossing mutants are options (Allaby et al. 2017). This technique is typically applied to produce novel genotypes when desired genes are either not present or have been exhausted in the gene pool. Using this technique, a large number of genotypes with abiotic stress tolerance have been produced. For instance, the mutant variety of barley, Diamant, which is high yielding, has good grain quality, a short stem, malting quality, and is resistant to lodging, was produced by X-ray irradiating its latent seeds (Meena et al. 2016). Similarly, Calrose76, a gamma-irradiated japonica rice cultivar with small stature, was released in 1976 and has increased lodging resistance. More recently, using EMS as the mutagen, mutation breeding was used to create the salt-tolerant rice variety Kaijin (Allard 1999).

(c) *Marker Aided Selection (MAS):* In MAS, a trait of interest is chosen based on a marker, which can be a morphological marker (for instance, the presence or absence of an awn in rice), a biochemical marker, or a DNA marker [Single Nucleotide Polymorphisms (SNPs), Simple Sequence Repeats (SSRs), Sequ DNA markers might be based on PCR (Polymerase Chain Reaction)]. Younis et al. (2020) examined many molecular markers and their use for QTL mapping in agronomic crops. In contrast to more complex traits like biotic and abiotic stress tolerance, where each plant must be tested individually for the presence or absence of the trait among a large number of progenies, screening of plants for clearly visible morphological traits (clearly observable phenotypes) like higher yield, bigger fruits, and seeds can be done easily (Melchinger and Gumber 1998). In order to avoid the necessity for phenotypic screening of plants lacking the marker of interest, MAS selects for markers, particularly DNA markers that are closely connected to a trait or gene of interest. Breeding for specific features requires arduous, expensive, and time-consuming traditional procedures. In general, more time and effort are required to accomplish desired results the more complicated the features (Pathirana 2011). Additionally, it is frequently necessary for the offspring to mature before the success of the cross can be assessed. Because it is growth-stage independent, unaffected by environmental factors, effective to use in early generations, and effective when field evaluation is very slow or expensive, DNA-based MAS is a useful technique for speeding up the breeding process (Ahloowalia et al. 2004).

Breeding for abiotic stress tolerance characteristics has proven particularly challenging due to certain significant abiotic stress-related restrictions, notwithstanding a variable success rate and the normal problems of compatibility and linkage drag (Rutger et al. 1976). First off, after homologous recombination, the abiotic stress response in plants consists of a variety of signal transduction pathways and a large number of genes, the majority of which are not tightly connected to one another. The introduction of many genes involved in a given abiotic stress becomes highly challenging and implausible, barring QTLs (Takagi et al. 2015). Second, various plants respond differently to the same stressors, making it difficult to select promising plants and develop a species for many resilient traits at once. Understanding epigenetic codes is crucial for breeding stress-tolerant crops because they affect chromatin structure at the transcriptional and posttranscriptional levels, where epigenetic modifications also play a significant role in stress responsiveness. Genotype performance frequently varies annually (Frisch and Melchinger 2005). Third, different abiotic stresses hardly ever occur alone and are frequently connected to one another. This is made worse by the fact that there are only a relatively few number of genetic markers that let crop plants tolerate numerous shocks (Ashraf and Foolad 2012).

5 Genetic Engineering

Genetic engineering techniques can be used to get around the majority of the problems with breeding procedures. When compared to breeding methods, genetic engineering or transgenic technology is more precise, unrestricted by interspecies incompatibility, free from linkage drag, and, most importantly, faster at transferring the desired gene(s) into crops regardless of their source (Neeraja et al. 2007). The pyramiding of genes for any desired trait or traits is also made possible through genetic engineering, allowing crops to be produced with a variety of desirable qualities. In reality, 24 years after their introduction, genetically altered crops have become the most widely used crops in modern history, making up around 191.7 million hectares in 2018. Herbicide-tolerant and BT crops make up a significant portion of this. However, newly created transgenic varieties of sugarcane and maize that are drought-tolerant also add to the total acreage of transgenic crops. Detailed knowledge of the genes involved in the stress response is needed when engineering for abiotic stress tolerance (Duque et al. 2013). A large number of candidate genes implicated in abiotic stress responses have been discovered by functional genomics and genetics, and stress-responsive cellular signal transduction pathways have also been mapped. The modification of gene expression is a common technique used in these researches. Most research has used one of two methods to manipulate gene expression: either overexpressing native or foreign genes in the background, or silencing the native genes (Chang et al. 2020). Using functional genomics, some of the genes encoding ion transporters, antioxidant enzymes, osmoregulatory metabolites, and transcription factors, etc., have been found to modulate the abiotic stress response in plants (Mittler 2006).

6 Gene Expression

By utilizing expression constructs that promote the gene's enhanced transcription, gene overexpression denotes the abundant creation of an important protein in a host. Epigenetic changes can occasionally cause an abundant synthesis of an endogenous gene transcript. The layout is quite basic. Depending on the situation, genes are cloned into plant expression vectors with constitutive, tissue-specific, or stress-responsive promoters (Wang et al. 2017). The various plant transformation techniques are then used to randomly integrate this cloned gene into the genome. Then, several properties connected to the gene of interest are examined between the transgenic plants overexpressing the gene and non-transformed plants. Numerous studies have demonstrated that transgenic plants are more tolerant than their wild-type counterparts due to the overexpression of genes, such as miRNAs, different transcription factors (TFs), and major and minor QTLs, which are triggered under abiotic stressors. To name a few, overexpression of miR169 in tomatoes has increased its ability to withstand drought stress because it causes fewer stomata to open, which leads to a low transpiration rate (ISAAA 2017). Similarly, rice grain production and drought stress tolerance have been improved by overexpressing the transcription factors OsNAC5 and OsNAC14. Rice now has increased resistance to salinity, drought, and high temperatures as a result of overexpression of the glyoxalase enzymes.

7 RNA Interference

Plants have an internal process called RNA interference that controls gene expression by means of tiny (20–25 bp) interfering RNA molecules. Small interfering RNAs (siRNAs) and microRNAs (miRNAs) are the main RNA types implicated in RNAi (Roy et al. 2011; Younis et al. 2014). The processing of these lengthy dsRNA into tiny RNAs, which can guide enzyme complexes to bind to mRNAs and cause mRNA degradation or translation inhibition (posttranscriptional gene silencing, or PTGS), reduces the expression of target proteins (Pradhan et al. 2015; Joshi et al. 2016). Small interfering RNAs are one of the techniques used by the cell to regulate the expression of a number of genes involved in growth and development, maintain the integrity of the genome, and act as an immune response against viruses and other foreign genetic material. By developing knockdown transgenic lines, RNA interference has emerged as a potent tool for the inhibition of the expression of target genes and for the functional investigation of genes (Anwar and Kim 2020).

There are two primary methods for knocking down genes: exogenously produced siRNAs that are introduced directly into the cells, or genetic engineering techniques that entail the stable integration of expression constructs that express antisense RNA, short hairpin RNAs, or long hairpin RNAs that are targeted specifically at a target RNA (Rao et al. 2020).

The enzyme dicer converts the hairpin RNA into a number of useful siRNAs that use the natural RNA Interference Silencing Complex (RISC) to target particular mRNAs. Numerous crops have been successfully subjected to increased abiotic stressors thanks to RNAi technology (Jeong et al. 2013). The RNAi method has improved rice's ability to withstand salt stress by downregulating OsCKX2. Even though there have been numerous instances of genetic engineering improving abiotic stress tolerance, relatively few of these transgenic crop plants have been effectively converted into marketable goods for cultivation and marketing. This is due to the fact that the majority of transgenic crops carry antibiotic marker and reporter genes, which are necessary for the effective selection of transgenic lines and molecular research. The testing of transgenic crops with reporters and markers in fields is currently prohibited by regulations (Shim et al. 2018). Additionally, it takes a lot longer and is much more difficult to create transgenic crops without markers and reporters while yet complying with present laws. Additionally, the cost of creating new transgenic crops has skyrocketed due to the several regulatory testing phases under laboratory, greenhouse, and field trials that are necessary for transgenic crops. Additionally, engineering abiotic stress-tolerant crops is quite challenging due to the complexity of abiotic stress tolerance features, which was already described before (Gupta et al. 2018).

8 Use of Genome Editing Tools for Crop Improvement

Three different types of chemically designed proteins—ZFNs, TALENs, and CRISPR/Cas—are now used for genome editing. The techniques used to turn SSNs into plants are the same ones currently used in genetic engineering (Joshi et al. 2018). The inserted transgene (like Cas) can be eliminated in later generations through segregation, which is an additional benefit of gene editing techniques once the genome has been altered in the transformed plants (Soda et al. 2018). Globally, the productivity of agriculture has been successfully increased through plant breeding and the usage of genetically modified crops. Conventional methods are inefficient, and despite GM plants' immense potential, their availability for human use is complicated by a convoluted regulatory process. Gene modification, gene knockout, or gene insertion are the three main effects of genome editing (Nutan et al. 2020). Genome-edited plants should not necessarily be categorized as transgenic plants, with the exception of those containing gene insertions. When compared to breeding methods, this allows for the considerably faster creation of better genotypes, which also makes the method more cost-effective. But it's important to remember that producing ZFNs and TALENs is expensive. Gene knockout, which is challenging through genetic engineering, can be accomplished by genome editing. Another benefit is that editing or altering gene regulatory areas, such as promoters, can be used to control gene expression. The infrequent risks associated with the transgenesis strategy, where random insertion of the transgene can alter gene functions or lead to the development of new Open Reading Frames (ORFs) encoding harmful proteins, are similarly reduced by targeted insertion of the transgene by HR in the case of gene editing. It is also possible to stack various transgenes at one location, which can then be transferred to other germplasm as a single locus (Khare et al. 2018). This is crucial for the development of numerous abiotic stress-tolerant plants. It is theoretically possible to simultaneously target several genes that control various abiotic challenges, which would allow the development of plants that are tolerant to a variety of abiotic pressures (Singh et al. 2018) (Table 2.1).

Table 2.1 Table summarizes the major methods for crop improvement/climate-resilient agriculture to tackle food and nutrition insecurity

Methods employed	Crossing of hundreds or thousands of genes to get the desirable combination in the progeny	Precise transfer of gene(s) (either native or alien) in an organism's background	Modification of existing genes or gene substitution or addition to get desirable outcomes
Working principle	Crossing and selecting individuals with a superior gene pool	Gene(s) are cloned in an expression vector and are transformed in the host via different transformation methods available	Using protein (ZFNs, TALENs) or RNA (CRISPR/Cas9) recognition of DNA, nonspecific endonuclease DSBs are introduced at a target site, which can be modified by exploiting endogenous repair pathways
Barrier across species	The gene pool can be derived from the same species or closely related species	Gene(s) from any source (plant, animal, microbe) can be introduced	Endogenous gene(s) can be modified or substituted, or gene(s) can be added irrespective of their source
Accuracy of gene transfer	Little or no guarantee of achieving the desirable combination of genes, even in hundreds and thousands of crosses. Often, undesirable genes along with the desirable ones get transferred, sometimes resulting in linkage drag	Only selected gene(s) are transferred	Selected gene(s) are modified or transferred
Integration in genome	Genes are transferred from parents to progeny in an ordered way; the natural mechanism of transfer of genes from parents to progeny	Genes are randomly integrated, which can activate or repress other genes	Precise modification or integration of gene(s)

(continued)

Table 2.1 (continued)

Methods employed	Crossing of hundreds or thousands of genes to get the desirable combination in the progeny	Precise transfer of gene(s) (either native or alien) in an organism's background	Modification of existing genes or gene substitution or addition to get desirable outcomes
Multiplexing	A complex process as it needs extensive crossing and selection	Easy to stack different genes in a single plant	Multiple genes can be targeted simultaneously
Time	Time-consuming; it takes several generations	Desirable changes can be achieved in few generations (three)	Desirable changes can be achieved in few generations (three)
Expenditure involved	Uses relatively inexpensive methods for phenotypic assessment	Current regulations and several levels of testing and field trials for the release of transgenic crops make it an expensive technology	Less costly alternative to GMOs. Not all products developed by this method can fit into the true definition of GMOs
Social acceptability	It is a traditional method of crop improvement	In some countries, these are acceptable	Rules and regulations are yet to be decided.
Final product	The final product is an improved genotype developed by the recombination of genes from parental lines	The final product is genetically engineered, usually consisting of a gene expression construct integrated into the genome	Except for gene insertions, the final product can be transgene-free

9 Challenges

Crop improvement, which involves developing new crop varieties with improved traits, faces several challenges. Crop genomes are often complex, with numerous genes contributing to various traits of interest. Understanding the genetic basis of desirable traits and identifying specific genes or genomic regions associated with them can be challenging. Genetic mapping and functional characterization of genes require extensive research and technological advancements (Raza et al. 2019). Crop breeding programs rely on genetic diversity to introduce new traits and improve the performance of crops. However, modern agricultural practices and the focus on a limited number of high-yielding varieties have led to a reduction in genetic diversity. Limited genetic diversity can hinder the development of new crop varieties with improved traits, as the gene pool for breeding becomes narrower. Crop improvement is a time-consuming process that requires significant resources (Yu and Li 2021). It takes years of breeding and selection to develop and test new varieties before they can be released to farmers. Conducting field trials, evaluating agronomic performance, and ensuring stability across diverse environments all require substantial investments of time, expertise, and financial resources. Trait complexity and interaction: Many desirable traits, such as yield, quality, and resistance to pests and diseases, are complex and influenced by multiple genes interacting with each other and the environment (Jorasch 2020). Understanding and manipulating these complex traits through breeding approaches can be challenging. Balancing multiple traits and achieving desired combinations is a complex task that requires careful selection and breeding strategies (Biswas et al. 2021). Crops face numerous challenges from abiotic stresses such as drought, heat, salinity, and floods, as well as biotic stresses from pests, diseases, and weeds. Developing crop varieties with improved tolerance or resistance to these stresses is a major challenge. Adapting crops to changing climate conditions and emerging pests and diseases requires continuous monitoring, research, and breeding efforts. The release and commercialization of new crop varieties are subject to regulatory frameworks and safety assessments to ensure their environmental safety and potential impacts on human health (Xiong et al. 2022). These regulatory processes can be complex, time-consuming, and expensive, posing challenges for breeders and companies seeking to bring new, improved varieties to the market. Intellectual property rights, including patents and plant breeders' rights, play a role in crop improvement. Balancing the need for innovation and investment with ensuring access to genetic resources and fair benefit-sharing can be challenging. Intellectual property issues can impact the availability and affordability of improved crop varieties, particularly for small-scale farmers in developing countries (Cerón-Souza et al. 2021). Addressing these challenges requires a combination of scientific advancements, collaboration among research institutions and breeding programs, supportive policies and regulations, adequate funding for research and development, and efforts to promote and preserve genetic diversity. Innovative breeding technologies, such as marker-assisted selection, genomic selection, and gene editing, hold promise for accelerating the crop improvement process, but they also raise ethical, social, and regulatory considerations that need to be addressed (Abdallah et al. 2015).

10 Future Perspective

Artificial intelligence (AI) has recently been a groundbreaking topic that offers a significant chance in influencing contemporary crop breeding and is widely applied indoors for plant science (Harfouche et al. 2019). Crop phenomics, environmental genomics, and other "omics" methods are advancing our understanding of the intricate biological mechanisms underlying how a crop responds to environmental stressors (Baduge et al. 2022). These "omics" approaches have given plant scientists the precise instruments they need to quickly assess the critical agronomic features for larger-sized germplasm in the early growth phases. Big data and the intricate relationships within it, however, make it difficult to comprehend the intricate mechanisms that underlie the genes that generate agronomic traits (Montesinos-López et al. 2018). AI provides enormous computational power as well as numerous innovative tools and breeding tactics. Considering the rapid and continuous change in the climate, it becomes more important to examine how applications of AI technology, used in current breeding practices, help to resolve the issue in high-throughput phenotyping and gene functional analysis, and how advancements in AI technologies bring new potential for future breeding, to make envirotyping data extensively used in breeding (Xu et al. 2021). Furthermore, the efficient implementation of high-throughput field phenotyping, genomics, and enviromics is hampered by the enormous problem of tying genotype to phenotype in the existing breeding approaches. Certainly, AI will be the go-to technique for improving the accuracy of high-throughput crop phenotyping, genotyping, and envirotyping data (Duan et al. 2019). Emerging methods and difficulties for integrating multiomics large data remain a challenge to adopting AI-based technologies in agriculture to tackle climate change impact (Cravero and Sepúlveda 2021). Therefore, the combination of "omics" techniques with AI can speed up crop development programs by enabling quick gene identification.

11 Conclusion

The long-term mean climate state has a significant impact on the nature of agriculture and farming practices in any given location; the infrastructure and experience of the local farming communities are typically appropriate for specific farming practices and a specific group of crops that are known to be productive under the current climate. Increased temperatures throughout the growing season have a considerable influence on farm income, food security, and agricultural output. In some circumstances, the best form of farming may alter as a result of mean climatic shifts away from existing conditions, necessitating modifications to current practices in order to sustain production. To address the challenges posed by climate change in agriculture, adaptation strategies are crucial. These include developing climate-resilient crop varieties, implementing efficient water management techniques, improving soil health, promoting sustainable farming practices, and supporting farmers with access to information, technology, and financial resources. Additionally, efforts to mitigate climate change by reducing greenhouse gas emissions are essential for long-term sustainability.

References

Abdallah NA, Prakash CS, McHughen AG (2015) Genome editing for crop improvement: challenges and opportunities. GM Crops Food 6(4):183–205. https://doi.org/10.1080/21645698.2015.1129937

Ahloowalia BS, Maluszynki M, Nichterlein K (2004) Global impact of mutation derived varieties. Euphytica 135:187–204. https://doi.org/10.1023/B:EUPH.0000014914.85465.4f

Ahmad S, Abbas G, Ahmed M, Fatima Z, Anjum MA, Rasul G et al (2019) Climate warming and management impact on the change of phenology of the rice-wheat cropping system in Punjab, Pakistan. Field Crop Res 230:46–61. https://doi.org/10.1016/j.fcr.2018.10.008

Allaby RG, Stevens C, Lucas L, Maeda O, Fuller DQ (2017) Geographic mosaics and changing rates of cereal domestication. Philos Trans R Soc Lond Ser B Biol Sci 372(1735):20160429. https://doi.org/10.1098/rstb.2016.0429

Allard RW (1999) Principles of plant breeding. J Agric Sci 135(3):331–334

Anwar A, Kim JK (2020) Transgenic breeding approaches for improving abiotic stress tolerance: recent progress and future perspectives. Int J Mol Sci 21(8):2695. https://doi.org/10.3390/ijms21082695

Arunrat N, Pumijumnong N, Hatano R (2018) Predicting local-scale impact of climate change on rice yield and soil organic carbon sequestration: a case study in Roi Et Province, Northeast Thailand. Agric Syst 164:58–70. https://doi.org/10.1016/j.agsy.2018.04.001

Aryal JP, Sapkota TB, Khurana R, Khatri-Chhetri A, Jat ML (2019) Climate change and agriculture in South Asia: adaptation options in smallholder production systems. Environ Dev Sustain 20:1–31. https://doi.org/10.1007/s10668-019-00414-4

Ashraf F, Foolad MR (2012) Crop breeding for salt tolerance in the era of molecular markers and marker-assisted selection. Plant Breed 132(1):10–20. https://doi.org/10.1111/pbr.12000

Asseng S, Martre P, Maiorano A, Rotter RP, O'leary GJ, Fitzgerald GJ et al (2019) Climate change impact and adaptation for wheat protein. Glob Chang Biol 64:155–173. https://doi.org/10.1111/gcb.14481

Azad N, Behmanesh J, Rezaverdinejad V, Tayfeh RH (2018) Climate change impacts modeling on winter wheat yield under full and deficit irrigation in Myandoab-Iran. Arch Agron Soil Sci 64:731–746. https://doi.org/10.1080/03650340.2017.1373187

Baduge SK, Thilakarathna S, Perera JS, Arashpour M, Sharafi P, Teodosio B, Shringi A, Mendis PJAIC (2022) Artificial intelligence and smart vision for building and construction 4.0: machine and deep learning methods and applications. Autom Constr 141:104440. https://doi.org/10.1016/j.autcon.2022.104440

Biswas S, Zhang D, Shi J (2021) CRISPR/Cas systems: opportunities and challenges for crop breeding. Plant Cell Rep 40(6):979–998. https://doi.org/10.1007/s00299-021-02708-2

Cai Y, Bandara JS, Newth DA (2016) Framework for integrated assessment of food production economics in South Asia under climate change. Environ Model Softw 75:459–497. https://doi.org/10.1016/j.envsoft.2015.10.024

Cerón-Souza I, Galeano CH, Tehelen K, Jiménez HR, González C (2021) Opportunities and challenges to improve a public research program in plant breeding and enhance underutilized plant genetic resources in the tropics. Gene 12(10):1584. https://doi.org/10.3390/genes12101584

Chang YN, Zhu C, Jiang J, Zhang H, Zhu JK, Duan CG (2020) Epigenetic regulation in plant abiotic stress responses. J Integr Plant Biol 62(5):563–580. https://doi.org/10.1111/jipb.12901

Chen H, Liang Z, Liu Y, Jiang Q, Xie S (2018) Effects of drought and flood on crop production in China across 1949–2015: spatial heterogeneity analysis with Bayesian hierarchical modeling. Nat Hazards 92:525–541. https://doi.org/10.1007/s11069-018-3216-0

Chun JA, Li S, Wang Q, Lee WS, Lee EJ, Horstmann N et al (2016) Assessing rice productivity and adaptation strategies for Southeast Asia under climate change through multi-scale crop modeling. Agric Syst 143:14–21. https://doi.org/10.1016/j.agsy.2015.12.001

Cravero A, Sepúlveda S (2021) Use and adaptations of machine learning in big data—applications in real cases in agriculture. Electronics 10:552. https://doi.org/10.3390/electronics10050552

Din MSU, Mubeen M, Hussain S, Ahmad A, Hussain N, Ali MA et al (2022) World nations priorities on climate change and food security. In: Building climate resilience in agriculture. Springer, Cham, pp 365–384. https://doi.org/10.1007/978-3-030-79408-8_22

Duan Y, Edwards JS, Dwivedi YK (2019) Artificial intelligence for decision making in the era of Big Data–evolution, challenges and research agenda. Int J Inf Manag 48:63–71. https://doi.org/10.1016/j.ijinfomgt.2019.01.021

Duque AS, de Almeida AM, da Silva AB, da Silva JM, Farinha AP, Santos D, Fevereiro P, de Sousa Araújo S (2013) Abiotic stress responses in plants: unravelling the complexity of genes and networks to survive. In: Abiotic stress-plant responses and applications in agriculture, vol 13. IntechOpen, pp 49–101

FAO IFAD, and WFP (2015) The state of food insecurity in the world: meeting the 2015 international hunger targets: taking stock of uneven progress. FAO, IFAD and WFP, Rome

FAO (2016) Food and Agriculture Organization of the United Nations. The State of Food and Agriculture 2016, 978-92-5-109374-0, https://www.fao.org/family-farming/detail/en/c/447860/

Frisch M, Melchinger AE (2005) Selection theory for marker-assisted backcrossing. Genetics 170(2):909–917. https://doi.org/10.1534/genetics.104.035451

Geng X, Wang F, Ren W, Hao Z (2019) Climate change impacts on winter wheat yield in Northern China. Adv Meteorol 4:11–12. https://doi.org/10.1155/2019/2767018

Ghaffar A, Habib MHR, Ahmad S, Ahmad I, Khan MA, Hussain J et al (2022) Adaptations in cropping system and pattern for sustainable crops production under climate change scenarios. CRC Press, Boca Raton, p 10. https://doi.org/10.1201/9781003286417-1

Gouldson A, Colenbrander S, Sudmant A, Papargyropoulou E, Kerr N, McAnulla F et al (2016) Cities and climate change mitigation: economic opportunities and governance challenges in Asia. Cities 54:11–19. https://doi.org/10.1016/j.cities.2015.10.010

Gupta BK, Sahoo KK, Ghosh A, Tripathi AK, Anwar K, Das P, Singh AK, Pareek A, Sopory SK, Singla-Pareek SL (2018) Manipulation of glyoxalase pathway confers tolerance to multiple stresses in rice. Plant Cell Environ 41(5):1186–1200. https://doi.org/10.1111/pce.12968

Harfouche AL, Jacobson DA, Kainer D, Romero JC, Harfouche AH, Mugnozza GS, Moshelion M, Tuskan GA, Keurentjes JJB, Altman A (2019) Accelerating climate resilient plant breeding by applying next-generation artificial intelligence. Trends Biotechnol 37:1217–1235. https://doi.org/10.1016/j.tibtech.2019.05.007

Hussain J, Khaliq T, Ahmad A, Akhter J, Asseng S (2018) Wheat responses to climate change and its adaptations: a focus on arid and semi-arid environment. Int J Environ Res 12:117–126. https://doi.org/10.1007/s41742-018-0074-2

Hussein HAA, Alshammari SO, Kenawy SK, Elkady FM, Badawy AA (2022) Grain-priming with L-arginine improves the growth performance of wheat (*Triticum aestivum* L.) plants under drought stress. Plan Theory 11:1219. https://doi.org/10.3390/plants11091219

IPCC (2019) Global warming of 1.5°C. Summary for policy makers. World Meteorological Organization, United Nations Environment Program, and Intergovernmental Panel on Climate Change, Bern

Iqbal N, Sehar Z, Fatma M, Umar S, Sofo A, Khan NA (2022) Nitric oxide and abscisic acid mediate heat stress tolerance through regulation of osmolytes and antioxidants to protect photosynthesis and growth in wheat plants. Antioxidants 11:372. https://doi.org/10.3390/antiox11020372

ISAAA Briefs (2017) Global status of commercialized biotech/GM crops in 2017: biotech crop adoption surges as economic benefits accumulate in 22 years, p 53. www.isaaa.org/resources/publications/briefs/53

Jeong JS, Kim YS, Redillas MC, Jang G, Jung H, Bang SW, Choi YD, Ha SH, Reuzeau C, Kim JK (2013) OsNAC5 overexpression enlarges root diameter in rice plants leading to enhanced drought tolerance and increased grain yield in the field. Plant Biotechnol J 11(1):101–114. https://doi.org/10.1111/pbi.12011

Jorasch P (2020) Potential, challenges, and threats for the application of new breeding techniques by the private plant breeding sector in the EU. Front Plant Sci 11:582011. https://doi.org/10.3389/fpls.2020.582011

Joshi R, Wani SH, Singh B, Bohra A, Dar ZA, Lone AA, Pareek A, Singla-Pareek SL (2016) Transcription factors and plants response to drought stress: current understanding and future directions. Front Plant Sci 7:1029. https://doi.org/10.3389/fpls.2016.01029

Joshi R, Sahoo KK, Tripathi AK, Kumar R, Gupta BK, Pareek A, Singla-Pareek SL (2018) Knockdown of an inflorescence meristem-specific cytokinin oxidase—OsCKX2 in rice reduces yield penalty under salinity stress condition. Plant Cell Environ 41(5):936–946. https://doi.org/10.1111/pce.12947

Khare T, Shriram V, Kumar V (2018) RNAi technology: the role in development of abiotic stress-tolerant crops. In: Biochemical, physiological and molecular avenues for combating abiotic

stress tolerance in plants. Academic, pp 117–133. https://doi.org/10.1016/B978-0-12-813066-7.00008-5
Kheir AM, El Baroudy A, Aiad MA, Zoghdan MG, El-Aziz MAA, Ali MG et al (2019) Impacts of rising temperature, carbon dioxide concentration and sea level on wheat production in North Nile delta. Sci Total Environ 651:3161–3173. https://doi.org/10.1016/j.scitotenv.2018.10.209
Liu Y, Chen Q, Ge Q, Dai J, Qin Y, Dai L et al (2018) Modelling the impacts of climate change and crop management on phenological trends of spring and winter wheat in China. Agric For Meteorol 248:518–526. https://doi.org/10.1016/j.agrformet.2017.09.008
Lv Z, Liu X, Cao W, Zhu Y (2013) Climate change impacts on regional winter wheat production in main wheat production regions of China. Agric For Meteorol 171:234–248. https://doi.org/10.1016/j.agrformet.2012.12.008
Meena HP, Bainsla NK, Yadav DK (2016) Recent advances in plant stress physiology, Breeding for abiotic stress tolerance in crop plants, vol 1. Daya Publishing House, New Delhi, pp 329–379
Melchinger AE, Gumber RK (1998) Overview of heterosis and heterotic groups in agronomic crops. Concept Breed Heterosis Crop Plants 1(25):29–44. https://doi.org/10.2135/cssaspecpub25.c3
Mittler R (2006) Abiotic stress, the field environment and stress combination. Trends Plant Sci 11(1):15–19. https://doi.org/10.1016/j.tplants.2005.11.002
Montesinos-López OA, Montesinos-López A, Crossa J, Gianola D, Hernández-Suárez CM, Martín-Vallejo J (2018) Multi-trait, multi-environment deep learning modeling for genomic-enabled prediction of plant traits. G3 Genes Genomes Genet 8:3829–3840. https://doi.org/10.1534/g3.118.200728
Myers SS, Smith MR, Guth S, Golden CD, Vaitla B, Mueller ND et al (2017) Climate change and global food systems: potential impacts on food security and undernutrition. Annu Rev Public Health 38:259–277. https://doi.org/10.1146/annurev-publhealth-031816-044356
Neeraja CN, Maghirang-Rodriguez R, Pamplona A et al (2007) A marker-assisted backcross approach for developing submergence-tolerant rice cultivars. Theor Appl Genet 115(6):767–776. https://doi.org/10.1007/s00122-007-0607-0
Noori MS, Taliman NA (2022) Effect of drought stress and chemical fertilizers on wheat productivity and grain and quality. Plant Prod Gene 3:43–56. https://doi.org/10.34785/J020.2022.800
Nutan KK, Singla-Pareek SL, Pareek A (2020) The Saltol QTL-localized transcription factor OsGATA8 plays an important role in stress tolerance and seed development in Arabidopsis and rice. J Exp Bot 71(2):684–698
Obermeier WA, Lehnert LW, Kammann CI, Muller C, Grunhage L, Luterbacher J et al (2017) Reduced CO_2 fertilization effect in temperate C_3 grasslands under more extreme weather conditions. Nat Clim Chang 7:137–148. https://doi.org/10.1038/nclimate3191
Pathirana R (2011) Plant mutation breeding in agriculture. Perspect Agric Vet Sci Nutr Nat Resour 6(32):107–126. https://doi.org/10.1079/PAVSNNR20116032
Pingali PL (2012) Green revolution: impacts, limits, and the path ahead. Proc Natl Acad Sci USA 109(31):12302–12308. https://doi.org/10.1073/pnas.0912953109
Pradhan A, Naik N, Sahoo KK (2015) RNAi mediated drought and salinity stress tolerance in plants. Am J Plant Sci 6:1990–2008. https://doi.org/10.4236/ajps.2015.612200
Rakszegi M, Darko E, Lovegrove A, Molnár I, Lang L, Bed OZ et al (2019) Drought stress affects the protein and dietary fiber content of wholemeal wheat flour in wheat/Aegilops addition lines. PLoS One 14:e0211892. https://doi.org/10.1371/journal.pone.0211892
Rao S, Balyan S, Jha S, Mathur S (2020) Novel insights into expansion and functional diversification of MIR169 family in tomato. Planta 251(2):55. https://doi.org/10.1007/s00425-020-03346-w
Rasul G, Neupane N, Hussain A, Pasakhala B (2019) Beyond hydropower: towards an integrated solution for water, energy and food security in South Asia. Int J Water Resour Develop 17:1–25. https://doi.org/10.1080/07900627.2019.1579705
Ray DK, Mueller ND, West PC, Foley JA (2013) Yield trends are insufficient to double global crop production by 2050. PLoS One 8(6):e66428. https://doi.org/10.1371/journal.pone.0066428

Raza A, Razzaq A, Mehmood SS, Zou X, Zhang X, Lv Y, Xu J (2019) Impact of climate change on crops adaptation and strategies to tackle its outcome: a review. Plan Theory 8(2):34. https://doi.org/10.3390/plants8020034

Ren S, Qin Q, Ren H (2019) Contrasting wheat phenological responses to climate change in global scale. Sci Total Environ 665:620–631. https://doi.org/10.1016/j.scitotenv.2019.01.394

Rezaei EE, Siebert S, Huging H, Ewert F (2018) Climate change effect on wheat phenology depends on cultivar change. Sci Rep 8:48–91. https://doi.org/10.1038/s41598-018-23101-2

Roy B, Noren SK, Mandal AB, Basu AK (2011) Genetic engineering for abiotic stress tolerance in agricultural crops. Biotechnol J 10(1):1–22

Rutger JN, Peterson ML, Hu CH, Lehman WF (1976) Induction of useful short stature and early maturing mutants in two japonica rice (*Oryza sativa* L.) cultivars. Crop Sci 16:631–635. https://doi.org/10.2135/cropsci1976.0011183X001600050008x

Shah L, Ali A, Yahya M, Zhu Y, Wang S, Si H et al (2018) Integrated control of fusarium head blight and deoxynivalenol mycotoxin in wheat. Plant Pathol 67:532–548. https://doi.org/10.1111/ppa.12785

Shim JS, Oh N, Chung PJ, Kim YS, Choi YD, Kim JK (2018) Overexpression of OsNAC14 improves drought tolerance in rice. Front Plant Sci 9:310. https://doi.org/10.3389/fpls.2018.00310

Singh AK, Kumar M, Choudhary D, Aher L, Rane J, Singh NP (2018) RNAi approach: a powerful technique for gene function studies and enhancing abiotic stress tolerance in crop plants. Biotech Crop Improv 2:113–127. https://doi.org/10.1007/978-3-319-90650-8_5

Soda N, Gupta BK, Anwar K, Sharan A, Singla-Pareek SL, Pareek A (2018) Rice intermediate filament, OsIF, stabilizes photosynthetic machinery and yield under salinity and heat stress. Sci Rep 8(1):1–3

Stevenson JR, Villoria N, Byerlee D, Kelley T, Maredia M (2013) Green revolution research saved an estimated 18 to 27 million hectares from being brought into agricultural production. Proc Natl Acad Sci USA 110(21):8363–8368. https://doi.org/10.1073/pnas.1208065110

Takagi H, Tamiru M, Abe A, Yoshida K, Uemura A et al (2015) MutMap accelerates breeding of a salt-tolerant rice cultivar. Nat Biotechnol 33(5):445–449. https://doi.org/10.1038/nbt.3188

Tian B, Yu Z, Pei Y, Zhang Z, Siemann E, Wan S et al (2019) Elevated temperature reduces wheat grain yield by increasing pests and decreasing soil mutualists. Pest Manag Sci 75:466–475. https://doi.org/10.1002/ps.5140

Wang M, Mao Y, Lu Y, Tao X, Zhu JK (2017) Multiplex gene editing in rice using the CRISPR-Cpf1 system. Mol Plant 10(7):1011–1013. https://doi.org/10.1016/j.molp.2017.03.001

Wang Y, Wang L, Zhou J, Hu S, Chen H, Xiang J et al (2019) Research progress on heat stress of rice at flowering stage. Rice Sci 26:1–10. https://doi.org/10.1016/j.rsci.2018.06.009

Wasaya A, Yasir TA, Sarwar N, Mubeen K, Rajendran K, Hadifa A et al (2022) Climate change and global rice security. In: Modern techniques of rice crop production. Springer, Singapore, pp 13–26. https://doi.org/10.1007/978-981-16-4955-4_2

Xiong W, Reynolds M, Xu Y (2022) Climate change challenges plant breeding. Curr Opin Plant Biol 70:102308. https://doi.org/10.1016/j.pbi.2022.102308

Xu Y, Liu X, Cao X, Huang C, Liu E, Qian S, Liu X, Wu Y, Dong F, Qiu C-W (2021) Artificial intelligence: a powerful paradigm for scientific research. Innovations 2:100179. https://doi.org/10.1016/j.xinn.2021.100179

Yang Z, Zhang Z, Zhang T, Fahad S, Cui K, Nie L et al (2017) The effect of season-long temperature increases on rice cultivars grown in the central and southern regions of China. Front Plant Sci 8:1908 1927. https://doi.org/10.3389/fpls.2017.01908

Younis A, Siddique MI, Kim CK, Lim KB (2014) RNA interference (RNAi) induced gene silencing: a promising approach of hi-tech plant breeding. Int J Biol Sci 10(10):1150–1158. https://doi.org/10.7150/ijbs.10452

Younis A, Ramzan F, Ramzan Y, Zulfiqar F, Ahsan M, Lim KB (2020) Molecular markers improve abiotic stress tolerance in crops: a review. Plan Theory 9(10):1374. https://doi.org/10.3390/plants9101374

Yu H, Li J (2021) Short- and long-term challenges in crop breeding. Natl Sci Rev 8(2):nwab002. https://doi.org/10.1093/nsr/nwab002

Yu H, Zhang Q, Sun P, Song C (2018) Impact of droughts on winter wheat yield in different growth stages during 2001–2016 in Eastern China. Int J Disaster Risk Sci 9:376–391. https://doi.org/10.1007/s13753-018-0187-4

Zafar SA, Hameed A, Nawaz MA, Wei MA, Noor MA, Hussain M (2018) Mechanisms and molecular approaches for heat tolerance in rice (*Oryza sativa* L.) under climate change scenario. J Integr Agric 17:726–738. https://doi.org/10.1016/S2095-3119(17)61718-0

Zahra N, Wahid A, Hafeez MB, Ullah A, Siddique KHM, Farooq M (2021) Grain development in wheat under combined heat and drought stress: plant responses and management. Environ Exp Bot 188:104517. https://doi.org/10.1016/j.envexpbot.2021.104517

Zhao K, Tao Y, Liu M, Yang D, Zhu M, Ding J et al (2022) Does temporary heat stress or low temperature stress similarly affect yield, starch, and protein of winter wheat grain during grain filling? J Cereal Sci 103:103408. https://doi.org/10.1016/j.jcs.2021.103408

Climate Change and Agroecosystems: The Unseen Consequences on Microbes and Soil Microbial Diversity

3

Suruchi Rai, Barsha Datta, Sayasta Ahmed, Nishma Dahal, and Rakshak Kumar

Abstract

Climate change is the most prevalent issue in the last decades and is likely to accelerate due to anthropogenic factors. One of the most prominent contributors to climate change is the agricultural sector, which includes the release of greenhouse gases, large volumes of organic agro-waste, and deforestation, to name a few. The industry itself is also heavily impacted, as plants are sensitive to slight environmental changes, which include increases in temperature and CO_2, drought, altered nutrient availability in the soil, and xenobiotics. The impact of climate change and unregulated agroecosystems can be categorized as direct and indirect, particularly concerning microbial diversity. The consequences include changes in environmental conditions that affect the soil microbiome of crops, ultimately decreasing productivity and yield. Microbes are essential for nutrient acquisition from the soil, maintaining soil health and biodiversity, and helping in biocontrol along with bioremediation of the soil. The microbes not only fulfil the criteria of a biofertilizer but also increase resistance against diseases by maintaining a healthy soil microbiome. Biofertilizers in agroecosystems are increasingly growing as they put less strain on the soil microbial community than synthetic fertilizers. Recent technological advancements like metaproteomics, metatranscriptomics, metabolomics, and novel gene-altering methods have significantly improved our understanding of the role of the microbiome in crop improvement and development. Rhizosphere engineering has shown promising

S. Rai · B. Datta · S. Ahmed · R. Kumar (✉)
Department of Molecular Biology & Bioinformatics, Tripura University (A Central University), Agartala, Tripura, India
e-mail: rakshakkumar@tripurauniv.ac.in

N. Dahal
Biotechnology Division, CSIR-Institute of Himalayan Bioresource Technology Palampur, Palampur, Himachal Pradesh, India

U. Pankaj et al. (eds.), *Plant-microbiome Interactions for Climate-resilient Agriculture*, https://doi.org/10.1007/978-981-96-3534-4_3

results in improving crop yields and productivity. However, a deeper understanding of dynamics in microbial diversity and plant–microbe interaction in agroecosystems is essential to keep up with the changing climate. Advancing research in this field, with a focus on the role of microbial diversity in enhancing plant growth and resistance against diseases using advanced technologies, is crucial in developing sustainable agricultural practices that are better equipped to adapt to an evolving climate.

Keywords

Climate change · Agroecosystem · Microbial diversity · Microbe · Biofertilizers

1 Introduction

Plants, as primary producers in an ecosystem, play a pivotal role in supplying energy to the entire ecosystem through the process of photosynthesis. Plants interact with various microorganisms, facilitating them with their growth, development, defence against pathogens, ecology, and diversity. However, the plant–microbe interaction is not only limited to natural ecosystems, but scientists have developed new ways to replicate this mechanism in agroecosystems (Adeleke et al. 2019). An agroecosystem can be defined as an ecosystem designed and managed manually, in which biotic and abiotic interactions occur within itself, primarily to produce food, fuel, or fibre (Fieding 2022). Microbes play a crucial role in the functioning of an ecosystem, though they are rarely considered to be individual models of ecosystems. Along with various methodological limitations, the overwhelming density of microbes is one of the main reasons for this gap. It is estimated that microbial diversity in a few grams of soil is estimated to range from thousands to millions of microbial species, and its impact on the ecosystem is widely unknown (Torsvik and Øvreås 2002). Recent research focuses on the role of microbial communities in soil fertility and crop productivity across various agroecosystems. Microbes can improve soil health through various biologically mediated processes by degrading organic residues, nutrient mineralization, and also the formation of soil aggregates (Haynes 1999).

Anthropogenic processes directly accelerate the impact of climate change, which affects plant and microbial diversity. For example, agricultural expansion and the use of pesticides and other intensive approaches have altered the main purpose of agriculture, which is the well-being of the people. Monoculture of agricultural products like rubber, tea, etc., is further known to alter soil–microbe interactions and reduce soil fertility by changing the key soil components like micronutrients, macronutrients, and organic matter. Climate change causes alterations in weather patterns and also increases the frequency of extreme weather events, which ultimately change the soil environment. Studies have shown how greenhouse gases such as CO_2, CH_4, NO_X, and water vapour, which are major contributors to greenhouse gas emissions and global temperature rise, impact the soil–microbe interactions, which in turn pose a serious threat to global food security. In a terrestrial ecosystem, plants and microbes are important biological components that undergo several types of interactions, including commensalism, competition, mutualism, and parasitism. Generally, microorganisms have a well-established symbiotic relationship with all the land-dwelling plants (Brundrett 2009). It is observed that a few alterations in the temperature and precipitation over a long period were small and steady, but even small variations in climatic conditions will have a significant influence on the soil health. Therefore, modern-day agricultural techniques currently face two important issues, which are mitigation and adaptation. These involve addressing the impacts of climate change while also developing sustainable methods to overcome harmful effects on food quantity and yields. These problems can be mitigated by responsibly using natural resources and reducing the volume of waste and pollutants (Singh et al. 2019).

Soil supports a variety of chemical, physical, and biological processes that are carried out in the soil only under very specific climatic and geographical conditions. It is well known for its ability to offer ecosystem services. In recent years, the concept of "soil health" has gained increasing attention (Guo 2021). The "soil microbiome" consists of bacteria, fungi, archaea, and protozoa, which predominantly exist in the topsoil and soil that makes up most of the ecosystems in the world. It also has abiotic components, such as water-saturated air, water-filled pores, and potential hotspots for bacteria (Kuzyakov and Blagodatskaya 2015; Kumar et al. 2020; Mukhia et al. 2022). Maintaining a healthy ecosystem in an agroecosystem is a complex process that not only depends on the availability of nutrients but also on various environmental factors. A healthy agroecosystem requires a well-balanced interaction between the biotic and abiotic factors. This chapter begins by highlighting the types and role of plant–microbe interaction in maintaining soil health and enhancing plant growth. It then explores the profound effects of climate change on plant–microbe interaction, including shifts in CO_2 levels, temperature, elevated drought, altered soil carbon and nitrogen availability in soil, and the influence of xenobiotics. Understanding these changes is necessary to reduce the effects of climate change on the soil microbiome. Further, it explains the impacts of harnessing the soil microbiome to enhance agroecosystems amidst climate change by promoting plant growth, managing nutrients in the soil, increasing plant resilience against pathogens, bioremediation, raising crop production and food security, and promoting sustainable ecosystem services. Finally, we discuss established approaches for the management of agroecosystems, enhancing plant–microbe interactions for climate resilience and sustainability, and its prospects in the future.

2 Plant–Soil Microbe Interactions and Their Susceptibility to Climate Change Effects

Numerous studies have emphasized the critical role of microbiomes in maintaining soil nutrient equilibrium and the assimilation of nutrients, making them easily accessible to plants. Additionally, these microbes also enhance the resilience of plants to various stressors (Singh et al. 2019). Climate change and shifting weather patterns have become pressing global concerns among researchers and agriculturalists today. This inevitable phenomenon is primarily driven by a combination of natural and anthropogenic activities over recent decades, including factors such as the greenhouse effect, deforestation, urbanization, and global warming. Climatic shifts lead to unpredictable events like excessive greenhouse gas emissions and rising air temperatures that dry out soil and reduce its moisture content, resulting in drought conditions and altered precipitation patterns across the globe (Singh et al. 2019). Such environmental changes pose a significant threat to agroecosystems by disrupting soil fertility, reducing water availability, and increasing the incidence of crop diseases and pest outbreaks, ultimately jeopardizing food security (Rosenzweig et al. 2013). The direct impact of climate change on plants encompasses morphological, physiological, and phenotypic alterations, along with variations in plant productivity.

The intricate relationship that exists between plants and the microorganisms that live in their rhizosphere and other plant parts is often referred to as plant–microbe interactions. Plant–soil microbe interactions are vital for maintaining soil health, plant growth, and ecosystem sustainability. These interactions can be neutral, antagonistic, or symbiotic. The structure of microbial communities affects plant growth, selection traits, and flowering timing (Lau and Lennon 2011). Nevertheless, little is known about how the environment influences plant–microbe interactions, especially at the molecular and mechanistic levels. This gap presents new opportunities for exploring the intricate relationship between plants and microbes in nature, which is vital for maintaining ecological resilience. Moreover, understanding how plant–soil microbe interactions respond to climate change is a research priority because of their significance in defining ecosystem properties and will help to clarify crucial ecosystem functions like soil carbon storage and net primary efficiency (Classen et al. 2015).

2.1 Types of Plant–Soil Microbe Interaction

Plant–soil microbe interactions can be categorized into symbiotic, antagonistic, and neutral associations. This section mainly focuses on symbiotic interactions, where both the plant and the microbe mutually benefit from their relationship. The symbiotic associations between plants and microbes are shown in Fig. 3.1.

Symbiotic associations are primarily classified into two types: mycorrhizal associations and rhizobial associations. Mycorrhizal associations are crucial symbiotic associations between plants and fungi. The term mycorrhiza, which means "fungus-root", was coined by Albert Bernhard Frank in 1885, who recognized that these interactions are essential for the nutrition of both partners. Any plant that harbours a fungus can be referred to as a "host". Mycorrhizal fungi within plants can be referred to by a variety of names, including symbiont, associate, mycobiont, inhabitant, etc., but it is typically enough to simply call them fungi (Brundrett 2009). Morphologically, two major types of mycorrhizal associations are there,

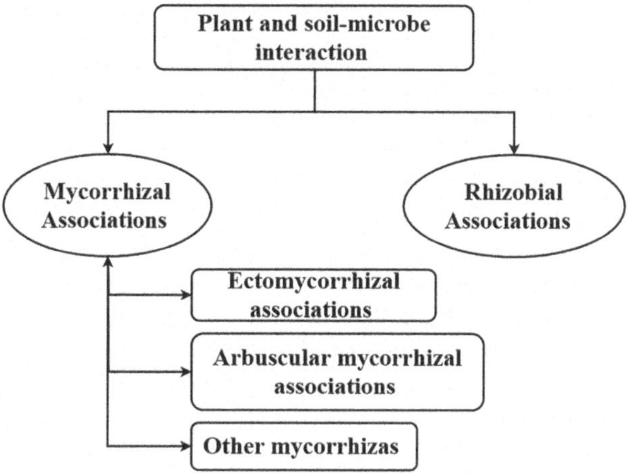

Fig. 3.1 Types of plant–soil microbe interaction: The figure highlights two major types of plant–microbe interactions: mycorrhizal associations (fungi-based) and rhizobial associations (bacteria-based). Mycorrhizal associations are further broken down into ectomycorrhizal associations, arbuscular mycorrhizal associations, and other mycorrhizas

i.e., vesicular-arbuscular mycorrhizal associations (VAM) and ectomycorrhizal associations (ECM). Vesicular-arbuscular mycorrhizal associations are also called arbuscular mycorrhizal associations (AM). The name arbuscular mycorrhiza has become increasingly popular because certain fungi do not generate vesicles in roots. Arbuscular mycorrhizas are widespread in both temperate and tropical ecosystems, including agricultural systems, and are composed of bryophytes, pteridophytes, gymnosperms, and angiosperms. Arbuscular mycorrhizas are sometimes referred to as "endomycorrhizas" due to the intraradical structures formed by the fungi inside the plant roots. Arbuscules, or coils, serve as the interface between plant and fungal tissues in AM associations, facilitating the exchange of materials between plant and fungal symbionts. Modifications of the fungal hyphae, such as arbuscules and coils, provide a larger surface area for the exchange of resources. Additionally, some fungi produce vesicles within roots that serve as storage (Johnson and Gehring 2007). Ectomycorrhizas are common in temperate and boreal forests and are found in certain woody plants, such as gymnosperms (e.g., Pinaceae) and angiosperms (e.g., Betulaceae). Ectomycorrhizas surround the roots of host plants with a hyphal network known as the Harting net, which helps in resource exchange and changes root architecture and root hair development (Johnson and Gehring 2007). In addition to these, other types of mycorrhizal associations, such as ericoid mycorrhizas, show distinct characteristics, though more research is needed to fully understand their structure and function, particularly in orchids (Brundrett 2009).

Rhizobial associations involve *Rhizobium*, a Gram-negative soil bacterium that forms an endosymbiotic association with the roots of legume plants. In this association, both organisms benefit from each other. *Rhizobium* colonizes within the root nodules of plants and helps in nitrogen fixation by converting the atmospheric nitrogen (N_2) into ammonia (NH_3) and other nitrogenous compounds with the aid of the enzyme nitrogenase (D'Souza et al. 2023). *Rhizobium* acts like a natural fertilizer in the soil, resulting in the stimulation of plant growth, called plant growth-promoting rhizobacteria (PGPR). PGPRs are categorized based on their beneficial properties. For example, biofertilizers fix nitrogen deficiency in soils, phyto-stimulators promote plant growth through hormone production, and biocontrol agents protect plants from pathogenic infections. Large-scale use of PGPRs can reduce reliance on chemical fertilizers and pesticides, minimizing environmental pollution. Recent studies also show evidence that rhizobia, in addition to nitrogen fixation in leguminous plants, can also promote the growth of non-leguminous plants, though further understanding of their effects on these plants is needed (Mehboob et al. 2009).

Soil acts as a reservoir for microbial populations, with plants interacting with a wide range of microbes and other entities present in their environment. Factors associated with global change, such as increased CO_2, rising temperatures, higher nitrogen levels, and changes in precipitation, as well as wetting-drying cycles and drought, significantly influence microbial communities by altering soil properties and conditions that affect plant performance (Toju et al. 2018). Because of the variations in the physiological responses, growth rates, and temperature sensitivity of the various members of soil microbial communities, climate change can also influence the makeup and operation of these communities. At the field level, these communities are crucial for supporting plant health and nutritional requirements, as well as for the efficient functioning of agroecosystems. The microbes also aid in the degradation of xenobiotics, the decomposition of organic materials, and the generation of soil aggregates (see Fig. 3.2). Additionally, the cycling of carbon and nitrogen in ecosystems depends on these communities (Microbiology Online 2015). Both biotic and abiotic parameters, such as the quality and the quantity of the waste, as well as temperature and moisture content, affect the activity of the microbial communities.

As an abiotic stressor, global warming can potentially modify the properties of soil and the diversity of soil microbial communities, thereby influencing the range of processes they facilitate (Shade et al. 2012). Research has revealed that the complex changes in soil bacterial composition and diversity along altitudinal gradients are influenced by various factors, including temperature. Any alterations in plant–microbe interactions might exert a major impact on the composition of plant communities and the functioning of an ecosystem because soil microbes are essential for nutrient transformation, providing nutrients to plants, facilitating neighbour coexistence, and controlling plant populations. By changing the distribution of soil microbial communities within the soil profile, plants can adapt to climate-induced stress. This reshuffling of soil communities and their interactions could further influence plant–microbe-process dynamics (Classen et al. 2015). There have been reports of beneficial as well as adverse impacts on soil microbial communities in response to variations in atmospheric carbon dioxide (CO_2) levels. Variations in root-shoot phenology can also result from changes in soil characteristics; these changes might impact the dynamics of the rhizosphere and even the periodic

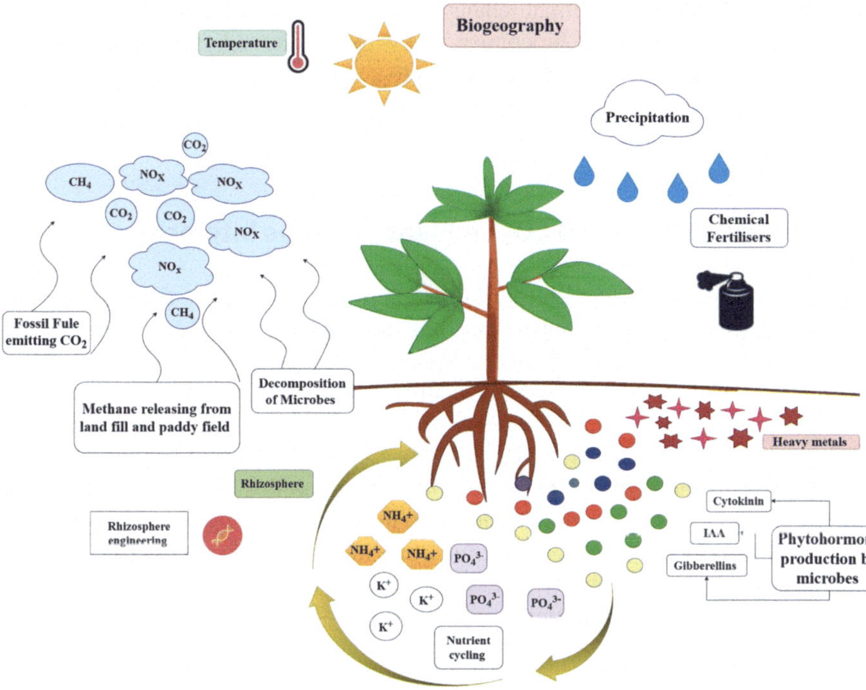

Fig. 3.2 Overview of climate change impacts on the rhizosphere and the role of beneficial microbes in mitigating the challenges: Microbes help in enhancing nutrient cycling (NPK), producing phytohormones, facilitating heavy metal bioremediation, degrading xenobiotics, and improving plant resilience against climate stress, such as elevated temperature, elevated CO_2, altered precipitation, and chemical stressors

composition of soil microbial groups. The microbial community shifts in favour of species that are better suited for warmer heat and faster growth rates as a result of elevated microbial activity, processing, and turnover brought on by rising temperatures. Furthermore, warming modifies the physiological makeup of decomposers, which affects soil CO_2 outflow, further contributing to climate change feedback loops (Mekala and Polepongu 2019).

The role of microorganisms in climate change is very critical as they produce and consume the three primary gases, such as methane (CH_4), carbon dioxide (CO_2), and nitrous oxide (N_2O), which are responsible for 98% of the increased global warming. The effects of abiotic variables on decomposition rates are influenced by microbial decomposers, which control important stages in the process of decomposition (Classen et al. 2015). Climate-induced changes may also have an impact on the makeup and organization of microbes, which may affect the frequency and severity of diseases occurring in plants. The continuous coevolution involving plants and disease-causing pathogens is impacted by these climate-induced alterations in one way or another. The ability of several microorganisms to provide crops resistance against both abiotic and biotic stress is well known. However, it has been observed that the ongoing climate changes have led to the outbreaks of new diseases, which may disrupt the biocontrol pathways mediated by microbes and can potentially reduce the ability of the microbes to control diseases (Singh et al. 2019). Given these challenges, it is crucial to thoroughly investigate the effects of the changing climate on various aspects of plant–microbe interactions.

2.2 Effects of Elevated CO_2 on Plant–Microbe Interaction

Elevated atmospheric CO_2 concentrations significantly affect the interactions between plants and microorganisms and the structure of microbial groups in the soil. In comparison to bulk soil, plants frequently undergo increased photosynthesis and growth as CO_2 levels rise, which results in increased root exudation and the discharge of organic compounds into the rhizosphere. This, in turn, raises microbial abundance and their action in the rhizosphere region (Kapoor and Mukerji 2006). The composition and activity of soil microbial communities are influenced by root exudates and hence are considered important in plant–microorganism interactions. However, the altered root exudate profiles may also favour the growth of certain pathogenic microbes, potentially increasing the risk of diseases in crops. The composition of soil microbial communities in the rhizosphere region may shift because of these changes, which could affect microbial functions such as decomposition of organic matter and nutrient cycling (Fierer et al. 2009). For example, microbial species that are more adept at utilizing elevated carbon inputs may become more dominant, altering soil processes like nutrient mineralization and organic matter turnover. As a result, microorganisms that are typically beneficial to plant growth and health may experience reduced efficacy in their desired functions and colonization capacity under specific conditions. Higher levels of atmospheric CO_2 can lead to soil microbes releasing greenhouse gases, including nitrous oxide and methane. Additionally, microbial respiration increases as a consequence of the enhanced plant productivity driven by elevated CO_2, which provides an increased carbon substrate for soil microbes (De Graaff et al. 2006).

High carbon availability can be advantageous for beneficial microorganisms like mycorrhizal fungi and nitrogen-fixing bacteria, which can potentially improve nutrient uptake by plants and enhance stress resistance. Under conditions of elevated CO_2, nutrients like nitrogen (N) may become limiting, which would require higher fertilizer inputs in agriculture. Microorganisms like mycorrhizae or nitrogen-fixing bacteria that promote plant development and help with nutrient acquisition may become more significant under these situations. These findings suggest that elevated CO_2 could influence plant–fungal symbiosis, potentially increasing the frequency of endophyte infections, but with less production of toxins (Brosi et al. 2009). Additionally, CO_2 affects both the host plant and its endosymbionts and might lead to changes in the plant's carbohydrate content. Most of the research has demonstrated that under high CO_2 concentrations, plants benefit from the presence of plant-associated microbes. The combined effects of elevated CO_2 and other environmental variables could lead to complex, often unpredictable changes in ecosystem functioning, emphasizing the need for a deeper understanding of how plant–microbe dynamics may evolve under future climate conditions (Van der Heijden et al. 2008).

2.3 Effects of Elevated Temperature on Plant–Microbe Interaction

Temperature variations are mostly accompanied by variations in soil moisture, which can help explain a few of the inconsistencies observed in research findings. Since soil microorganisms and the processes they support are temperature-sensitive, therefore global factors like rising temperatures have a direct impact on the respiration rates of soil microbes (Mekala and Polepongu 2019). For instance, in a temperate forest, a 5 °C increase in temperature caused changes in the distribution and abundance of soil bacteria, which raised the microbial community's bacterial-to-fungal ratio (DeAngelis et al. 2015). It is anticipated that higher temperatures will hasten the decomposition of fungi, increasing the amount of CO_2 released from the soil. Higher temperatures do, however, also increase nitrogen concentration in soil, which may slow the rate of fungal decomposition. Microbial diversity and their activity are likely to decline with increased nitrogen availability (American Society for Microbiology 2008). Conversely, biochemical reactions in bacteria under the stress of warmer climates become less efficient, and as a result, microbes emit more CO_2 rather than converting a significant portion of it into biomass. The increased levels of CO_2 released through this and other processes stimulate the release of nitrous oxide and methane from plants, which in turn contributes to the feedback loop of climate change (see Fig. 3.3). Studies have shown that rising temperatures generally speed up metabolic processes in microorganisms, which increases microbial activity as well as faster nutrient cycling (see Fig. 3.3). Enhanced microbial activity can lead to an increased rate of organic decomposition, which can affect nutrient availability for plants (see Fig. 3.3). While elevated temperature might benefit plant growth in the short term, long-term variations in the content of soil organic matter and nutrient dynamics leads to nutrient imbalances or soil degradation (DeAngelis et al. 2015).

Changes in microbial activity can subsequently affect decomposition, storage of organic carbon, and nitrogen mineralization. Deposition of nitrogen is expected to reduce mycorrhizal fungal biomass while elevating the biomass of bacteria and saprotrophic fungi. This process could enhance carbon cycling by stimulating microbial enzymes involved in carbon transformation. Furthermore, mycorrhizal associations may change from symbiotic to parasitic as mycorrhizal hyphae are expected to receive more carbon when temperatures increase. These alterations in mycorrhizal-plant associations can trigger changes in soil microbial composition and microbial activity (Nazir et al. 2009).

The rate of microbial activity, processing, and turnover rises with temperature. Consequently, microbial communities shift to accommodate species that are more quickly adapted to higher temperatures (Castro et al. 2009). For instance, the

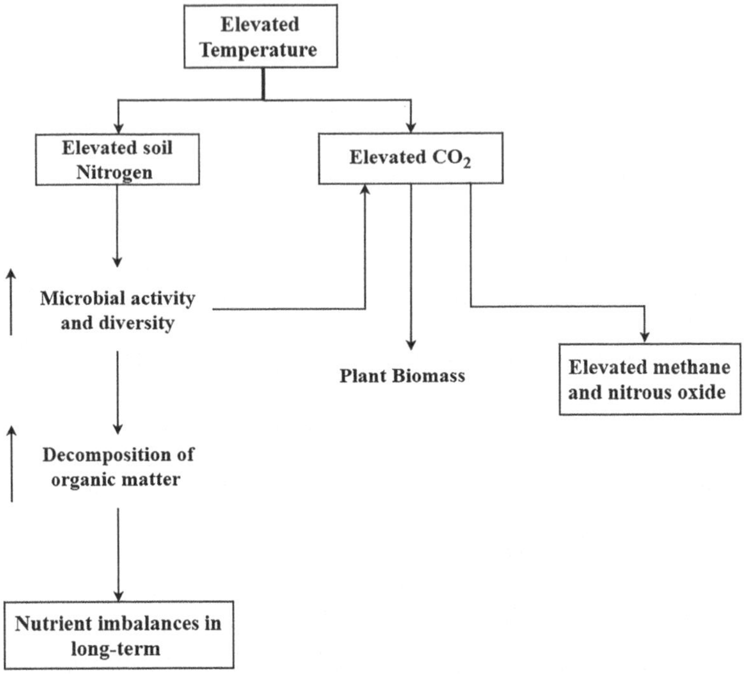

Fig. 3.3 Impact of elevated temperature on soil Nitrogen dynamics, Microbial activity, and Greenhouse gas emissions: Elevated temperatures increase soil nitrogen levels and CO_2, which in turn enhance microbial activity and diversity, leading to faster decomposition of organic matter but can eventually cause nutrient imbalances in the soil. Similarly, elevated CO_2 levels promote plant biomass growth along with increasing methane and nitrous oxide emissions

effects of climatic changes on two important cyanobacteria species in the arid topsoil of the western USA: *Microcoleus vaginatus* and *Microcoleus steenstrupii*. The thermo-tolerant *M. steenstrupii* has been observed to outcompete and even replace the psychrotolerant *M. vaginatus* as global temperatures rise. These microorganisms are essential for preserving the topsoil's microbial population, which is necessary to prevent soil erosion (DiGregorio 2013). Thus, variations in soil microbial populations can be seen as a result of prolonged warming. It should be noted in this context that warming is likely to have an impact on the populations of bacteria as well as fungi. This is crucial because certain microbes control vital ecological processes, including nitrification, nitrogen fixation, denitrification, and methanogenesis, and shifts in their relative abundances have a direct impact on how quickly these processes take place. Additionally, warmer soils can alter the dynamics of soil pathogens, with some studies suggesting that rising temperatures may increase the severity of plant diseases, as certain pathogens become more virulent at higher temperatures (Garrett et al. 2006). Overall, elevated temperature shows complex and multifaceted effects on soil–microbe interactions, which require careful consideration of both direct microbial responses and broader ecosystem implications.

2.4 Effects of Drought on Plant–Microbe Interaction

Drought stress is often a byproduct of climate change, with soil desiccation enhancing O_2 availability and accelerating carbon cycling, ultimately increasing carbon dioxide emissions. Additionally, drought conditions exacerbate the differing temperature sensitivities of bacterial as well as fungal populations (Briones et al. 2014). A significant outcome of drought is the suppression of plant growth in both root systems and aerial structures, leaving plants more vulnerable to attacks by other pathogens. This can alter the distribution of photosynthates within the rhizosphere and affect the development of ectomycorrhizal mycelium (ECM). Soil microbes, particularly those residing in the rhizosphere region, are also directly affected by drought, as reduced root exudation limits the carbon and energy available to microbial communities, thereby impairing beneficial plant–microbe interactions (Mekala and Polepongu 2019).

Meta-analyses have revealed that drought significantly reduces soil microbial diversity on a global scale. Drought negatively impacts plant productivity, ground cover, and litter quality, which further diminishes the decomposition of organic substances by microorganisms. This results in the retention of nitrogen and carbon in the soil, ultimately affecting nutrient availability (Bouskill et al. 2012). Studies show that drought stress often disrupts symbiotic relationships, including those between plants and arbuscular mycorrhizal fungi (AMF) or nitrogen-fixing bacteria. However, drought-tolerant microorganisms, such as specific endophytes, can help enhance plant drought resistance through mechanisms like osmotic adjustment and hormone regulation. Drought can impose selective pressure on microbial communities, favouring drought-resistant species like certain Actinobacteria and *Bacillus* spp., while diminishing the populations of water-dependent microbes such as fungi and gram-negative bacteria (Chowdhury et al. 2019). This microbial shift contributes to a reduction in microbial diversity, potentially affecting soil resilience and its ability to recover after drought events, thus affecting plant health, soil fertility, and overall agroecosystem productivity.

2.5 Impact of Altered Soil Carbon (C) and Nitrogen (N) Availability on Plant–Microbe Interactions

Altered C and N availability in soil can have significant consequences on microbial diversity and the functioning of agroecosystems. Both C and N are key elements for metabolic processes required for microbial growth, and their availability in the soil has a critical role in forming the diversity and composition of microbial groups. The microbial world plays a vital role in carbon cycling and other biogeochemical cycles, and their varying interactions due to changing climatic conditions make it a critical area of study. Projections suggest that atmospheric nitrogen (N) deposition could rise by two to three times over the next century (Yu et al. 2019). A growing body of research indicates that elevated nitrogen levels can directly impact soil microbial biomass and community composition or indirectly influence them by lowering soil pH or reducing the availability of organic carbon. High nitrogen concentrations can exert toxic effects on certain saprophytic microorganisms, leading to a reduction in microbial diversity, which, in turn, decreases soil microbial diversity and biomass on both regional and global scales. Excessive carbon input can lead to an imbalance in microbial communities, with the overrepresentation of fast-growing microbes that outcompete other species, potentially reducing overall microbial diversity (Fierer et al. 2009). Increases in nitrogen availability, often due to anthropogenic activities such as fertilizer application or atmospheric deposition, can lead to an increase in nitrogen-fixing bacteria and nitrifying microorganisms. However, excessive nitrogen input can also cause a shift in microbial communities, favouring nitrogen-utilizing bacteria and fungi while suppressing species that rely on other nutrient sources (Högberg 1997) and can lead to soil acidification, which may harm soil microbial communities, particularly fungi and mycorrhizal symbionts that are crucial for nutrient uptake in plants.

Climate change-induced variations in the soil microbial community may have enduring impacts on plant growth, establishment, and soil carbon balance. The relationships between soil microbes and plants, especially plant–soil feedback processes, are important but are not widely recognized aspects that influence the cycling of carbon and nitrogen in the soil. Significant releases of carbon from soil into the atmosphere can result from even minor temperature changes because carbon cycling mechanisms are highly temperature-sensitive. Carbon derived from plants plays a key role in moderating the temperature sensitivity of soil carbon breakdown.

An important factor in reducing the temperature sensitivity of carbon decomposition in the soil is carbon inputs derived from plants. However, the impact of climatic alterations on soil carbon dynamics is still unclear, particularly in ecosystems undergoing transitions from one state to another (Classen et al. 2015). Several studies have shown that nitrogen-degrading enzymes exhibit lower temperature sensitivity compared to carbon-degrading enzymes. As soils warm, this difference may lead to increased nitrogen limitation, prompting microbes to ramp up the production of nitrogen-degrading enzymes while reducing the production of carbon-degrading enzymes. Although microbes naturally produce greenhouse gases as part of their metabolic processes, recent increases in these gases can be attributed to human activities, which have provided microbes with greater access to carbon and nitrogen, subsequently leading to higher greenhouse gas emissions into the atmosphere (Wallenstein et al. 2011). In agroecosystems, the balance of carbon and nitrogen availability influences soil fertility, plant growth, and overall ecosystem productivity. Changes in microbial diversity due to altered C and N availability can cause shifts in microbial diversity and can further affect the resilience of agroecosystems to stressors such as drought or pest infestations. Therefore, maintaining a balance of these nutrients is necessary for maintaining soil health, optimizing crop production, and promoting environmental sustainability in agricultural systems.

2.6 Xenobiotic Compounds in Agroecosystems: A Climate-Driven Threat

Changing climatic factors, including high atmospheric temperatures, altered rainfall patterns, and the frequent occurrence of extreme weather events, cause significant effects on soil microbial communities within agroecosystems. The result of these shifts and the associated consequences, such as drought, flood, and warming, along with the emergence of new pests, has compelled farmers to rely more heavily on chemical pesticides and herbicides to ensure higher crop yields (González-Alcaraz and van Gestel 2016). Additionally, elevated CO_2 levels can lead to nutrient limitations, particularly nitrogen, which in turn increases the use of fertilizers in agriculture. The alterations in soil microbial processes, properties, and enzymatic activities resulting from pesticide use are critical factors that affect overall soil productivity (Dhuldhaj et al. 2023). The widespread application of chemical pesticides and other human activities has led to the release of xenobiotic compounds into the environment, including the soil (Singh et al. 2019). These pollutants, which result primarily from rapid urbanization, are pervasive in ecosystems, with agriculture being a significant contributor. Xenobiotics are characterized by their increased production, environmental persistence, and their potential biological impacts (Punia et al. 2023). These compounds, which include pesticides, solvents, fuels, alkanes, and polycyclic aromatic hydrocarbons (PAHs), can accumulate in living organisms due to their complex chemical composition, making them difficult to break down. The degradation of xenobiotics can lead to even greater environmental problems, as incomplete breakdown by soil bacteria can produce toxic metabolites that accumulate in the soil (Pande et al. 2020). These pollutants are often not recognized by the metabolic processes of microbes and plants, allowing them to infiltrate ecosystems either in their original forms or as hazardous metabolites. Once introduced into agroecosystems, the movement and persistence of pesticides depend on their chemical nature, the characteristics of the soil, and the microbial and ecological communities present (Punia et al. 2023). Some common sources of pesticides used in agriculture include organochlorine pesticides, polycyclic aromatic hydrocarbons (PAHs), hexachlorocyclohexanes (HCHs), dichlorodiphenyltrichloroethanes (DDTs), and endosulfan. Environmental scientists particularly express concern about organochlorines, as they have a higher potential for causing pollution (Punia et al. 2023). Studies have shown that the overuse of xenobiotics can lead to several unintended consequences, including animal fatalities, degradation of the environment, and disruptions to ecological balance (Li and Jennings 2018). Excessive application of environmentally harmful pesticides, such as organochlorine pesticides (OCPs), can have negative impacts on the soil's chemical, biological, and physical properties. The chemicals particularly regulate the functions of various microbial communities and other organisms within the soil ecosystem, ultimately influencing the productivity of agroecosystems (Pang et al. 2020). Under uncontrolled conditions, xenobiotics can completely disrupt biological nitrogen processes by inhibiting the initial stages of nitrification. Insecticides have been found to interfere with the molecular signals required for the functioning of nitrogen-fixing bacteria, thus reducing their ability to fix nitrogen. Consequently, soils exposed to high pesticide levels may become nitrogen-deficient, leading to a considerable decline in crop productivity over time (Fox et al. 2007).

In summary, the reports that are currently available indicate that the physicochemical characteristics of xenobiotics, their relationship with soil matrices, and their persistence in the soil environment all play a significant role in how they affect the biological properties of soil, including the composition and activities of microbial communities. However, excessive application, improper handling, insufficient understanding of safety measures, and neglect of regulatory guidelines can lead to the accumulation of pesticides in crops, resulting in food contamination and potential risks to consumer health. Overall, the effects of pesticides can be influenced by soil microorganisms, which create opportunities for exploring bio-based remediation of pesticides in agroecosystems (Dhuldhaj et al. 2023). Additionally, environmental stresses such as drought, heat stress, and others can reduce microbial biomass and activity, impairing the soil's capacity to process xenobiotics effectively. As a result, the persistence of these harmful chemicals in the environment may lead to bioaccumulation in crops, posing potential risks to food safety and human health (Fatta-Kassinos et al. 2010).

3 Impacts of Utilizing the Soil Microbiome to Improve Agroecosystem in the Context of Climate Change

Microbes, in the short term, have a minimal effect on a plant's adaptation to climate change, but plants and microbiomes have a long-term relationship in developing adaptation strategies to climate change (Trivedi et al. 2022). Some of the strategies by which the soil microbiome helps improve agroecosystems in terms of climate change include promoting plant growth, raising crop production and food security, managing nutrients in the soil, increasing plant resilience against pathogens, bioremediation, and promoting sustainable ecosystem services (see Fig. 3.4). These strategies are mentioned below.

3.1 Role of Microbes in Improving Crop Yield and Food Security

The population on a global scale is increasing significantly at a concerning rate, and food security has been a huge challenge. Food produced should improve significantly from the available farming land to meet its increasing demands. Moreover, the injudicious use of chemical fertilizers, intensive farming, and technological inputs has led to soil erosion, ultimately decreasing soil fertility. Developing countries lack food security, which puts tremendous pressure on agricultural land (Chen et al. 2011). To build sustainable food production, various mechanisms associated with the interaction between the rhizospheric microbial

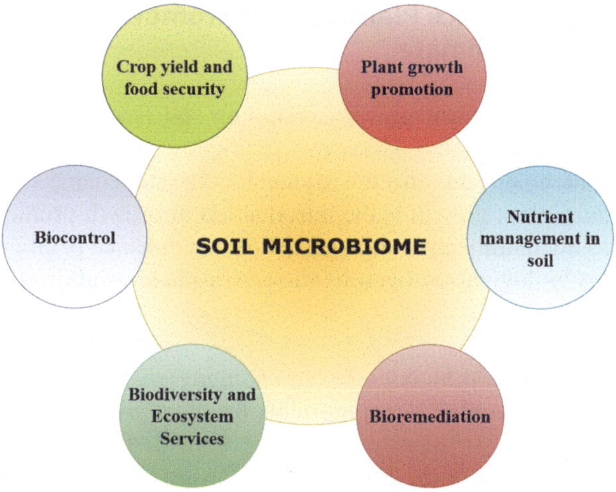

Fig. 3.4 Impacts of harnessing the soil microbiome to enhance agroecosystems amidst climate change: promoting plant growth, managing nutrients in the soil, increasing plant resilience against pathogens, bioremediation, raising crop production and food security, and promoting sustainable ecosystem services

community and plants need to be studied thoroughly. The rate of greenhouse emissions is likely to continue in the future and is predicted to cause changes in the climatic conditions and weather patterns. This will have a detrimental effect on food production due to changes in growth behaviour, disturbing nutrient uptake, and mineralization processes. This is where the microbes have an important role in mitigating climate change through various biogeochemical cycles and regulating mechanisms, improving food production (Sharma et al. 2022). Biological fertility restoration techniques of soil in agroecosystems and other management practices can provide efficient solutions for improving the reducing world food production (Raimi et al. 2017). Recently, there has been a substantial increase in the implementation of products that have microbes, such as biofertilizers, biostimulants in agroecosystems, and also biocontrol agents, but this system lacks proper management practices. In the future, the improvement in bioinoculants can be done only through consistent research and innovations in this field. One of the few ways this can be achieved is by using microbial consortia, endophytic microbes, synthetic microbial communities, and more refined formulation techniques. Ultimately, microbial studies in the future are in the hands of the scientists dealing with the extraction of potential microbiota from indigenous phytomicrobiomes. Though the ongoing research is in steady progress, there is more to understand regarding the vital processes of microbial colonizations in plants, including knowing the molecular signals for engaging with the phytomicrobiome as well as identifying plant-associated microbes concerning the agronomy and climate change (Astapati and Nath 2023).

3.2 Microbes Enhance Plant Growth Promotion in Changing Climate

Several factors influence plant growth, including soil health, atmospheric conditions, and microbial diversity. However, predicting plant growth has become increasingly challenging due to the dynamic climate changes. One effective approach to induce plant growth is the introduction of growth-promoting microbes that can adapt to fluctuating environments. Microbes enhance plant growth through functions such as biological nitrogen fixation, phosphate solubilization, and potassium solubilization.

Biological nitrogen fixation by microbes in changing climatic scenarios: Nitrogen is one of the most essential macronutrients for plant growth, but it is not accessible to most plants. In most cases, Plant Growth Promoting Rhizobacteria (PGPR) fix nitrogen from the atmosphere, converting it into ammonia through the process of nitrogen fixation. Ammonia is a form of nitrogen available to the plant for utilization (Ahemad and Kibret 2013). One of the important applications of nitrogen-fixing microbes is that they act as a better replacement for commercial fertilizer; approximately 50–70 teragrams (Tg) of N per year are used as biological nitrogen fixers (BNF) in agroecosystems (Herridge et al. 2008). In general, biofertilization methods are of two types: one being the utilization of endophytic microorganisms to increase nitrogen fixation in the host plants, and the other is the implementation of free-living microorganisms to make nitrogen available in the soil for the plant. Nitrogen-fixing microbes can be grouped based on five evaluation methods, including acetylene-reduction assays, xylem sap urides, ^{15}N isotope methods, the ratio of the amount of inoculated and uninoculated plants, and changes in the abundance of the *nifH.* gene (a gene that is responsible for nitrogen fixation and a genetic marker for diazotrophs) (Zehr et al. 2003).

In the case of arbuscular mycorrhizal fungi, which spread in the soil with the help of mycelium to the parts where the plant roots cannot access them, these mycorrhizal fungi can synchronize the NO_2 production in the soil with the plant's need for nitrogen. For example, arbuscular mycorrhizal fungi can supply nitrogen to the plant by producing ammonium and reducing NO_2 (Zhang et al. 2022). Nitrogen-fixing bacteria are also present; the most common are *Mesorhizobium*, *Azotobacter*, *Rhizobium*, *Azospirillum*, and *Sinorhizobium*, which are widely recognized for their ability to increase plant growth (Emtiazi et al. 2007). The relevance of biofertilizers has been growing in recent years in contrast with chemical fertilizers, which can cause environmental stress. These measures are a necessary approach in combating climate change and its effect on microbial diversity.

Phosphate solubilization by microbes during climate-induced stress conditions: The plant can mostly uptake soluble forms of phosphorus from the soil. In soil, the phosphate present is inorganically or/and organically bound insoluble forms. Some specialized microbes solubilize inorganically bound phosphate by organic acids (viz., lactic, gluconic, oxalic, malic, etc.). These organic acids decrease the pH of the soil, causing a release of phosphate. A few examples of bacteria responsible for solubilizing phosphate are *Klebsiella, Pseudomonas*, *Enterobacter*, *Azotobacter*, and *Arthrobacter* sp., etc., and some of the fungi that solubilize phosphate are *Aspergillus* sp. (*A. awamori*, *A. niger*, and *A. flavus*), *Penicillium*, and also Vesicular-Arbuscular Mycorrhizal fungi (VAM) to a certain extent. The inorganically bound phosphate is solubilized enzymatically by extracellular enzymes (alkaline and acid phosphatases) released by special microbes. Dissimilar to inorganically bound phosphate, the organically bounded is solubilized enzymatically by extracellular enzymes from specialized microbes (e.g., *Colletotrichum graminicola* and *Aspergillus niger*). These enzymes release phosphate from organic compounds such as inositol phosphates like phytic acid ($C_6H_{18}O_{24}P_6$) (Khan et al. 2014). One of the other mechanisms by which microbes solubilize phosphate is reported in a study where the microbes release enzymes like phytase. This enzyme can release phosphorus (P) bound with organic materials in the soil, particularly from the phytate. As a result, this degradation process of phytate releases P in a form that is available to the plant, as phytate is not directly absorbed. However, the microbes convert it into soluble forms, making phosphorus available to the plant for uptake (Richardson and Simpson 2011).

Microbial contribution to potassium solubilization in changing climatic conditions: After Nitrogen and Phosphorus, Potassium is the third most essential nutrient for the growth of plants, as it maintains the cellular osmotic pressure of the plant and acts as a transport for active compounds inside the cell. Moreover, there are many other functions of Potassium (K), including an increase in the photosynthetic rate that triggers the activation of essential enzymes, enhances nitrogen utilization efficiency, and facilitates the production of sugar and proteins inside the cell (Sparks 1987). Changes in temperature, humidity, and other abiotic stresses have enabled plants to undergo various physiological changes that make it difficult to uptake nutrition from the soil. Therefore, microorganisms become a medium for plants to utilize the available nutrients in the soil. Though potassium is available in the soil, only a portion of it is present in a form that plants can use. Usually, it is found in forms associated with other elements that plants are unable to absorb. Since Potassium is not readily available in soil, it primarily exists in the form of minerals like biotite, muscovite, feldspar, orthoclase, illite, and mica. Some microorganisms can convert these minerals into soluble forms, and they are known as Potassium-solubilizing microorganisms (KSM), which are located in the rhizosphere. These microbes located in the soil produce organic acids that facilitate the dissolution of rocks; in doing so, it enhances the chelation of silicon ions from minerals and the extraction of free potassium ions from the soil (Friedrich et al. 1991).

Some of the microbes that can solubilize potassium include bacteria and fungi such as *Bacillus*, *Burkholderia*, *Pseudomonas, Aspergillus,* and *Acidithiobacillus* spp. (Raghavendra et al. 2016). Potassium is a fundamental mineral responsible for plant growth and is a moderator for signalling, aiding plants to protect themselves from various biotic and abiotic stresses. Therefore, potassium-solubilizing microorganisms are used for plant growth, as they act as biofertilizers for better plant growth and development.

3.3 Microbial Contributions to Phytohormone Production in a Changing Climate

Phytohormones are hormones produced by plants in a very limited amount; phytohormones function as signalling molecules responsible for a major part of the regulatory processes in the plant's growth and development. Cytokinin, gibberellins, and indole acetic acid (IAA) are some of the phytohormones produced by the plants. Many microorganisms help in producing these hormones for the plant, and PGPRs (Plant Growth Promoting Rhizobia) being one of them, such as *Enterobacter cloacae*, *Pseudomonas fluorescens*, and *Paenibacillus polymyxa*, produce Cytokinin. In contrast, *Enterobacter cloacae* and *Azospirillum brasilense* are known to produce growth-promoting hormones like IAA, while gibberellins by other bacteria like *Bacillus pumilus*, *Bacillus cereus*, and *Acetobacter diazotropicus* (Kaymak 2010). The phytohormones produced by the plants are not only for growth but also help them to overcome abiotic stresses. This theory is supported by various findings, one of them being Zaheer et al. (2016), where it was found that cucumber biomass was increased by synthesizing IAA placed in greenhouse conditions. These IAA-producing bacteria were also associated with the roots, which help the plant under drought stress. This type of experiment was also performed on a chickpea plant, inoculated with IAA-producing bacterial species from the genus *Serratia,* which proliferated its grain production. In a similar case, inoculating *Arabidopsis thaliana* with the fungus *Trichoderma virens* showed significant growth in the root hair, lateral roots, and overall root organization. The microbes associated with the roots use hormonal balancing, osmoregulation, and various biochemical processes; this will enhance the metabolic interactions of the host plant with other beneficial microbes and also increase the plant growth. In the case of salt stress, Chickpea was inoculated with bacteria that produce IAA, including *Mesorhizobium ciceri* and *Bacillus subtilis*, which elevated nodule formation and shoot biomass. Similarly, cytokinin-producing bacteria, which include *Azospirillum*, *Arthrobacter*, *Bacillus*, and *Pseudomonas*, increased root and shoot biomass along with cytokinin and proline production in soybeans (Egamberdieva et al. 2017).

Therefore, these findings demonstrate that the production of Phytohormones by microbial colonies will improve plant growth and development even under certain biotic and abiotic stresses with changing climate.

3.4 Significance of Microbes in Soil Nutrient Management Amid Changing Climate

The effects of climate change on soil fertility and nutrient management in soil are poorly understood, which makes it difficult to understand the potential effects of climate change and nutrient acquisition by plants from the soil. The agricultural sector heavily depends upon the application of new and improved fertilizers to maximize plant productivity worldwide; this practice is not only costly, but also the consistent and injudicious use of synthetic fertilizers puts a huge pressure on the soil, decreasing its fertility (Gyaneshwar et al. 2002). It is impossible to imagine a world where microbes are non-existent, and all the nutrients are locked inside the organic wastes. After a particular duration, none of the nutrients will be available to living organisms, including human beings, which will make the world inhabitable for us. Microorganisms have their uses from soil to our plates. Due to the presence of these minute living beings, soil is considered to be living and has a vital role in maintaining nutrient and carbon cycles to prevent them from getting locked up. Microorganisms are involved in nearly every chemical reaction that occurs in soil (Rakshit et al. 2020).

The extensive farming practices have drastically changed agroecosystem functioning, affecting microbial diversity. The disturbances in microbial diversity have led to a decrease in biofertility, mainly caused by improper management of nutrients in soil. This lack of strategy in nutrient management is a major problem for plant growth and ideal soil health (Fernández-Romero et al. 2016). In an experiment, a plant's response to a single microbial culture and a microbial community was evaluated by calculating stable carbon isotope ratios ($\delta^{13}C$) in leaves and moisture content in the soil. The stable carbon isotope was calculated because it helps determine the plant's efficiency in water uptake and its response to climate change. It was observed that species-rich soil had higher moisture content in the top layer of soil near the roots, which suggests the positive role of microbial species diversity and nutrient availability that leads to higher crop yield (Caldeira et al. 2001). Bender and van der Heijden (2014) studied the results of soil microbial communities and their impact on sustainable agricultural practices. It was observed that the microbial community increased the efficiency of nutrient uptake and reduced nutrient leaching in 2 years. The experiment was conducted using two Lysimeters, which were filled with sterilized soil; one was inoculated with beneficial microbial communities, and the other had reduced microbial communities. After 2 years of crop rotation, microbial-rich soil had increased uptake of N and P by plants, also there was reduced nutrient leaching. Therefore, the soil microbiome containing AMF (Arbuscular mycorrhizal fungi) increases nutrient uptake as well as limits nutrient leakage in the soil. Moreover, several microorganisms play a vital role in enhancing the nutrient content in the soil through various biological processes. The nutrient availability in the soil increases the plant health and the adaptation ability of the plant in changing environments. Apart from NPK, there are other organic sources of nutrients that need maintenance to preserve soil fertility. Without the help of beneficial microbes in the soil, it becomes almost impossible to predict the adaptability of plants in a changing climate.

3.5 Role of Microbes in Bioremediation with Changing Climate

Microbes that can volatilize, transform, or immobilize the heavy metals from the soil are known as "bioremediation of heavy metals". Certain types of bacteria are known to eradicate heavy metals from polluted soils, including lead, mercury, and cadmium (Tarfeen et al. 2022). Microorganisms exhibit a remarkable ability to adapt themselves under changing climatic conditions. The microbes are exhibited as agents for bioremediation that can successfully regulate numerous xenobiotic compounds from the soil. Genetic modification through modern high-throughput molecular tools and techniques has enhanced their bioremediation potential (Singh et al. 2019).

Rhizobacteria, in particular, are significantly responsible for heavy metal resistance in soil. For example, the rhizospheric bacteria *Enterobacter cloacae* isolated from *Capsicum annum*, this bacterium exhibited resistance against heavy metals such as Cd^{2+}, Cr^{6+}, Pb^{2+}, and Ni^{2+}. The synthesis of siderophore, 2,3-dihydroxybenzoic acid, salicylic acid, ACC deaminase, and EPS enhanced the bacteria's tolerance. However, this defence mechanism weakens as the concentration of heavy metals increases. This takes place till the tolerance limit has reached its maximum point, beyond which the viability of the cell is lost because of respiratory inhibition. With an increase in heavy metal content, the IAA production of rhizobacteria was significantly reduced (Syed et al. 2021).

Biodiversity and ecosystem functioning (BDEF) research has encountered the exploration of various microbiome diversity to restore contaminated sites. It was found that an increase in the bacterial colony increased the bacterial productivity in a petroleum-contaminated ecosystem, demonstrating the effects of manipulating microbes to improve ecosystem health in contaminated sites (Venail and Vives 2013). However, the success rate of bioremediation also depends on the prior environmental conditions of the contaminated area. Factors such as biotic interactions (e.g., competition, predation) and abiotic factors (e.g., resource availability, physicochemical properties) affect the microbiome diversity–bioremediation relationship. Additionally, manipulating microbiome diversity–bioremediation interactions could significantly enhance the development of effective phytoremediation technologies (Saleem 2015).

3.6 Microbes and Their Importance in Biodiversity and Ecosystem Services in the Face of Climate Change

Microbes are present everywhere and play a crucial role in our existence. They are tiny organisms that are not visible to the naked eye; they range from free-living bacteria to a fungus that can occupy several square kilometres. They are an essential part of our ecosystem; their functions range from supporting sustainable plant productivity by regulating carbon dioxide and nitrogen levels to producing oxygen (50%) and nitrogen (75%) in the atmosphere on a global scale. Some bacteria are also known to remove nitrogen from the atmosphere (Kling 2010). The microbes are part of a larger network in the ecosystem in the wild, and anthropogenic factors that have caused major climatic changes pose a threat to disbalance this equilibrium. Soil microbial diversity is threatened due to the exploitation of chemical fertilizers in the agricultural industries, which has affected its productivity and yield. This has caused a shift in agricultural practices, where the use of biofertilizers and biopesticides is increasing. This is a necessary measure to maintain the soil health. The combination of microbial symbiosis along with sustainable agroecological principles offers a promising opportunity to shift more towards environmentally sound and adaptable agricultural methods. With the effective utilization of the untapped potential of microbes in sustainable farming strategies, agricultural stakeholders can improve soil health, reduce ecological disruptions, and enhance overall agricultural productivity (Nicholls et al. 2017). Therefore, various environmental factors that are responsible for the depletion of microbial diversity have a major impact on biodiversity and ecosystem functioning. Thus, focusing on microbes as a tool for plant growth and development is essential to maintain a balance between sustainable agricultural practices and ecosystem services.

3.7 Role of Microbes in Biocontrol Under Climate Change

Most of the beneficial microbes play a vital role in the process of biocontrol, providing eco-friendly strategies for managing plant diseases and serving as an alternative to harmful chemical pesticides. However, global warming is connected to increasing CO_2, temperatures, and ozone levels, in addition to drought. These factors can affect microbial diversity in the soil, which can lead to the severity of plant diseases. These circumstances can have a direct or indirect effect on the coevolution of the host plant and its pathogen (Garrett et al. 2006). The microbes that are responsible for the process of biocontrol are known as Biocontrol agents (BCAs); BCAs produce various compounds that contribute to their antagonistic properties. These include Knosamine, Mycosubtilin, Zwittermicin, Iturin A, and Bacillomycin D, which are produced by various *Bacillus* species. Other examples are Herbicolin and Gliotoxin, produced by *Pantoea agglomerans* and *Trichoderma virens*, respectively, among many others (Ram et al. 2018). The strategy by which the biocontrol agents use against plant diseases involves applying antagonistic microorganisms before or after the plants get infected by pests or diseases. The infected part and type of infection determine how BCAs can be applied to manage plant disease. Biological control agents can be bacteria, fungi, viruses, or actinomycetes that kill or reduce the microbe's pathogenicity through various modes of action. Numerous bacterial and fungal microbes are known to be utilized as biocontrol agents, along with their growth-promoting and disease management. Several bacterial genera, including *Azoarcus*, *Acetobacter*, *Achromobacter*, *Vibrio*, *Anabaena*, *Arthrobacter*,

Azospirillum, *Azotobacter*, *Bacillus*, *Burkholderia*, *Enterobacter*, *Flavobacterium*, *Phyllobacterium*, *Frankia*, *Hydrogenophaga*, *Microcoleus*, *Clostridium*, *Kluyvera*, *Pseudomonas*, *Serratia*, *Streptomyces*, and *Staphylococcus*, have demonstrated their biocontrol and plant growth potential. The ability of plant growth-promoting rhizobacteria (PGPR) to produce antibiotics that function as a biocontrol trait has been shown in research. *Bacillus subtilis*, *Bacillus amyloliquefaciens*, and *Pseudomonas stutzeri* have significantly suppressed the pathogen *Phytophthora capsici* by colonizing the roots of the cucumber plant during growth (Islam et al. 2016). The fungal biocontrol agents have also demonstrated a significant potential for their beneficial contribution to enhancing crop yield and productivity. These fungal biocontrol agents, like bacteria, are environmentally friendly and do not harm the ecosystem; to date, no report of resistance has developed against fungal BCAs in any of the target pests like insects, weeds, and pathogens. This is because of their complex method of functioning. Thus, they have been proven to be an alternative solution against other chemical pesticides. Examples of fungi exhibiting this potential are *Beauveria bassiana*, *Aspergillus* sp., *Culicinomyces*, *Verticillium*, *Hirsutella*, *Metarhizium*, *Trichoderma*, *Metarhizium anisopliae*, *Nomuraea*, *Paecilomyces*, and *Tolypocladium* (Jyoti and Singh 2016). Climate change is a worldwide phenomenon and has affected the agricultural sector on a large scale. To overcome these challenges, the use of biofertilizers has increased every year. However, various environmental changes have reduced the biocontrol ability of the microbes, making it even more important to explore the plant–microbe interactions through strategic research and technologically advanced demonstrations.

4 Advances in Genetic Engineering and Emerging Technologies: Enhancing Plant–Microbe Interactions for Climate Resilience and Sustainability

Soil health maintenance is one of the key factors for agricultural sustainability and is essential for the productivity of agroecosystems. However, soil resources are currently threatened by anthropogenic activities, including climate change, changes in climatic conditions, added variabilities, and difficulties in agroecosystems, which will compromise their sustainability. Beneficial microbes that are beneficial to plants will enhance plant growth and resistance to various biotic and abiotic stresses. Establishing a link between ecosystem functioning and microbial diversity is essential in understanding how ecosystems respond to environmental changes. Therefore, soil microbial taxa are crucial in the context of global climate change, as they play a pivotal role in biogeochemical cycling, carbon sequestration, and plant growth.

Traditional studies focused on understanding the dynamics of microbial communities and their functioning have employed various techniques, including Denaturing Gel Gradient Electrophoresis (DGGE), Terminal Restriction Fragment Length Polymorphism (TRFLP), Amplified Ribosomal DNA Restriction Analysis (ARDRA), Phospholipid Fatty Acid Analysis (PLFA), etc. These molecular techniques have enabled researchers to comprehend the compositional patterns of microbial communities at a broader level (Gray et al. 2011; Bell-Dereske et al. 2017). Modern genomics shows great potential for identifying uncultured soil microbial diversity. Genomics has revealed that changes in bacterial communities are associated with both sensitive and disease-tolerant plants, which help us to understand how plant–microbial interactions are affected by climate change. Consequently, exploring microbial diversity is the top priority for its beneficial roles in helping plants grow and overcome diseases. Agricultural practices heavily influence the bulk of these beneficial microbes present in the soil. Therefore, by utilizing microbiome engineering techniques like metaproteomics, metabolomics, metatranscriptomics, and novel gene-altering methods, heterogeneous and stable microbial populations in the soil can be effectively filtered, enhancing its vitality and efficiency. Subsequently, cultivation strategies may be enhanced to achieve sustainability. Changing our ways to adopt efficient natural and organic farming practices is a necessity, where the microbes can be effectively utilized to improve soil health, reduce pests and diseases, lower greenhouse emissions, and ultimately prevent environmental degradation (Sudheer et al. 2020). Successful implementation of genetic engineering has been observed in a study done by Wang et al. (2022) when a plant was exposed to salt stress. It was observed that there was a significant change in rhizosphere metabolites and rhizobacterial diversity of Sea Rice 86 (SR86) using transcriptome and genome sequencing analysis. Application of correlation analysis and co-occurrence, the researchers identified key taxa that support the application of particular plant growth-promoting rhizobacteria in managing salty soil. Rhizosphere engineering has surfaced as a practical approach to enhance the cellular stress resistance of microorganisms for sustainable agriculture. Advances in metagenomic analysis have enabled quicker identification of various rhizospheric microorganisms (Babalola et al. 2021). With improving sequencing technologies and the rapidly evolving field of bioinformatics, metagenomic analysis has reshaped our understanding of the structure as well as the function of microbial communities, which is often referred to as the "black box" of nature; by exploring the astounding genetic diversity, it also uncovers new pathways for bioremediation and biogeochemical cycles (Lewis et al. 2010).

As stated earlier, microbes inhabit every part of our environment and play pivotal roles in our agriculture, food, ecosystem functioning, industry, and in cleaning our environment. Therefore, the microbial diversity, functioning, community structure, and stability of these invisible microorganisms are essential for enhancing our comprehension of community dynamics, stability, and the resulting evolution of life on Earth (refer Table 3.1).

5 Future Prospects

There are exciting prospects for the future of plant and soil microbes in terms of agriculture, ecological health, and environmental stability. The importance of microbes in soil health in a changing climate will only increase as our knowledge of plant–soil microbe interactions expands. More potent microbial inoculants and

Table 3.1 Table showcasing various microbial inoculants, their application conditions, and corresponding beneficial roles

Microbial inoculant	Microbiological medium (plant/ substrate)	Condition	Role of the inoculant	References
Micromonospora sp. (PGPB)	*Marrubium vulgare*	Elevated CO_2	It increased biomass and photosynthetic capacity by improving primary and secondary metabolite biosynthesis and promoting antioxidant capacity.	AbdElgawad et al. (2023)
Actinobacterium sp. (PGPB)	*Brassica napus* L.	Elevated CO_2	Higher accumulation of antioxidant enzymes, β-tocopherols, and flavonoids was improved. Additionally, glucosinolates and amino acids were enhanced.	Hagagy and AbdElgawad (2023)
Rhizophagus irregularis (AMF)	*Helianthus annuus* L.	Elevated CO_2	Improved the absorption of N, while also decreasing the plant's oxidative stress	Bellido et al. (2021)
Azolla sp.	*Oryza sativa* (BRRI dhan)	Heat stress	Replacement of 20–40% of inorganic fertilizers (NPKS) and increased rice yield.	Ali et al. (2024)
Pseudomonas sp. and *Bacillus* sp. (PG-F)	*Triticum aestivum* L.	Heat stress	Improved Chlorophyll fluorescence parameters and increased yield	Eisvand et al. (2018)
Priestia megaterium (PGPR)	*Glycine max*	Heat stress	Stimulates plant growth reduces heat stress and improves physiological parameters by regulating phytohormones, antioxidant protection system, and genetic expression	Shaffique et al. (2024)

(continued)

Table 3.1 (continued)

Microbial inoculant	Microbiological medium (plant/substrate)	Condition	Role of the inoculant	References
Rhizoglomus irregulare, *Glomus* sp., *Sclerocystis* sp. and *Acaulospora* sp. (AMF)	*Phoenix dactylifera* L.	Drought stress	Enhanced plant growth, leaf water potential, electrical, and conductivity parameters	Anli et al. (2020)
Funelliformis mosseae	*Cucurbita* pepo *convar.* Pepo var. *Styriaca*	Drought stress	Improve the agronomical traits and will help to enhance the yield under drought stress conditions.	Najafi et al. (2021)
Pseudomonas aeruginosa and *Bacillus firmus*	Soil	Heavy metal stress	Enhance soil health and remediate heavy metal-contaminated soils	Haroun et al. (2023)

biofertilizers will be filtered, which can facilitate nutrient cycling, strengthen soil structure, and boost plant growth, and are anticipated as a result of advances in microbiology. These bio-based solutions can decrease the negative effects on the environment as a better alternative to synthetic pesticides and fertilizers.

Future research can concentrate on the following concepts to better understand and control soil microbial populations, including their possible applications to improve soil health. Additionally, further research is required to investigate the diversity of soil microbial communities and their functions in various habitats. Advanced molecular techniques like metagenomics and metatranscriptomics may be used to characterize and identify specific microbial taxa and their active genes. This will enable farmers to optimize soil health and plant performance based on microbial populations.

Microbes increase resistance in plants to better overcome abiotic stresses like heat, salinity, and drought through the synthesis of protective compounds, inducing systemic resistance, or improving water retention. This will be essential for maintaining food security in the face of climate change. In the future, crop breeding might entail choosing plants that get along better with helpful microorganisms. This may lead to crops that become more resistant to environmental stress and require fewer chemical inputs. Agricultural soils release three potent gases: carbon dioxide, nitrous oxide, and methane. Through sustainable agricultural practices, these emissions can be reduced. Additionally, this approach could stabilize organic carbon and contribute to help lock away carbon.

Scientists might be able to create early diagnostic instruments for controlling plant pathogens by locating microbial biomarkers linked to soilborne illnesses. This would lessen the need for broad-spectrum pesticides by enabling targeted interventions. In contrast to conventional agricultural inputs, researchers may be able to create synthetic microbial consortia communities that cooperate to carry out particular functions or mechanisms that promote plant growth, nutrient cycling, or pest resistance in more effective and sustainable ways. To address particular environmental issues, such as decomposing pollutants in contaminated soils or offering biological control against invasive plant pests and diseases, engineered microbes may be created. Climate change is a global issue that demands international collaboration for effective solutions. Thus, to mitigate its impact on agroecosystems worldwide, it is essential to emerge in cooperative research focused on climate-resilient crops, ecosystem restoration, and sustainable farming practices.

6 Conclusion

Climate change is a worldwide phenomenon that has disrupted various natural processes, including microbial diversity within agroecosystems. The use of chemical fertilizers and various abiotic stressors has significantly damaged soil health. Anthropogenic factors include the emission of Greenhouse gases (GHS), which is a major contributor to the increase in global temperatures, which has altered the plant's natural mechanism, making it more susceptible to pathogens and diseases. These changes in plant dynamics have negatively affected the crop yield and productivity. With the increasing demand for food, maintaining an efficient agroecosystem is a must. Different soils have specific chemical compositions where crops are grown. The addition of chemical fertilizers has allowed it to replicate a pseudo-ecosystem for the plant. However, this method is not only expensive but it also requires extensive maintenance and is time-consuming. This is where biofertilizers play a crucial role. The implantation of binoculars into the soil will not only cater to the plant's specific needs but also increase resistance against diseases. There are naturally occurring beneficial fungi and bacteria that help in various biogeochemical processes. These include mycorrhizal associations and also rhizobial associations. These plant–microbe interactions are affected due to the changing climatic conditions like elevated CO_2, drought, and increasing temperatures. This indirectly causes a change in the availability of nutrients in the soil, along with its impact on xenobiotics. The ways in which the microbes can help the plant overcome these challenges are by increasing the nutrient availability in the soil through nutrient cycling, phytohormone production, bioremediation, and biocontrol. Exploring the microbial populations will enable us to understand the plant–microbe interaction as a more sustainable method to overcome climatic-related challenges, which is crucial for agroecosystems. Advances in genetic engineering and emerging omic technologies have made it assess the impacts of various stressors and improve the microbe's efficiency in enhancing plant growth and resistance. Since climate change is an irreversible phenomenon, it is necessary to implement measures that incorporate microbes to maintain sustainable agricultural practices and improve agroecosystems.

References

AbdElgawad H, Korany SM, Hagagy N, Khanghahi MY, Reyad AM, Crecchio C, Zakri AM, Alsherif EA, Bakkar MR (2023) Biochemical and pharmaceutical traits of *Marrubium vulgare* L. plants treated with plant growth-promoting bacteria and elevated CO_2. 3 Biotech 13(12). https://doi.org/10.1007/s13205-023-03836-0

Adeleke RA, Nunthkumar B, Roopnarain A, Obi L (2019) Applications of plant-microbe interactions in agro-ecosystems. In: Kumar V, Prasad R, Kumar M, Choudhary D (eds) Microbiome in plant health and disease. Springer, Singapore. https://doi.org/10.1007/978-981-13-8495-0_1

Ahemad M, Kibret M (2013) Mechanisms and applications of plant growth promoting rhizobacteria: current perspective. J King Saud Univ Sci 26(1):1–20. https://doi.org/10.1016/j.jksus.2013.05.001

Ali MA, Rabbi SKMF, Baten MA, Hiya HJ, Auyon ST, Rahman MS, Kundu D, Amin K, Barman SC, Hossain T, Nobi FB (2024) Response of rice cultivars to elevated air temperature and soil amendments: implications towards climate change adaptations and mitigating global warming potentials. Am J Clim Chang 13(03):406–426. https://doi.org/10.4236/ajcc.2024.133019

American Society for Microbiology (2008, June 8) Climate change could impact vital functions of microbes. ScienceDaily. Retrieved December 6, 2024, from www.sciencedaily.com/releases/2008/06/080603085922.htm

Anli M, Baslam M, Tahiri A, Raklami A, Symanczik S, Boutasknit A, Ait-El-Mokhtar M, Ben-Laouane R, Toubali S, Rahou YA, Chitt MA, Oufdou K, Mitsui T, Hafidi M, Meddich A (2020) Biofertilizers as strategies to improve photosynthetic apparatus, growth, and drought stress tolerance in the date palm. Front Plant Sci 11. https://doi.org/10.3389/fpls.2020.516818

Astapati AD, Nath S (2023) The complex interplay between plant-microbe and virus interactions in sustainable agriculture: harnessing phytomicrobiomes for enhanced soil health, designer plants, resource use efficiency, and food security. Crop Design 2(1):100028. https://doi.org/10.1016/j.cropd.2023.100028

Babalola OO, Emmanuel OC, Adeleke BS, Odelade KA, Nwachukwu BC, Ayiti OE, Adegboyega TT, Igiehon NO (2021) Rhizosphere microbiome cooperations: strategies for sustainable crop production. Curr Microbiol 78(4):1069–1085. https://doi.org/10.1007/s00284-021-02375-2

Bell-Dereske L, Takacs-Vesbach C, Kivlin SN, Emery SM, Rudgers JA (2017) Leaf endophytic fungus interacts with precipitation to alter belowground microbial communities in primary successional dunes. FEMS Microbiol Ecol 93(6). https://doi.org/10.1093/femsec/fix036

Bellido E, De La Haba P, Agüera E (2021) Physiological alteration in sunflower plants (*Helianthus annuus* L.) exposed to high CO_2 and arbuscular mycorrhizal fungi. Plan Theory 10(5):937. https://doi.org/10.3390/plants10050937

Bender SF, Van Der Heijden MG (2014) Soil biota enhance agricultural sustainability by improving crop yield, nutrient uptake and reducing nitrogen leaching losses. J Appl Ecol 52(1):228–239. https://doi.org/10.1111/1365-2664.12351

Bouskill NJ, Lim HC, Borglin S, Salve R, Wood TE, Silver WL, Brodie EL (2012) Pre-exposure to drought increases the resistance of tropical forest soil bacterial communities to extended drought. ISME J 7(2):384–394. https://doi.org/10.1038/ismej.2012.113

Briones MJI, McNamara NP, Poskitt J, Crow SE, Ostle NJ (2014) Interactive biotic and abiotic regulators of soil carbon cycling: evidence from controlled climate experiments on peatland and boreal soils. Glob Chang Biol 20(9):2971–2982. https://doi.org/10.1111/gcb.12585

Brosi GB, Nelson JA, McCulley RL, Classen AT, Norby R (2009, August) PS 45–40: global change factors interact with fungal endophyte symbiosis to determine tall fescue litter chemistry. In: The 94th ESA annual meeting

Brundrett MC (2009) Mycorrhizal associations and other means of nutrition of vascular plants: understanding the global diversity of host plants by resolving conflicting information and developing reliable means of diagnosis. Plant Soil 320(1–2):37–77. https://doi.org/10.1007/s11104-008-9877-9

Caldeira MC, Ryel RJ, Lawton JH, Pereira JS (2001) Mechanisms of positive biodiversity–production relationships: insights provided by δ13C analysis in experimental Mediterranean grassland plots. Ecol Lett 4(5):439–443. https://doi.org/10.1046/j.1461-0248.2001.00238.x

Castro HF, Classen AT, Austin EE, Norby RJ, Schadt CW (2009) Soil microbial community responses to multiple experimental climate change drivers. Appl Environ Microbiol 76(4):999–1007. https://doi.org/10.1128/aem.02874-09

Chen C, Baethgen WE, Wang E, Yu Q (2011) Characterizing spatial and temporal variability of crop yield caused by climate and irrigation in the North China Plain. Theor Appl Climatol 106(3–4):365–381. https://doi.org/10.1007/s00704-011-0440-x

Chowdhury SP, Babin D, Sandmann M, Jacquiod S, Sommermann L, Sørensen SJ, Fliessbach A, Mäder P, Geistlinger J, Smalla K, Rothballer M, Grosch R (2019) Effect of long-term organic and mineral fertilization strategies on rhizosphere microbiota assemblage and performance of lettuce. Environ Microbiol 21(7):2426–2439. https://doi.org/10.1111/1462-2920.14631

Classen AT, Sundqvist MK, Henning JA, Newman GS, Moore JAM, Cregger MA, Moorhead LC, Patterson CM (2015) Direct and indirect effects of climate change on soil microbial and soil microbial-plant interactions: what lies ahead? Ecosphere 6(8):1–21. https://doi.org/10.1890/es15-00217.1

D'Souza SR, Singh S, Ravi L (2023) Secondary metabolites produced from symbiotic microbes. In: Microbial symbionts: functions and molecular interactions on host. Elsevier eBooks, pp 803–830. https://doi.org/10.1016/b978-0-323-99334-0.00015-3

DeAngelis KM, Pold G, Topçuoğlu BD, Van Diepen LTA, Varney RM, Blanchard JL, Melillo J, Frey SD (2015) Long-term forest soil warming alters microbial communities in temperate forest soils. Frontiers in Microbiology, 6. https://doi.org/10.3389/fmicb.2015.00104

De Graaff M, Van Groenigen K, Six J, Hungate B, Van Kessel C (2006) Interactions between plant growth and soil nutrient cycling under elevated CO_2: a meta-analysis. Glob Chang Biol 12(11):2077–2091. https://doi.org/10.1111/j.1365-2486.2006.01240.x

Dhuldhaj UP, Singh R, Singh VK (2023) Pesticide contamination in agro-ecosystems: toxicity, impacts, and bio-based management strategies. Environ Sci Pollut Res 30(4):9243–9270. https://doi.org/10.1007/s11356-022-24381-y

DiGregorio BE (2013) Climate change affecting microbes in North America soils. Microbe Mag 8(11):436–437. https://doi.org/10.1128/microbe.8.436.1

Egamberdieva D, Wirth SJ, Shurigin VV, Hashem A, Abd-Allah EF (2017) Endophytic bacteria improve plant growth, symbiotic performance of chickpea (*Cicer arietinum* L.) and induce suppression of root rot caused by *Fusarium solani* under salt stress. Front Microbiol 8. https://doi.org/10.3389/fmicb.2017.01887

Eisvand HR, Kamaei H, Nazarian F (2018) Chlorophyll fluorescence, yield and yield components of bread wheat affected by phosphate bio-fertilizer, zinc and boron under late-season heat stress. Photosynthetica 56(4):1287–1296. https://doi.org/10.1007/s11099-018-0829-1

Emtiazi G, Pooyan M, Shamalnasab M (2007) Cellulase activities in nitrogen-fixing Paenibacillus isolated from soil in N-free media. World J Agric Sci 3(5):602–608

Fatta-Kassinos D, Kalavrouziotis I, Koukoulakis P, Vasquez M (2010) The risks associated with wastewater reuse and xenobiotics in the agroecological environment. Sci Total Environ 409(19):3555–3563. https://doi.org/10.1016/j.scitotenv.2010.03.036

Fernández-Romero ML, Parras-Alcántara L, Lozano-García B, Clark JM, Collins CD (2016) Soil quality assessment based on carbon stratification index in different olive grove management practices in Mediterranean areas. Catena 137:449–458. https://doi.org/10.1016/j.catena.2015.10.019

Fieding J (2022) A brief note on agroecosystems. Glob J Agric Res Rev 10(1):1–2. https://creativecommons.org/licenses/by-nc-nd/4.0/

Fierer N, Carney KM, Horner-Devine MC, Megonigal JP (2009) The biogeography of ammonia-oxidizing bacterial communities in soil. Microb Ecol 58(2):435–445. https://doi.org/10.1007/s00248-009-9517-9

Fox JE, Gulledge J, Engelhaupt E, Burow ME, McLachlan JA (2007) Pesticides reduce symbiotic efficiency of nitrogen-fixing rhizobia and host plants. Proc Natl Acad Sci 104(24):10282–10287. https://doi.org/10.1073/pnas.0611710104

Friedrich S, Platonova NP, Karavaiko GI, Stichel E, Glombitza F (1991) Chemical and microbiological solubilization of silicates. Acta Biotechnol 11:187–196. https://doi.org/10.1002/ABIO.370110302

Garrett KA, Dendy SP, Frank EE, Rouse MN, Travers SE (2006) Climate change effects on plant disease: genomes to ecosystems. Annu Rev Phytopathol 44(1):489–509. https://doi.org/10.1146/annurev.phyto.44.070505.143420

González-Alcaraz MN, Van Gestel CA (2016) Toxicity of a metal(loid)-polluted agricultural soil to *Enchytraeus crypticus* changes under a global warming perspective: variations in air temperature and soil moisture content. Sci Total Environ 573:203–211. https://doi.org/10.1016/j.scitotenv.2016.08.061

Gray SB, Classen AT, Kardol P, Yermakov Z, Mille RM (2011) Multiple climate change factors interact to alter soil microbial community structure in an old-field ecosystem. Soil Sci Soc Am J 75(6):2217–2226. https://doi.org/10.2136/sssaj2011.0135

Guo M (2021) Soil health assessment and management: recent development in science and practices. Soil Syst 5(4):61. https://doi.org/10.3390/soilsystems5040061

Gyaneshwar P, Naresh Kumar G, Parekh LJ, Poole PS (2002) Role of soil microorganisms in improving P nutrition of plants. Plant Soil 245:83–93. https://doi.org/10.1023/A:1020663916259

Hagagy N, AbdElgawad H (2023) Rapeseed plant: biostimulation effects of plant growth-promoting Actinobacteria on metabolites and antioxidant defense system under elevated CO_2 conditions. J Sci Food Agric 104(1):51–62. https://doi.org/10.1002/jsfa.12909

Haroun M, Xie S, Awadelkareem W, Wang J, Qian X (2023) Influence of biofertilizer on heavy metal bioremediation and enzyme activities in the soil to revealing the potential for sustainable soil restoration. Sci Rep 13(1). https://doi.org/10.1038/s41598-023-44986-8

Haynes RJ (1999) Size and activity of the soil microbial biomass under grass and arable management. Biol Fertil Soils 30(3):210–216. https://doi.org/10.1007/s003740050610

Herridge DF, Peoples MB, Boddey RM (2008) Global inputs of biological nitrogen fixation in agricultural systems. Plant Soil 311(1–2):1–18. https://doi.org/10.1007/s11104-008-9668-3

Högberg P (1997) 15 N natural abundance in soil-plant systems. New Phytol 137(2):179–203

Islam S, Akanda AM, Prova A, Islam MT, Hossain MM (2016) Isolation and identification of plant growth promoting rhizobacteria from cucumber rhizosphere and their effect on plant growth promotion and disease suppression. Front Microbiol 6. https://doi.org/10.3389/fmicb.2015.01360

Johnson NC, Gehring CA (2007) Mycorrhizas: symbiotic mediators of rhizosphere and ecosystem processes. In: The rhizosphere. Elsevier eBooks, pp 73–100. https://doi.org/10.1016/b978-012088775-0/50006-9

Jyoti S, Singh DP (2016) Fungi as biocontrol agents in sustainable agriculture. In: Microbes and environmental management. Studium Press, New Delhi, pp 172–194

Kapoor R, Mukerji KG (2006) Rhizosphere microbial community dynamics. In: Microbial activity in the rhizosphere, Soil biology. Springer eBooks, pp 55–69. https://doi.org/10.1007/3-540-29420-1_4

Kaymak HC (2010) Potential of PGPR in agricultural innovations. In: Microbiology monographs, pp 45–79. https://doi.org/10.1007/978-3-642-13612-2_3

Khan MS, Zaidi A, Ahmad E (2014) Mechanism of phosphate solubilization and physiological functions of phosphate-solubilizing microorganisms. In: Khan M, Zaidi A, Musarrat J (eds) Phosphate solubilizing microorganisms. Springer, Cham. https://doi.org/10.1007/978-3-319-08216-5_2

Kling G (2010) Microbes: transformers of matter and material. University of Michigan, Ecology and Evolutionary Biology, Ann Arbor. Retrieved from https://sites.lsa.umich.edu/globalchange/global-change-1/lectures/microbes-transformers-of-matter-and-material/

Kumar A, Mukhia S, Kumar N, Acharya V, Kumar S, Kumar R (2020) A broad temperature active lipase purified from a psychrotrophic bacterium of Sikkim Himalaya with potential application in detergent formulation. Front Bioeng Biotechnol 8. https://doi.org/10.3389/fbioe.2020.00642

Kuzyakov Y, Blagodatskaya E (2015) Microbial hotspots and hot moments in soil: concept & review. Soil Biol Biochem 83:184–199. https://doi.org/10.1016/j.soilbio.2015.01.025
Lau JA, Lennon JT (2011) Evolutionary ecology of plant–microbe interactions: soil microbial structure alters selection on plant traits. New Phytol 192(1):215–224. https://doi.org/10.1111/j.1469-8137.2011.03790.x
Lewis K, Epstein S, D'Onofrio A, Ling LL (2010) Uncultured microorganisms as a source of secondary metabolites. J Antibiot 63(8):468–476. https://doi.org/10.1038/ja.2010.87
Li Z, Jennings A (2018) Global variations in pesticide regulations and health risk assessment of maximum concentration levels in drinking water. J Environ Manage, 212, 384–394. https://doi.org/10.1016/j.jenvman.2017.12.083
Mehboob I, Naveed M, Zahir ZA (2009) Rhizobial association with non-legumes: mechanisms and applications. Crit Rev Plant Sci 28(6):432–456. https://doi.org/10.1080/07352680903187753
Mekala S, Polepongu S (2019) Impact of climate change on soil microbial community. In: Plant biotic interactions. Springer eBooks, pp 31–41. https://doi.org/10.1007/978-3-030-26657-8_3
Microbiology Online (2015) Microbes and climate change. http://www.microbiologyonline.org.uk/aboutmicrobiology/microbesandclimatechange. Accessed 15 Dec 2015
Mukhia S, Kumar A, Kumari P, Kumar R (2022) Psychrotrophic plant beneficial bacteria from the glacial ecosystem of Sikkim Himalaya: genomic evidence for the cold adaptation and plant growth promotion. Microbiol Res 260:127049. https://doi.org/10.1016/j.micres.2022.127049
Najafi S, Nasi HN, Tuncturk R, Tuncturk M, Sayyed RZ, Amirnia R (2021) Biofertilizer application enhances drought stress tolerance and alters the antioxidant enzymes in medicinal pumpkin (*Cucurbita pepo* convar. pepo var. *Styriaca*). Horticulturae 7(12):588. https://doi.org/10.3390/horticulturae7120588
Nazir R, Warmink JA, Boersma H, Van Elsas JD (2009) Mechanisms that promote bacterial fitness in fungal-affected soil microhabitats. FEMS Microbiology Ecology, 71(2):169–185. https://doi.org/10.1111/j.1574-6941.2009.00807.x
Nicholls CI, Altieri MA, Vazquez L (2017) Agroecological principles for the conversion of farming systems. In: Agroecological practices for sustainable agriculture: principles, applications, and making the transition. World Scientific, pp 1–18
Pande V, Pandey SC, Sati D, Pande V, Samant M (2020) Bioremediation: an emerging effective approach towards environment restoration. Environ Sustain 3(1):91–103. https://doi.org/10.1007/s42398-020-00099-w
Pang S, Lin Z, Zhang W, Mishra S, Bhatt P, Chen S (2020) Insights into the microbial degradation and biochemical mechanisms of neonicotinoids. Front Microbiol 11. https://doi.org/10.3389/fmicb.2020.00868
Punia A, Singh RP, Chauhan NS (2023) Impact of Xenobiotics under changing climate scenario. In: Xenobiotics in urban ecosystems. Springer eBooks, pp 3–26. https://doi.org/10.1007/978-3-031-35775-6_1
Raghavendra MP, Nayaka SC, Nuthan BR (2016) Role of rhizosphere microflora in potassium solubilization. In: Potassium solubilizing microorganisms for sustainable agriculture. Springer eBooks, pp 43–59. https://doi.org/10.1007/978-81-322-2776-2_4
Raimi A, Adeleke R, Roopnarain A (2017) Soil fertility challenges and biofertiliser as a viable alternative for increasing smallholder farmer crop productivity in sub-Saharan Africa. Cogent Food Agric 3(1):1400933. https://doi.org/10.1080/23311932.2017.1400933
Rakshit A, Singh HB, Singh AK, Singh US, Fraceto L (2020) New frontiers in stress management for durable agriculture. Springer eBooks. https://doi.org/10.1007/978-981-15-1322-0
Ram RM, Keswani C, Bisen K, Tripathi R, Singh SP, Singh HB (2018) Biocontrol technology. In: Omics technologies and bio-engineering. Elsevier eBooks, pp 177–190. https://doi.org/10.1016/b978-0-12-815870-8.00010-3
Richardson AE, Simpson RJ (2011) Soil microorganisms mediating phosphorus availability update on microbial phosphorus. Plant Physiol 156(3):989–996. https://doi.org/10.1104/pp.111.175448
Rosenzweig C, Elliott J, Deryng D, Ruane AC, Müller C, Arneth A, Boote KJ, Folberth C, Glotter M, Khabarov N, Neumann K, Piontek F, Pugh TAM, Schmid E, Stehfest E, Yang H, Jones JW

(2013) Assessing agricultural risks of climate change in the 21st century in a global gridded crop model intercomparison. Proc Natl Acad Sci 111(9):3268–3273. https://doi.org/10.1073/pnas.1222463110

Saleem M (2015) Global microbiome for agroecology, industry, and human well-being: opportunities and challenges in climate change. In: Microbiome community ecology, SpringerBriefs in ecology. Springer, Cham, pp 125–152. https://doi.org/10.1007/978-3-319-11665-5_6

Shade A, Peter H, Allison SD, Baho DL, Berga M, Bürgmann H, Huber DH, Langenheder S, Lennon JT, Martiny JBH, Matulich KL, Schmidt TM, Handelsman J (2012) Fundamentals of microbial community resistance and resilience. Front Microbiol 3. https://doi.org/10.3389/fmicb.2012.00417

Shaffique S, Shah AA, Peter O, Injamum-Ul-Hoque M, Elansary HO, Kang S, Azzawi TNIA, Yun B, Lee I (2024) The rhizobacterial *Priestia megaterium* strain SH-19 mitigates the hazardous effects of heat stress via an endogenous secondary metabolite elucidation network and molecular regulation signalling. BMC Plant Biol 24(1). https://doi.org/10.1186/s12870-024-05534-2

Sharma B, Singh BN, Dwivedi P, Rajawat MVS (2022) Interference of climate change on plant-microbe interaction: present and future prospects. Front Agron 3. https://doi.org/10.3389/fagro.2021.725804

Singh VK, Shukla AK, Singh AK (2019) Impact of climate change on plant–microbe interactions under agroecosystems. In: Climate change and agricultural ecosystems: current challenges and adaptation. Elsevier eBooks, pp 153–179. https://doi.org/10.1016/b978-0-12-816483-9.00007-4

Sparks DL (1987) Potassium dynamics in soils, pp 1–63. https://doi.org/10.1007/978-1-4612-4682-4_1

Sudheer S, Bai RG, Usmani Z, Sharma M (2020) Insights on engineered microbes in sustainable agriculture: biotechnological developments and future prospects. Curr Genomics 21(5):321–333. https://doi.org/10.2174/1389202921999200603165934

Syed A, Zeyad MT, Shahid M, Elgorban AM, Alkhulaifi MM, Ansari IA (2021) Heavy metals induced modulations in growth, physiology, cellular viability, and biofilm formation of an identified bacterial isolate. ACS Omega 6(38):25076–25088. https://doi.org/10.1021/acsomega.1c04396

Tarfeen N, Nisa KU, Hamid B, Bashir Z, Yatoo AM, Dar MA, Mohiddin FA, Amin Z, Ahmad RA, Sayyed RZ (2022) Microbial remediation: a promising tool for reclamation of contaminated sites with special emphasis on heavy metal and pesticide pollution: a review. PRO 10(7):1358. https://doi.org/10.3390/pr10071358

Toju H, Peay KG, Yamamichi M, Narisawa K, Hiruma K, Naito K, Fukuda S, Ushio M, Nakaoka S, Onoda Y, Yoshida K, Schlaeppi K, Bai Y, Sugiura R, Ichihashi Y, Minamisawa K, Kiers ET (2018) Core microbiomes for sustainable agroecosystems. Nat Plants 4(5):247–257. https://doi.org/10.1038/s41477-018-0139-4

Torsvik V, Øvreås L (2002) Microbial diversity and function in soil: from genes to ecosystems. Curr Opin Microbiol 5(3):240–245. https://doi.org/10.1016/s1369-5274(02)00324-7

Trivedi P, Batista BD, Bazany KE, Singh BK (2022) Plant–microbiome interactions under a changing world: responses, consequences and perspectives. New Phytol 234(6):1951–1959. https://doi.org/10.1111/nph.18016

Van Der Heijden MG, Bardgett RD, Van Straalen NM (2008) The unseen majority: soil microbes as drivers of plant diversity and productivity in terrestrial ecosystems. Ecol Lett 11:296–310. https://doi.org/10.1111/j.1461-0248.2007.01139.x

Venail PA, Vives MJ (2013) Positive effects of bacterial diversity on ecosystem functioning driven by complementarity effects in a bioremediation context. PLoS One 8(9):e72561. https://doi.org/10.1371/journal.pone.0072561

Wallenstein MD, Haddix ML, Lee DD, Conant RT, Paul EA (2011) A litter-slurry technique elucidates the key role of enzyme production and microbial dynamics in temperature sensitivity of organic matter decomposition. Soil Biol Biochem 47:18–26. https://doi.org/10.1016/j.soilbio.2011.12.009

Wang G, Weng L, Huang Y, Ling Y, Zhen Z, Lin Z, Hu H, Li C, Guo J, Zhou JL, Chen S, Jia Y, Ren L (2022) Microbiome-metabolome analysis directed isolation of rhizobacteria capable of enhancing salt tolerance of Sea Rice 86. Sci Total Environ 843:156817. https://doi.org/10.1016/j.scitotenv.2022.156817

Yu G, Jia Y, He N, Zhu J, Chen Z, Wang Q, Piao S, Liu X, He H, Guo X, Wen Z, Li P, Ding G, Goulding K (2019) Stabilization of atmospheric nitrogen deposition in China over the past decade. Nat Geosci 12(6):424–429. https://doi.org/10.1038/s41561-019-0352-4

Zaheer A, Mirza BS, Mclean JE, Yasmin S, Shah TM, Malik KA, Mirza MS (2016) Association of plant growth-promoting Serratia spp. with the root nodules of chickpea. Res Microbiol 167(6):510–520. https://doi.org/10.1016/j.resmic.2016.04.001

Zehr JP, Jenkins BD, Short SM, Steward GF (2003) Nitrogenase gene diversity and microbial community structure: a cross-system comparison. Environ Microbiol 5(7):539–554. https://doi.org/10.1046/j.1462-2920.2003.00451.x

Zhang Y, Wang W, Yao H (2022) Urea-based nitrogen fertilization in agriculture: a key source of N2O emissions and recent development in mitigating strategies. Arch Agron Soil Sci 69(5):663–678. https://doi.org/10.1080/03650340.2022.2025588

4 Climate Change and Agroecosystem: How Entomopathogenic Fungi Influence Soil Microbial Diversity and Plant Growth?

Gulafsha Parveen, Anuj Chaudhary, Harjeet Tomar, Priyanka Mishra, Navin Kumar, Vishal Tripathi, Parul Chaudhary, Ram Krishna, and Durgesh Kumar Jaiswal

Abstract

Climate change significantly threatens agricultural productivity, necessitating sustainable pest management strategies. EPFs, such as *Beauveria bassiana* and *Metarhizium anisopliae*, serve as effective biological control agents against insect pests, enhancing soil health and promoting plant growth through improved nutrient cycling and stress tolerance. The chapter highlights how EPF contributes to maintaining soil microbial diversity, which is crucial for ecosystem functionality and resilience against climatic shifts. EPF plays a vital role in sustainable agriculture by suppressing pest populations and facilitating nutrient uptake. The findings underscore the importance of integrating EPF into agricultural practices to address food security challenges while minimizing environmental impacts. Further research is essential to fully understand the ecological roles of EPF in adapting agroecosystems to changing climate conditions.

Keywords

Agriculture · Climate change · Fungi · Microbes · Plant growth

---

G. Parveen · A. Chaudhary · H. Tomar · P. Mishra · N. Kumar · V. Tripathi · D. K. Jaiswal (✉)
Department of Biotechnology, Graphic Era Deemed to be University, Dehradun, Uttarakhand, India

P. Chaudhary
School of Agriculture, Graphic Era Hill University, Dehradun, Uttarakhand, India

R. Krishna
ICAR-Indian Institute of Vegetable Research, Varanasi, Uttar Pradesh, India

U. Pankaj et al. (eds.), *Plant-microbiome Interactions for Climate-resilient Agriculture*, https://doi.org/10.1007/978-981-96-3534-4_4

1 Introduction

Climate change represents the most significant natural threat to agricultural systems globally, fundamentally altering the dynamics of food production and ecosystem health (Praveen and Sharma 2019). The interconnection between climate change and agriculture is critical, particularly in light of the increasing global population and the corresponding demand for food (US EPA 2017). Since the mid-1940s, the application of pesticides has escalated dramatically, primarily due to the rise of commercial farming practices (Elbialy et al. 2021; Umapathi et al. 2022). Over the past 50 years, the release of synthetic chemicals into the environment has outpaced other contributors to environmental change, including greenhouse gases, with pesticides being the most prevalent (Bernhardt et al. 2017). While these chemicals have temporarily boosted agricultural output, their long-term environmental impacts are increasingly concerning (Seibold et al. 2019). Modern agriculture's challenge is to meet the rising food demands while minimizing negative environmental effects (Tilman et al. 2011). Soil hosts diverse microorganisms—fungi, bacteria, and actinobacteria—essential for biogeochemical cycles and ecosystem health (Alves et al. 2020; Frouz et al. 2020). The ecological roles of these microorganisms are increasingly recognized, particularly in soil dominated by fungi (Brandan et al. 2017). The diversity of soil fungi is crucial for maintaining soil health; without it, soil ecosystems would be severely compromised (Chu et al. 2020). Studies indicate that fungal diversity positively correlates with air temperature, suggesting that climate-induced changes in fungal community composition could enhance nutrient cycling and ecosystem productivity (Bryant et al. 2008; Newsham et al. 2016). As agricultural systems grapple with climate change adaptation and mitigation, authorities and farmers must develop integrated pest management strategies that promote sustainable agriculture to ensure food security (Hillocks 2012). Climate change and globalization have significantly increased arthropod pests in agricultural and forestry areas, with insects alone accounting for 10–16% of global crop losses annually (Savary et al. 2019). Climate change may also affect the relationship between insect pests and their natural enemies, which can influence pest distribution and insecticide efficacy (Skendzic et al. 2021). Chemical pesticides have traditionally been employed to protect crops from pests; however, their use raises significant health and environmental concerns. Prolonged exposure can lead to insect pesticide resistance, necessitating stronger chemicals that may adversely affect wildlife and beneficial insects (Chaudhary et al. 2022a). Furthermore, some pesticides persist in the environment, causing long-term ecological harm (Swathy et al. 2024). Biological control methods are emerging as effective alternatives for managing insect pests. Entomopathogenic fungi (EPF), which specifically infect arthropods, play a critical role in pest management within forestry and agriculture. These fungi attack insects through their cuticles—differentiating them from bacterial and viral pathogens that enter through mouthparts or intestines (Dara et al. 2019; de Faria and Wraight 2007). EPF produce secondary metabolites that can serve as nutrients or bioactive compounds with potential medicinal properties (Hu et al. 2016). Notable EPFs, such as *Beauveria bassiana*, *Purpureocillium platinum*, *Metarhizium anisopliae*, and *Isaria fumosorosea*, have been developed as biopesticides for controlling agricultural pests like locusts, aphids, whiteflies, moths, and phytopathogenic nematodes (Fang et al. 2014). Entomopathogenic fungi thrive natively in soil ecosystems, contributing significantly to insect population control (Hallouti et al. 2020). They are typically isolated from soil using selective media or bait-trapping techniques due to their natural occurrence in these environments (Bravo et al. 2016; Bihal et al. 2023). Human reliance on ecosystem services—benefits derived from healthy ecosystems—has been profound over the past century (De Groot et al. 2010). With projections indicating a global population increase of approximately 70% by 2050, understanding how climate change affects agroecosystem productivity becomes increasingly urgent (Sahana et al. 2023). Addressing food security while ensuring safety and quality throughout production and distribution is paramount in this context (Ibrahim et al. 2023). Insect diversity is vast, with nearly one million identified species; however, many negatively impact food production and human health. Herbivorous insects alone account for an estimated 18% loss in global agricultural production (Jankielsohn 2018). Recent studies by Swathy et al. (2024) have isolated and characterized entomopathogenic fungi's pathogenicity against key insect populations while assessing their nontarget effects on soil bioindicator species. Their findings indicate that conidia from *B. bassiana* effectively killed *Halyomorpha halys* and *Tenebrio molitor* within 10 days posttreatment. The study also revealed significant reductions in insect lifespans after treatment with *B. bassiana*, highlighting its potential as a biological control agent. Since climate change increasingly alters agricultural landscapes worldwide, comprehending the function of entomopathogenic fungi in these systems is essential for formulating sustainable strategies that guarantee food security while reducing environmental repercussions.

2 Climate Change and Agroecosystem Dynamic

Climate change adaptation and mitigation represent two of modern agriculture's most pressing challenges (Fig. 4.1). The increasing variability in climate patterns, including temperature fluctuations and altered precipitation regimes, directly impacts agricultural productivity and sustainability. This necessitates the development of innovative solutions by authorities and farmers alike to ensure food security and promote sustainable agricultural practices through integrated pest management (Hillocks 2012). Temperature is a critical factor influencing entomopathogenic fungi (EPF)'s effectiveness in managing microbial pests. Research indicates that thermal stress can significantly affect the entire infection process of EPF, which includes well-known species such as *Beauveria* and *Metarhizium* (Paixao et al. 2019). Both high and low temperatures can inhibit the growth of these fungi, limiting their potential as biological control agents in agricultural systems (Muniz et al. 2020). The relationship between soil organic matter and EPF dynamics is complex. Soils with high organic matter content often harbor a diverse community of microorganisms, including antagonistic species that may compete with EPF for resources. However, these same soils also support a higher density and diversity of insect

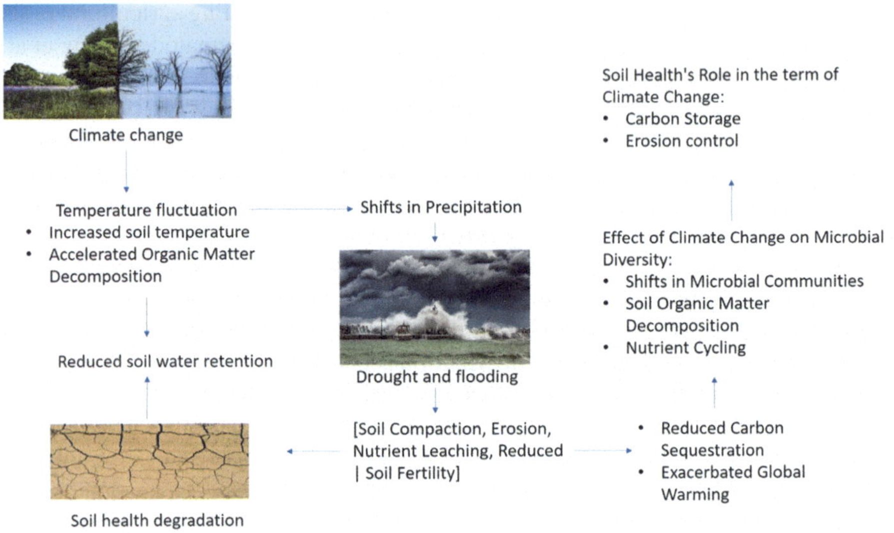

Fig. 4.1 Effect of climate change on agroecosystem

populations, providing a broader range of hosts for EPF to infect (Uzman et al. 2019). The carbon-to-nitrogen ratio in the soil can further influence EPF activity; an optimal balance is crucial for maintaining microbial health and promoting effective pest control. In natural ecosystems, organic matter decomposes through the action of various microorganisms, returning essential nutrients—such as carbon, nitrogen, potassium, phosphorus, calcium, magnesium, sulfur, and other minerals—back into the soil (Chaudhary et al. 2021a, b, 2024). This nutrient cycling is facilitated by a diverse community of soil microorganisms, including actinomycetes, bacteria, fungi, protozoa, and algae (Chaudhary et al. 2021a, b, 2022b). These organisms play vital roles in maintaining soil health and fertility by breaking down organic materials and recycling nutrients (Nannipieri and Smalla 2006; Chaudhary et al. 2022c, 2023).

Despite the recognized importance of soil biological diversity in sustaining productivity within natural and managed agroecosystems, our understanding of microbial community dynamics remains limited. Soil serves as a substrate for root development and a critical source of nutrients and water for plants (Adeleke et al. 2024). The biotic activities of microorganisms contribute to the breakdown of rocks and minerals into soil—a process essential for soil formation. Historically, soil science focused primarily on chemical and physical properties; however, contemporary research acknowledges that biological activity interacts closely with these properties. Soil biodiversity is now understood to be crucial in preserving fertility (Manlay et al. 2007; Khatoon et al. 2024; Bhutani et al. 2024). Indicators encompassing chemical, physical, and biological characteristics, including microbial diversity, are essential for evaluating soil health (Srivastava et al. 2024; Bisht and Chaudhary 2024). Cultivation practices often lead to decreased levels of organic soil matter over time. This reduction negatively impacts crop nutrient availability (Schlesinger and Bernhardt 2013). Sustainable development hinges on robust soil systems vital for producing biomass—including food, feed, fiber, and fuel—and preserving biodiversity by maintaining habitats for various organisms. Healthy soils also contribute to water purification processes while mitigating greenhouse gas emissions through carbon sequestration. Climate change exacerbates existing challenges within agroecosystems. Increased temperatures can shift pest populations and their interactions with natural enemies. For instance, warmer conditions may expand certain insect pests' range while affecting their natural predators' efficacy (Skendzic et al. 2021). This dynamic can create imbalances within ecosystems that further threaten agricultural productivity. Moreover, climate change influences soil moisture levels; droughts can reduce microbial activity, while excessive rainfall can lead to nutrient leaching. Both scenarios compromise soil health and its ability to support plant growth effectively. As agricultural systems adapt to these changes, integrating biological control methods—such as EPF—into pest management strategies becomes increasingly important.

Briefly, tackling the combined difficulties of climate change necessitates new strategies that include biological control techniques, such as entomopathogenic fungi, in agricultural operations. These technologies facilitate sustainable insect population management and improve soil health, thereby ensuring the productivity of agroecosystems in the face of changing climatic conditions. As we advance into an unpredictable future marked by climate variability and population expansion, comprehending the complex interconnections between climate change, agroecosystem dynamics, and soil health will be essential for creating resilient agricultural systems that can satisfy global food requirements while maintaining environmental integrity. Sustainable practices must prioritize enhancing soil biological diversity and functionality to ensure food supply in an era increasingly characterized by climate change problems.

3 Role of Entomopathogenic Fungi in Pest Management

The world's largest engineering environment, the agroecosystem, occupies 38% of the planet's land surface (Faostat 2011) and provides ecosystem services that facilitate human life and growth (Mea 2005). Agroecosystems are vital to the survival of human life, and crop production is one of its primary functions (Li et al. 2022). Human existence has been directly and inherently dependent on the ecosystem's ability to provide products and services, collectively called ecosystem services, for the past century (De Groot et al. 2010). The flow of the ecosystem is made possible by the world's stock of natural resources, such as biodiversity, air, water, and soil, known as natural capital (Mace et al. 2015; Costanza 2020). An agroecosystem is a portion of natural capital directly managed by humans, with biodiversity selected and simplified to maintain the ecosystem by producing food, feed, fiber, lumber, and other goods for markets (Moonen and Bàrberi 2008). Climate change, the primary cause of damage to ecosystem functionality, has seriously compromised the ability of agroecosystems to provide ecosystem services (Leal Filho et al. 2023; Kamboj et al. 2023). For instance, high temperatures brought on by a lack of water can lower primary production and increase plant evapotranspiration, which is essential (Semeraro et al. 2019; Semeraro et al. 2020). According to estimates, the effects of climate change, such as drought and intense rainfall, will hasten soil erosion by a factor of 10–100 compared to the time it takes to form (Masson-Delmotte 2018). Climate change will further disturb the microbial population and interfere with their ability to perform enzymatic tasks in the soil (Chia et al. 2023). Food security mostly refers to the potential to guarantee that everyone has equal access to food and that there will always be enough food for a healthy existence. There is food security when each person has physical and financial access to enough nutritious food that satisfies his or her food preferences and dietary requirements for a healthy and active life (Rinaldi 2023). Thus, if economic activity is the driving force behind climate change, then climate change may diminish the ability of natural capital to sustain the economy, with detrimental consequences, especially for food security, which is critical to human survival (Mirón et al. 2023; Azadi et al. 2022). Since it has been predicted that the global population will expand by 70% by 2050, the consequences of climate change on agroecosystem productivity will be a significant concern (Sahana et al. 2023).

Entomopathogenic fungi, with their potential as biological agents for sustainable pest control, offer a ray of hope in the growing need for healthier agricultural products. These fungi, crucial in controlling the arthropod population, have shown great promise as biological control agents. Their reproductive traits and nutritional needs make them easily mass-produced and prepared for use in crops, further enhancing their potential (Bucarei et al. 2019). The coexistence of insects and fungi for the past 400 million years has led to the evolution of distinct interactions (Araujo and Hughes 2016).

Entomopathogenic fungi play a crucial role in naturally regulating numerous insect pests and pest populations. Various entomopathogenic fungal species are grown commercially as biological controls against insect pests in many regions of the world. These microorganisms infect insects and use them as hosts to complete a stage of their life cycle. The process involves the fungi landing on the host and penetrating their cuticles, which are vital barriers that the fungi must detect and overcome to enter the body cavity (Delgado and Murcia 2011; Zhang et al. 2021). Extensive research has identified the enzymes and proteins involved in breaking down the chitin-rich insect cuticles, with *Metarhizium* and *Beauveria* spp. as a model (Hong et al. 2024).

4 Impact of Entomopathogenic Fungi on Plant Growth

The potential application of EPF in plant management is further enhanced by its capacity to colonize growing plants and perform other roles, such as promoting stress tolerance, plant growth, and improving water uptake, in addition to suppressing pest populations (Sui et al. 2021) (Fig. 4.2). Scientists are becoming

Fig. 4.2 Mechanism of EPF for plant growth and crop production

increasingly interested in studying fungal endophytes because of their functions in biocontrol, promoting plant growth, and possibly serving as a near-term substitute for synthetic pesticides and inorganic fertilizers (Shen et al. 2019). Fungal endophytes are an abundant source of biologically active chemicals, and their secondary products, processing unique structures and possible applications, hold significant value in agriculture, medicine, and industry (Rajamanikyam et al. 2017). They yield a wide spectrum of secondary metabolites necessary for use as antifungal, antibacterial, and insecticidal (Gibson et al. 2014). It has been discovered that mutual relationships among fungal endophytes and their colonized hosts are crucial for host plant nutrition and defense against disease-causing pathogens and insect pests (Tyagi et al. 2022a, b). In addition to receiving nutrition and defense from the host tissue, endophytes actively intake nutrients and shield the host from stress conditions. In general, fungal endophytes affect their hosts' growth, development, and well-being. Additionally, the actions of the current fungal endophytes impact population dynamics, the plant community, and the functioning of the ecosystem (Hardoim et al. 2015) (Table 4.1).

Numerous investigations have demonstrated that endophytic fungus can improve plant nutritional status and increase plant development by influencing hormone synthesis (Bamisile et al. 2021). For instance, *M. robertsii* can establish colonies in plant roots and development and root hair density (Sasan and Bidochka 2012). Additionally, the plant length and newly harvested weight of the wide bean roots and shoots were improved by endophytic colonization by strains of *M. brunneum*, *B. bassiana*, and *Beauveria brongniartii* (Jaber and Enkerli 2017). Likewise, through a mechanism reliant on the auxin hormone, *M. robertsii* (EPF) enhanced the growth parameters of *Arabidopsis* seedlings by promoting the production of lateral root hairs and roots (Harman et al. 2021). Additionally, this fungus reduced the root hair defects in the mutants lacking auxin by triggering an expression of auxin-regulated genes.

Table 4.1 Impact of entomopathogenic fungi on soil microbes in different agroecosystems

Sr. no.	Entomopathogenic fungi	Impact on soil microbes/role of EPF	Agroecosystem/host plant	Study	References
1.	*Beauveria bassiana* and *Metarhizium brunneum*	Report a 75–80% fatality rate in larvae of the leaf-chewing insect (*Spodoptera littoralis*).	Alfalfa, tomato, and melon plants	Temporary endophytic colonizations of plants enhance the efficacy of foliar treatments of mycoinsecticides against plant-eating insects.	Resquín-Romero et al. (2016)
2.	*B. bassiana*	Function as an antifungal against *B. cinerea.*	Chili pepper and Tomato	*B. bassiana* antifungal activity against *Botrytis cinerea* in two solanaceae crops.	Barra-Bucarei et al. (2019a, b)
3.	*L. attenuatum* and *L. longisporum*	Inhibit *Sphaerotheca fuliginea*'s growth and spore generation, which is the cause of cucumber powdery mildew.	Cucumber	*Lecanicillium* species' potential for controlling nematodes, insects, and plant illnesses.	Goettel et al. (2008)
4.	*Metarhizium* spp. and *B. bassiana*	Create plant resistance to the fungus *Fusarium graminearum* and the larvae of *Costelytra giveni.*	Maze	Impact of entomopathogenic fungal coating on maize seed on plant growth and resistance to *Costelytra giveni* and *Fusarium graminearum.*	Rivas-Franco et al. (2019)
5.	*B. bassiana*	Potentially employed as a biocontrol agent to combat a variety of bacteria, particularly G+ve ones.	Microbes	The antimicrobial properties of *Beauveria bassiana*'s volatile and diffusible metabolites.	Camele et al. (2023)
6.	*S. strictum*, *S. zeae*, and *L. lecanii*	Produce antibacterial, insecticidal, and poisonous substances.	Maze seed	Maize seed-associated *Sarocladium* and *Lecanicillium* and their capability to produce specific secondary metabolites.	Błaszczyk et al. (2021)

7.	*I. javanica*	In addition to increasing plant height, they also reduce plant viruses, which lowers the quantity of TYLCV in the insect vector.	Tomato	The functions of viruliferous whitefly entomopathogenic fungal infection in regulating tomato yellow leaf curl virus	Sun et al. (2021)
8.	*M. robertsii* and *M. brunneum*	Protect coffee plants from CLMs and encourage their growth.	Coffee	*Metarhizium*-inoculated coffee seeds encourage plant development and coffee leaf miner biocontrol.	Martins et al. (2024)
9.	*M. rileyi* and *M. anisopliae*	Utilized as a pesticide to combat Fall armyworm (*Spodoptera frugiperda*) larvae.	Maze	The condition of autumn armyworms, *Spodoptera frugiperda*, and their natural enemies in Haveri district's main corn-growing regions.	Darshan and Prasanna (2024)
10.	*B. bassiana* and *M. brunneum*	Encourage plant development, enhance nutrient and water absorption, resist drought or salinity, and boost harvests.	Cabbage	Effect of entomopathogenic fungi on the development, growth, and Well-being of cabbage grown under water stress conditions.	Dara et al. (2017)
11.	*B. bassiana* and *M. brunneum*	Boost the concentrations of P and Na in grains and promote plant development.	Durum wheat	Entomopathogenic fungi's effects on the growth and nutrition of durum wheat in the field.	Gonzalez-Guzman et al. (2021)
12.	*B. bassiana*	Improves the quality of the host plant and eliminates the potato tuber moth.	Potato	The endophytic *Beauveria bassiana* promotes the growth of potato plants while suppressing *Phthorimaea operculella* activity.	Zhang et al. (2022)

(continued)

Table 4.1 (continued)

Sr. no.	Entomopathogenic fungi	Impact on soil microbes/role of EPF	Agroecosystem/host plant	Study	References
13.	*M. anisopliae*, *B. bassiana*, and *V. lecanii*	High mortality rates of *Bemisia tabaci* (whiteflies) were from 80% to 100%.	Tomato	Effects of entomopathogenic fungus on the Egyptian tomato crop's whitefly, *Bemisia tabaci*.	Reyad (2017)
14.	*B. bassiana* and *M. anisopliae*	Decrease the quantity of *Thrips tabaci* Lindeman (onion thrips) eggs and larvae.	Onion	Laboratory, greenhouse, and field studies on the combined effects of entomopathogens on *Thrips tabaci* Lindeman (Thysanoptera: Thripidae).	Gulzar et al. (2021)
15.	*M. robertsii*	Act as a pesticide and use fungal mycelia to supply nitrogen to plant hosts.	Haricot bean and switchgrass	Nitrogen is directly transferred from insects to plants by endophytic insect-parasitic fungi.	Behie et al. (2012)
16.	*B. bassiana*	Demonstrate harmful activity against Bihar hairy caterpillars (*Spilosoma obliqua*) and semiloopers (*Anomis sabulifera*), two jute pests.	Jute plant	*Beauveria bassiana*, a fungal entomopathogen, has been identified as a season-long endophyte in jute (*Corchorus olitorius*).	Biswas et al. (2012)
17.	*V. lecanii*	Provide defense mechanisms in plants against root pathogens (*Pythium ultimum*).	Cucumber	*Verticillium lecanii*, a mycoparasite, is pre-injected into Ri T-DNA-transformed cucumber roots to trigger host defense responses against *Pythium ultimum* infection.	Benhamou and Brodeur (2001)
18.	*B. bassiana*	Defend against *Cosmopolites sordidus*, a banana weevil.	Banana	The endophytic *Beauveria bassiana* on bananas (*Musa* spp.) decreases the damage and fitness of banana weevils (*Cosmopolites sordidus*).	Akello et al. (2008)

19.	*B. bassiana* and *M. anisopliae*	Possess the ability to kill *Rhynchophorus ferrugineus*.	Date palm orchards	*Beauveria bassiana* and *Metarhizium anisopliae* are entomopathogenic fungi that are virulent against *Rhynchophorus ferrugineus* (Olivier), the red palm weevil.	Yasin et al. (2019)
20.	*I. fumosorosea*	Utilized as a biocontrol agent against insect species that feed on plant sap.	Pear	Using the entomopathogenic fungus *Isaria fumosorosea* strain (Ifu13a) to control insects that feed on plant sap.	BuGtI et al. (2018)
21.	*B. bassiana*	Lower the survival rate of *Bemisia tabaci* on plant.	Tomato	Evaluation of the impact of endophytic colonization with *Beauveria bassiana* on the incidence of the illness and its vector, *Bemisia tabaci*, as well as the response of certain tomato cultivars to the tomato leaf curl virus.	El-Deeb et al. (2012)
22.	*M. anisopliae* and *B. bassiana*	Applied as a biocontrol to prevent the larvae of *Agrotis ipsilon*.	Wheat and soybeans	*Agrotis ipsilon* pathogenicity induced by the entomopathogenic fungus *Beauveria bassiana* and *Metarhizium anisopliae*.	Gabarty et al. (2014)
23.	*C. tennuisimum* and *B. bassiana*	Infect larvae, neonates, and eggs of *S. frugiperda* during a pest invasion.	Maize	Efficiency of entomopathogenic fungi on fall armyworm, *Spodoptera frugiperda* larvae feeding performance and immature stages.	Idrees et al. (2021)

(continued)

Table 4.1 (continued)

Sr. no.	Entomopathogenic fungi	Impact on soil microbes/role of EPF	Agroecosystem/host plant	Study	References
24.	*I. fumosorosea*	Serve as a preventative measure against the Asian citrus psyllid, which causes citrus greening disease (*Diaphorina citri*).	Citrus plants	*Isaria fumosorosea* is being locally autoinoculated and dispersed in South Texas to control Asian citrus psyllids.	Moran et al. (2011)
25.	*B. bassiana*	Encouraged the development and production of soybean plants.	Soybean	Impact of endophytic entomopathogenic fungi on the growth and yield of soybeans.	Russo et al. (2019)

Furthermore, by transforming harmful reactive oxygen species toward nontoxic substances, root EPF such as *Trichoderma* spp. could control biochemical pathways and improve plant development and productivity in stressed crops (Harman et al. 2021). It has been demonstrated that endophytic EPF increase plants' resistance to several abiotic and biotic stressors (Sui et al. 2023). Plant illnesses are less common, and the body reacts better to dietary deficits. Endophytic fungi-mediated improvements in the health of plants also, although not directly, boost plants' ability to resist insect pests (Song et al. 2024). (Sui et al. 2021) conducted an experiment on tomato plants under pot and field conditions using *B. bassiana* and concluded that it improved plant growth and resistance toward gray mold.

Liu et al. (2022) studied the fact that entomopathogenic fungi *B. bassiana* and *M. anisopliae* enhanced the growth of maize (*Zea mays*). Plant growth indicators, such as fresh weight, root length, and plant height, were tracked and documented for 35 days to demonstrate that EPF inoculation promoted growth. The ability of *M. anisopliae* and *B. bassiana* to endophytically colonize maize tissue was confirmed by the fungal recovery rate (FRR). By the seventh day, the regaining rates of *B. bassiana* were 100% in leaves and stems.

Wilberts et al. (2023) demonstrated that the growth of sweet pepper plants is influenced by crop cultivar and fungal strain via root injection with entomopathogenic fungi. In addition to promoting plant development and health, entomopathogenic fungi can shield plants from abiotic and biotic stressors. According to the results, the EPF might increase plant development, especially plant weight and canopy area.

Song et al. (2024) studied the effect of endophytic fungi against *Rhopalosiphum padi* and concluded that the application of *R. padi* improved wheat resistance. The findings demonstrated that regardless of *R. padi* infestation, the quantity of elevated differentially expressed genes (DEGs) was greater than that of downregulated DEGs in wheat plants treated with fungal strains. In wheat plants, fungal strains markedly raised the levels of flavone, jasmonic acid, and tannins while decreasing the levels of salicylic acid. The findings indicated that by modifying the phytohormone pathway and secondary metabolite production in wheat plants, endophytic fungus considerably reduced the eating effectiveness and community fitness of *R. padi*.

Endophytic fungi also increase the activity of nutrient transport genes in plants that they colonize, such as *M. robertsii*, which transfers nitrogen from the *Galleria mellonella* (L.) larvae to plants (Behie et al. 2012). In the same way, it was discovered that *B. bassiana*, *M. brunneum*, *M. robertsii*, and *M. guizhouense* enhanced the development of switchgrass (*Panicum* spp.), green beans, wheat, and soybean by supplying these host plants with nitrogen derived from insects (Behie and Bidochka 2014).

5 Effect of EPF on Soil Microbial Diversity

Soil is a crucial reservoir of terrestrial biodiversity, hosting many microorganisms that play essential roles in various biological processes (Jansson and Hofmockel 2020; Guerra et al. 2021). Estimates suggest that a single gram of surface soil can contain up to one billion bacterial cells, approximately 200 million fungal hyphae, trillions of viruses, and tens of thousands of protists (Sokol et al. 2022). This immense microbial diversity supports plant growth and contributes to ecosystems' resilience and functionality. The microbial seed bank present in soil serves as a reservoir for microbial communities that can colonize crops, enhancing their growth and health (Trivedi et al. 2020). Plants provide various microhabitats—such as the phyllosphere, endosphere, and rhizosphere—that facilitate the establishment and proliferation of microbial communities (Knights et al. 2021). Specific members of soil microbiomes can migrate from soil to these plant compartments, forming intricate co-associations with crop hosts (Muller et al. 2016; Cordovez et al. 2019). Over the past 400 million years, plants and their associated microbiomes have co-evolved, leading to complex interactions that enhance stability and functionality within terrestrial ecosystems. This relationship has resulted in the "holobiont," where plants and their microbiomes function as a single ecological unit with evolutionary advantages linked to their interactions (Hassani et al. 2018). The crop microbiome encompasses diverse microorganisms, including beneficial mutualistic organisms like arbuscular mycorrhizal fungi and nitrogen-fixing bacteria, as well as neutral and pathogenic microbes (Cordovez et al. 2019). Understanding these microbial communities' diversity, composition, and functional roles throughout the soil-plant continuum is essential for advancing sustainable agricultural practices.

Soil bacteria play a vital role in ecosystem functionality, particularly in nutrient cycling and organic matter decomposition. Various studies have highlighted the distinct ecological roles of microbial groups within soil environments dominated by fungi (Bradan et al. 2017). The diversity of soil fungi is crucial for maintaining soil health; their activities underpin essential processes that sustain life on Earth (Chu et al. 2020). Entomopathogenic fungi (EPF) are notable for their ability to infect and kill pest arthropods, including numerous insect species. Their versatility as biological control agents stems from their capacity to target various hosts. Historically viewed primarily as agents for pest management (Deka et al. 2021), emerging research indicates that EPF may also enhance soil-plant systems' productivity through various mechanisms. Mechanisms by which EPF affects soil microbial diversity:

1. *Soil Aggregation:* Entomopathogenic fungi contribute to soil structure by producing glycoproteins and polysaccharides resembling glomalin. These compounds enhance soil aggregation, improving water retention and aeration while creating favorable conditions for microbial communities (Li et al. 2022).
2. *Nutrient Cycling:* EPF plays a role in nutrient recycling by breaking down organic matter and facilitating plant nutrient availability. Their interactions with other soil microorganisms can lead to enhanced nutrient cycling processes.
3. *Plant Growth Promotion:* EPF has been shown to support plant growth by inducing systemic resistance in host plants. This induced resistance helps mitigate stress from herbivory and reduces disease incidence, thereby contributing to overall plant health.

4. *Microbial Dynamics:* The application of EPF, such as *Beauveria brongniartii* and *B. bassiana*, has been observed to temporarily influence the structure of soil microbial communities under specific cropping conditions, such as strawberry plantations. These fungi can alter the dynamics and abundance of various soil microbial operational taxonomic units (OTUs) (Canfora et al. 2023).
5. *Management of Soil-Dwelling Pests:* As Brunner et al. (2024) reported, EPF could serve as a promising management strategy for controlling soil-dwelling pests. Their ability to infect multiple insect hosts effectively integrates them into pest management programs.

Entomopathogenic fungi profoundly influence soil microbial diversity via many pathways that improve soil health and foster sustainable agricultural practices. EPF enhances soil structure, promotes nutrient cycling, supports plant growth, and regulates insect populations, thus bolstering the resilience and productivity of agroecosystems. Comprehending these interactions is essential for maximizing the potential of entomopathogenic fungi in sustainable agriculture. In addressing climate change and rising food demand, incorporating biological control technologies such as EPF into agricultural systems will safeguard food security and preserve environmental integrity.

6 Relevant Sustainable Development Goals (SDGs)

Climate change poses unprecedented challenges to global agricultural systems, significantly impacting food security and sustainability. With the world's population projected to reach nearly 10 billion by 2050, the demand for food is escalating at an alarming rate. Research indicates that rising greenhouse gas emissions may lead to a 35% reduction in green vegetable yields (Kumar et al. 2022; Pereira et al. 2022). This decline is exacerbated by extreme weather events, such as droughts and floods, which disrupt traditional farming practices and threaten the stability of food supply chains. The effects of climate change extend beyond mere yield reductions; they also compromise the nutritional quality of crops. For instance, elevated carbon dioxide levels have been linked to decreased protein and mineral content in staple foods like wheat and rice, further aggravating malnutrition issues globally (Smith et al. 2019). This scenario is particularly dire in developing nations, where food insecurity is already prevalent. In response to these challenges, the agricultural sector increasingly turns to sustainable practices that minimize environmental impact while maximizing productivity. One promising avenue is the use of entomopathogenic fungi (EPFs) as biocontrol agents. These fungi offer a natural alternative to chemical pesticides, which have been widely used but often lead to detrimental effects on nontarget species and contribute to environmental degradation (Sinha et al. 2016). EPFs are particularly effective in targeting agricultural pests without harming beneficial insects, thus preserving biodiversity within agroecosystems. By enhancing soil microbial diversity, EPFs also contribute to healthier soil ecosystems, which are crucial for sustainable agriculture (Baron et al. 2019).

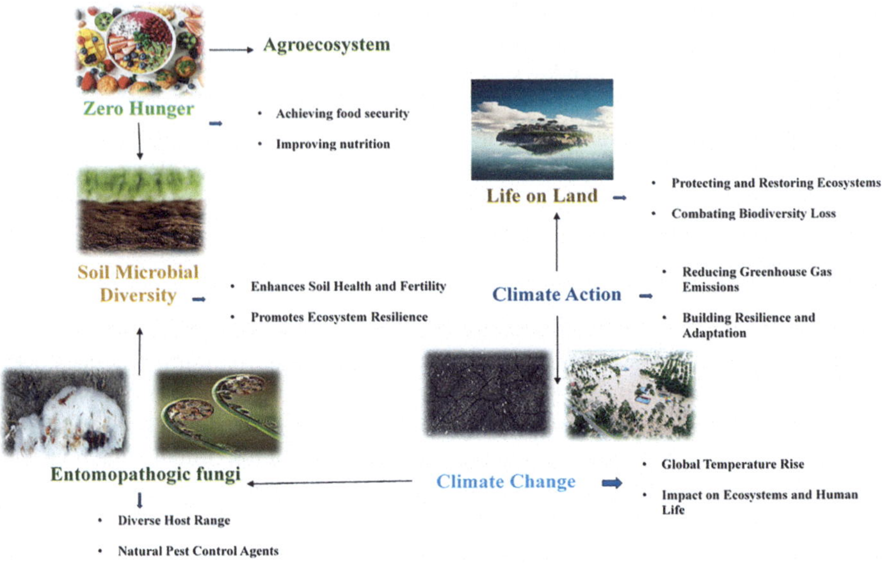

Fig. 4.3 Relationship between climate, EPF, and agroecosystem for sustainable development goals

Moreover, the symbiotic relationship between EPFs and plants can enhance nutrient uptake and improve plant resilience against environmental stressors, making them a valuable tool for farmers facing the pressures of climate change (Litwin et al. 2020). Integrating EPFs into agricultural practices aligns with several Sustainable Development Goals (SDGs), particularly SDG 2: Zero Hunger and SDG 12: Responsible Consumption and Production. By adopting biocontrol agents like EPFs, farmers can reduce their reliance on harmful chemicals while improving crop yields and soil health. This approach addresses immediate pest management issues and contributes to long-term sustainability by fostering resilient agricultural systems capable of withstanding climate variability (Fig. 4.3).

Developing climate-resilient crop varieties that can thrive under extreme weather conditions is critical for maintaining productivity in a changing environment (Sahoo et al., 2024). Moreover, implementing agroecological practices that enhance biodiversity and reduce chemical inputs will support long-term agricultural sustainability. These strategies are vital for ensuring food security while mitigating the adverse effects of climate change on agroecosystems. As we grapple with the complexities of climate change and its impact on agriculture, we must prioritize innovative solutions that promote ecological balance and resilience. The adoption of EPFs represents not just a shift in pest management, but a holistic approach to building sustainable agricultural systems capable of feeding future generations while safeguarding our planet's health. In conclusion, addressing the intertwined challenges of climate change and food security requires a multifaceted strategy that leverages natural solutions like entomopathogenic fungi, alongside technological advancements in crop development and sustainable farming practices. Through collaborative efforts across sectors and disciplines, we can pave the way toward a more resilient agricultural future that meets the needs of a growing population without compromising environmental integrity.

7 Future Research Directions

Recent advancements in microbial biotechnology have paved the way for enhancing the efficacy of entomopathogenic fungi (EPF) in greenhouse and field applications. Various protocols have been established to optimize the use of EPF as biological control agents in agricultural systems (Cuthbertson et al. 2011). One promising approach involves augmentation trials, where crops are treated with fungal conidia suspended in water-based solutions combined with synthetic additives. This method has been shown to significantly improve the effectiveness of EPF against pest populations (Shah and Pell 2003).

Another critical area of research is the exploration of various EPF strains for potential synergistic effects when used alongside chemical pesticides. Studies indicate that combining EPF with conventional chemical pesticides could enhance pest resistance management strategies and significantly reduce the environmental degradation caused by excessive pesticide use (Dara 2019; Vanderberg et al. 1998). This integrated approach holds great promise for more sustainable agricultural practices, reducing reliance on chemical inputs and preserving beneficial soil microbiota. Recent scientific investigations have delved into the ecology of fungal endophytes, focusing on their interactions with plants, herbivores, and other trophic levels within ecosystems (Hartley and Gange 2009).

Understanding these interactions is crucial for harnessing the full potential of EPF in agricultural settings. However, it is also important to consider the ethical implications of using EPF as a pest management tool, particularly regarding its potential impact on nontarget organisms and ecosystem balance. Moreover, advancements in genetic engineering techniques, such as CRISPR and RNA interference, enable researchers to enhance the virulence and overall performance of specific fungal strains, tailoring them for specific pest challenges (Chen et al. 2015). Modern research increasingly employs proteomics, metagenomics, genomics, and other omics technologies to uncover various cultural and noncultural fungal endophytes in various environments (Fadiji and Babalola 2020). These cutting-edge technologies are instrumental in discovering and classifying previously uncharacterized taxa, providing crucial insights into their ecological roles.

Furthermore, they allow for identifying whether isolated endophytes are beneficial, nonpathogenic, or harmful (Brader et al. 2017). This capability is vital for developing effective disease management strategies and producing high-quality crops. Advancements in molecular biology now enable researchers to investigate the molecular mechanisms underlying host plant responses to interactions with both beneficial and pathogenic fungi (Wang et al. 2020). Understanding these mechanisms is essential for developing targeted strategies that enhance plant resilience against pests while promoting beneficial interactions with microbial communities.

Recent developments suggest that improved formulations of EPF may be safer for nontarget organisms, exhibit longer shelf lives, and minimize allergenic potential in humans. This is crucial for increasing public acceptance and regulatory approval of biocontrol products. As such, specialized research focusing on host–environment–pathogen interactions is necessary to explore new mycoinsecticide formulations capable of addressing a broader spectrum of insect pests while maximizing crop production and nutritional value (Sharma et al. 2023).

Despite the growing literature on endophytic EPF, significant data gaps remain regarding their ecological roles under varying environmental conditions. These gaps underscore the urgent need for further research. Current studies often yield inconsistent results due to differing methodologies or environmental contexts. Future research should focus on standardizing experimental protocols and expanding field trials to generate robust datasets that inform best practices for EPF application in diverse agroecosystems.

8 Conclusion

The potential of entomopathogenic fungi (EPF) as biological agents for managing insect pests is substantial. These fungi enhance plant growth, bolster pathogen resistance, and provide a comprehensive approach to crop protection. Applying EPF can lead to sustainable agricultural practices, as they are environmentally friendly, help mitigate pesticide resistance, and pose minimal risks to animal and human health. Further field-level research is essential to commercialize biocontrol agents based on endophytic entomopathogenic fungi. Understanding the ecological and physiological dynamics of these fungi in various environmental conditions and host plants is crucial for ensuring their consistent effectiveness in real-world applications. Integrating EPF into integrated pest management (IPM) programs holds considerable promise as we strive to create a sustainable environment and secure the global food supply. Investigating how EPF can synergize with other sustainable pest management strategies—such as cultural practices, natural enemies, and host plant resistance—can lead to more comprehensive and reliable IPM strategies. This holistic approach can significantly facilitate the transition from conventional to sustainable agriculture. By focusing on the interactions between entophytic EPF and other components of agroecosystems, we can develop innovative solutions that enhance pest control while promoting ecological balance and resilience. In summary, further exploration of entomopathogenic fungi will advance our understanding of their role in pest management and contribute to the broader goals of sustainable agriculture, ultimately supporting food security in an increasingly challenging environmental landscape.

Acknowledgments Gulafsha Praveen is thankful to the Department of Biotechnology, Graphic Era (Deemed to be University) for providing financial support in the form of a Doctoral Research Fellowship (No.241172057). The authors are also thankful to the Department of Biotechnology, Graphic Era (Deemed to be University), Dehradun, Uttarakhand; the School of Agriculture, Graphic Era Hill University, Dehradun; and ICAR-Indian Institute of Vegetable Research for their gracious assistance with this review paper.

Declaration of Competing Interests The authors declare that they have no known competing financial interests or personal relationships that could have appeared to influence the work reported in this chapter.

Data Availability The authors are unable or have chosen not to specify which data has been used.

References

Adeleke BS, Chaudhary P, Ayilara MS, Ojo FM, Erinoso SM, Upadhayay VK, Adeyemo AI, Akinola SA (2024) Rhizosphere microbiomes mediating abiotic stress mitigation for improved plant nutrition. Ecologies 5(3):375–401. https://doi.org/10.3390/ecologies5030024

Akello J, Dubois T, Coyne D, Kyamanywa S (2008) Endophytic *Beauveria bassiana* in banana (Musa spp.) reduces banana weevil (*Cosmopolites sordidus*) fitness and damage. Crop Prot 27(11):1437–1441

Araújo JP, Hughes DP (2016) Diversity of entomopathogenic fungi: which groups conquered the insect body? Adv Genet 94:1–39

Azadi H, Barati AA, Nazari Nooghabi S, Scheffran J (2022) Climate-related disasters and agricultural land conversion: Towards prevention policies. Climate and Development, 14(9):814–828.

Bamisile BS, Akutse KS, Siddiqui JA, Xu Y (2021) Model application of entomopathogenic fungi as alternatives to chemical pesticides: prospects, challenges, and insights for next-generation sustainable agriculture. Front Plant Sci 12:741804

Baron NC, Rigobelo EC, Zied DC (2019) Filamentous fungi in biological control: current status and future perspectives. Chil J Agric Res 79(2):307–315

Barra-Bucarei L, France Iglesias A, Pino Torres C (2019a) Entomopathogenic fungi. In: Natural enemies of insect pests in neotropical agroecosystems: biological control and functional biodiversity. Springer, pp 123–136

Barra-Bucarei L, France Iglesias A, Gerding González M, Silva Aguayo G, Carrasco-Fernández J, Castro JF, Ortiz Campos J (2019b) Antifungal activity of *Beauveria bassiana* endophyte against *Botrytis cinerea* in two solanaceae crops. Microorganisms 8(1):65

Behie SW, Bidochka MJ (2014) Ubiquity of insect-derived nitrogen transfer to plants by endophytic insect-pathogenic fungi: an additional branch of the soil nitrogen cycle. Appl Environ Microbiol 80(5):1553–1560

Behie SW, Zelisko PM, Bidochka MJ (2012) Endophytic insect-parasitic fungi translocate nitrogen directly from insects to plants. Science 336(6088):1576–1577

Benhamou N, Brodeur J (2001) Pre-inoculation of Ri T-DNA transformed cucumber roots with the mycoparasite, *Verticillium lecanii*, induces host defense reactions against *Pythium ultimum* infection. Physiol Mol Plant Pathol 58(3):133–146

Bernhardt ES, Rosi EJ, Gessner MO (2017) Synthetic chemicals as agents of global change. Front Ecol Environ 15(2):84–90

Bhutani S, Mehra A, Mishra A, Suyal DC, Bhasin S, Dinkar N, Srivastava N, Fayssal SA, Chaudhary P, Singh M (2024) Importance of microbial inoculants in maintaining plant growth and productivity. In: Chaudhary P, Chaudhary A (eds) Microbial inoculants, Interdisciplinary biotechnological advances. Springer, Singapore. https://doi.org/10.1007/978-981-97-0633-4_9

Bihal R, Al-Khayri JM, Banu AN, Kudesia N, Ahmed FK, Sarkar R, Abd-Elsalam KA (2023) Entomopathogenic fungi: an eco-friendly synthesis of sustainable nanoparticles and their nanopesticide properties. Microorganisms 11(6):1617

Bisht N, Chaudhary P (2024) Plant-microbe interactions and crop production. In: Chaudhary P, Chaudhary A (eds) Microbial inoculants, Interdisciplinary biotechnological advances. Springer, Singapore. https://doi.org/10.1007/978-981-97-0633-4_4

Biswas C, Dey P, Satpathy S, Satya P (2012) Establishment of the fungal entomopathogen Beauveria bassiana as a season long endophyte in jute (*Corchorus olitorius*) and its rapid detection using SCAR marker. BioControl 57(4):565–571

Błaszczyk L, Waśkiewicz A, Gromadzka K, Mikołajczak K, Chełkowski J (2021) Sarocladium and Lecanicillium associated with maize seeds and their potential to form selected secondary metabolites. Biomol Ther 11(1):98

Brader G, Compant S, Vescio K, Mitter B, Trognitz F, Ma LJ, Sessitsch A (2017) Ecology and genomic insights into plant-pathogenic and plant-nonpathogenic endophytes. Annu Rev Phytopathol 55(1):61–83

Brandan CP, Chavarría D, Huidobro J, Meriles JM, Brandan CP, Gil SV (2017) Influence of a tropical grass (*Brachiaria brizantha* cv. Mulato) as cover crop on soil biochemical properties in a degraded agricultural soil. Eur J Soil Biol 83:84–90

Brunner M, Zeisler C, Neu D, Rotondo C, Rubbmark OR, Reinbacher L, Traugott M (2024) Trap crops enhance the control efficacy of *Metarhizium brunneum* against a soil-dwelling pest. J Pest Sci 97:1–13

Bryant JA, Lamanna C, Morlon H, Kerkhoff AJ, Enquist BJ, Green JL (2008) Microbes on mountainsides: contrasting elevational patterns of bacterial and plant diversity. Proc Natl Acad Sci 105(Suppl 1):11505–11511

Barra-Bucarei L, France Iglesias A, Pino Torres C (2019) Entomopathogenic fungi. Natural Enemies of Insect Pests in Neotropical Agroecosystems: Biological Control and Functional Biodiversity, 123–136.

BuGtI GA, Na C, Bin W, FeNG LH (2018) Control of plant sap-sucking insects using entomopathogenic fungi *Isaria fumosorosea* strain (Ifu13a). Plant Prot Sci 54(4):258–264

Camele I, Sadeek SA, Racioppi R, Elshafie HS (2023) Antimicrobial activity of diffusible and volatile metabolites emitted by *Beauveria bassiana*: chemical profile of volatile organic compounds (VOCs) using SPME-GC/MS analysis. Plan Theory 12(15):2854

Canfora L, Tartanus M, Manfredini A, Tkaczuk C, Majchrowska-Safaryan A, Malusà E (2023) The impact of *Beauveria* species bioinocula on the soil microbial community structure in organic strawberry plantations. Front Microbiol 13:1073386

Chia RW, Lee JY, Lee M, Lee GS, Jeong CD (2023) Role of soil microplastic pollution in climate change. Science of the Total Environment, 887, 164112.

Chaudhary A, Parveen H, Chaudhary P, Bhatt P, Khatoon H (2021a) Rhizospheric microbes and their mechanism. In: Bhatt P, Gangola S, Udayanga D, Kumar G (eds) Microbial technology for sustainable environment. Springer, Singapore. https://doi.org/10.1007/978-981-16-3840-4_6

Chaudhary P, Parveen H, Gangola S, Kumar G, Bhatt P, Chaudhary A (2021b) Plant growth promoting rhizobacteria and their application in sustainable crop production. In: Bhatt P, Gangola S, Udayanga D, Kumar G (eds) Microbial technology for sustainable environment. Springer, Singapore. https://doi.org/10.1007/978-981-16-3840-4_13

Chaudhary P, Singh S, Chaudhary A, Agri U, Bhandari G (2022a) Chapter 3: agrochemicals and their effects on soil microbial population. In: Soni R, Suyal D, Goel R (eds) Plant protection: from chemicals to biologicals. De Gruyter, Berlin/Boston, pp 45–58. https://doi.org/10.1515/9783110771558-003

Chaudhary P, Agri U, Chaudhary A, Kumar A, Kumar G (2022b) Endophytes and their potential in biotic stress management and crop production. Front Microbiol 13:933017. https://doi.org/10.3389/fmicb.2022.933017

Chaudhary P, Singh S, Chaudhary A, Sharma A, Kumar G (2022c) Overview of biofertilizers in crop production and stress management for sustainable agriculture. Front Plant Sci 13:930340. https://doi.org/10.3389/fpls.2022.930340

Chaudhary P, Xu M, Ahamad L, Chaudhary A, Kumar G, Adeleke BS, Verma KK, Hu DM, Širić I, Kumar P, Popescu SM, AbouFayssal S (2023) Application of synthetic consortia for improvement of soil fertility, pollution remediation, and agricultural productivity: a review. Agronomy 13:643. https://doi.org/10.3390/agronomy13030643

Chaudhary A, Chaudhary P, Fayssal SA, Singh S, Jaiswal DK, Tripathi V, Kumar J (2024) Exploring beneficial microbes and their multifaceted applications: an overview. In: Chaudhary P, Chaudhary A (eds) Microbial inoculants, Interdisciplinary biotechnological advances. Springer, Singapore. https://doi.org/10.1007/978-981-97-0633-4_1

Chen X, Li L, Hu Q, Zhang B, Wu W, Jin F, Jiang J (2015) Expression of dsRNA in recombinant *Isaria fumosorosea* strain targets the TLR7 gene in *Bemisia tabaci*. BMC Biotechnol 15:1–8

Chu H, Gao GF, Ma Y, Fan K, Delgado-Baquerizo M (2020) Soil microbial biogeography in a changing world: recent advances and future perspectives. MSystems 5(2):10–1128

Cordovez V, Dini-Andreote F, Carrión VJ, Raaijmakers JM (2019) Ecology and evolution of plant microbiomes. Ann Rev Microbiol 73(1):69–88

Costanza R (2020) Valuing natural capital and ecosystem services toward the goals of efficiency, fairness, and sustainability. Ecosyst Serv 43:101096

Cuthbertson AG, Blackburn LF, Eyre DP, Cannon RJ, Miller J, Northing P (2011) *Bemisia tabaci*: the current situation in the UK and the prospect of developing strategies for eradication using entomopathogens. Insect Sci 18(1):1–10

Dara SK (2019) Non-entomopathogenic roles of entomopathogenic fungi in promoting plant health and growth. Insects 10(9):277

Dara SK, Dara SS, Dara SS (2017) Impact of entomopathogenic fungi on the growth, development, and health of cabbage growing under water stress. Am J Plant Sci 8(06):1224

Dara SK, Montalva C, Barta M (2019) Microbial control of invasive forest pests with entomopathogenic fungi: a review of the current situation. Insects 10(10):341

Darshan R, Prasanna P (2024) Status of fall armyworm, *Spodoptera frugiperda* (JE Smith) (Lepidoptera: Noctuidae) and its natural enemies at major maize growing areas of Haveri district. J Farm Sci 37(01):29–31

de Faria MR, Wraight SP (2007) Mycoinsecticides and mycoacaricides: a comprehensive list with worldwide coverage and international classification of formulation types. Biol Control 43(3):237–256

De Groot RS, Alkemade R, Braat L, Hein L, Willemen L (2010) Challenges in integrating the concept of ecosystem services and values in landscape planning, management and decision making. Ecol Complex 7(3):260–272

de Jesus Alves L, Nunes FC, da Silva Santos I, Loureiro DM, Casaes PA, Gross E, Prasad MNV (2020) Microbial approach for alleviation of potentially toxic elements in agricultural soils. In: Climate change and soil interactions. Elsevier, pp 271–303

Deka B, Baruah C, Babu A (2021) Entomopathogenic microorganisms: their role in insect pest management. Egypt J Biol Pest Control 31:1–8

Delgado PAM, Murcia OP (2011) Hongos entomopatógenos: uma alternativa para la obtención de Biopesticidas. Ambi-Agua 6:77–90

Elbialy ZI, Assar DH, Abdelnaby A, Asa SA, Abdelhiee EY, Ibrahim SS et al (2021) Healing potential of Spirulina platensis for skin wounds by modulating bFGF, VEGF, TGF-ß1 and α-SMA genes expression targeting angiogenesis and scar tissue formation in the rat model. Biomed Pharmacother 137(111349):10–1016

El-Deeb HM, Lashin SM, Arab YAS (2012) Reaction of some tomato cultivars to tomato leaf curl virus and evaluation of the endophytic colonisation with *Beauveria bassiana* on the disease incidence and its vector, *Bemisia tabaci*. Arch Phytopathol Plant Protect 45(13):1538–1545

Epa U (2017) Climate impacts on agriculture and food supply. United States Environmental Protection Agency, Washington, DC

Fadiji AE, Babalola OO (2020) Exploring the potentialities of beneficial endophytes for improved plant growth. Saudi J Biol Sci 27(12):3622–3633

Fang W, Lu HL, King GF, St. Leger RJ (2014) Construction of a hypervirulent and specific mycoinsecticide for locust control. Sci Rep 4(1):7345

FAO, F (2011) Available online at http://faostat.fao.org/site/291/default.aspx. Food and Agriculture Organization

Fernández-Bravo M, Garrido-Jurado I, Valverde-García P, Enkerli J, Quesada-Moraga E (2016) Responses to abiotic environmental stresses among phylloplane and soil isolates of *Beauveria bassiana* from two holm oak ecosystems. J Invertebr Pathol 141:6–17

Frouz J, Novotná K, Čermáková L, Pivokonský M (2020) Soil fauna reduce soil respiration by supporting N leaching from litter. Appl Soil Ecol 153:103585

Gabarty A, Salem HM, Fouda MA, Abas AA, Ibrahim AA (2014) Pathogenicity induced by the entomopathogenic fungi *Beauveria bassiana* and *Metarhizium anisopliae* in *Agrotis ipsilon* (Hufn.). J Radiat Res Appl Sci 7(1):95–100

Gibson DM, Donzelli BG, Krasnoff SB, Keyhani NO (2014) Discovering the secondary metabolite potential encoded within entomopathogenic fungi. Nat Prod Rep 31(10):1287–1305

Goettel MS, Koike M, Kim JJ, Aiuchi D, Shinya R, Brodeur J (2008) Potential of *Lecanicillium* spp. for management of insects, nematodes and plant diseases. J Invertebr Pathol 98(3):256–261

Gonzalez-Guzman A, Raya-Diaz S, Sacristán D, Yousef M, Sánchez-Rodríguez AR, Barrón V, Torrent J (2021) Effects of entomopathogenic fungi on durum wheat nutrition and growth in the field. Eur J Agron 128:126282

Guerra CA, Bardgett RD, Caon L, Crowther TW, Delgado-Baquerizo M, Montanarella L, Eisenhauer N (2021) Tracking, targeting, and conserving soil biodiversity. Science 371(6526):239–241

Gulzar S, Wakil W, Shapiro-Ilan DI (2021) Combined effect of entomopathogens against *Thrips tabaci* Lindeman (Thysanoptera: Thripidae): laboratory, greenhouse and field trials. Insects 12(5):456

Hallouti A, Ait Hamza M, Zahidi A, Ait Hammou R, Bouharroud R, Ait Ben Aoumar A, Boubaker H (2020) Diversity of entomopathogenic fungi associated with the Mediterranean fruit fly (Ceratitis capitata (Diptera: Tephritidae) in Moroccan Argan forests and nearby area: impact of soil factors on their distribution. BMC ecology, 20:1–13.

Hardoim PR, Van Overbeek LS, Berg G, Pirttilä AM, Compant S, Campisano A, Sessitsch A (2015) The hidden world within plants: ecological and evolutionary considerations for defining functioning of microbial endophytes. Microbiol Mol Biol Rev 79(3):293–320

Harman GE, Doni F, Khadka RB, Uphoff N (2021) Endophytic strains of *Trichoderma* increase plants' photosynthetic capability. J Appl Microbiol 130(2):529–546

Hartley SE, Gange AC (2009) Impacts of plant symbiotic fungi on insect herbivores: mutualism in a multitrophic context. Annu Rev Entomol 54(1):323–342

Hassani MA, Durán P, Hacquard S (2018) Microbial interactions within the plant holobiont. Microbiome 6:1–17

Hillocks RJ (2012) Farming with fewer pesticides: EU pesticide review and resulting challenges for UK agriculture. Crop Prot 31(1):85–93

Hong S, Shang J, Sun Y, Tang G, Wang C (2024) Fungal infection of insects: molecular insights and prospects. Trends Microbiol 32(3):302–316

Hu Q, Li F, Zhang Y (2016) Risks of mycotoxins from mycoinsecticides to humans. Biomed Res Int 2016(1):3194321

Ibrahim RL, Al-Mulali U, Ajide KB, Mohammed A, Al-Faryan MAS (2023) The implications of food security on sustainability: do trade facilitation, population growth, and institutional quality make or mar the target for SSA? Sustain For 15(3):2089

Idrees A, Qadir ZA, Akutse KS, Afzal A, Hussain M, Islam W, Li J (2021) Effectiveness of entomopathogenic fungi on immature stages and feeding performance of Fall Armyworm, *Spodoptera frugiperda* (Lepidoptera: Noctuidae) Larvae. Insects 12(11):1044

Jaber LR, Enkerli J (2017) Fungal entomopathogens as endophytes: can they promote plant growth? Biocontrol Sci Tech 27(1):28–41

Jankielsohn A (2018) The importance of insects in agricultural ecosystems. Adv Entomol 6(2):62–73

Jansson JK, Hofmockel KS (2020) Soil microbiomes and climate change. Nat Rev Microbiol 18(1):35–46

Kamboj R, Kamboj S, Kamboj S, Kriplani P, Dutt R, Guarve K, Gautam SP (2023) Climate uncertainties and biodiversity: an overview. In: Visualization techniques for climate change with machine learning and artificial intelligence. Elsevier, Cambridge, MA, pp 1–14

Khatoon H, Chaudhary P, Chaudhary A (2024) Microbial inoculants and soil microbial population. In: Chaudhary P, Chaudhary A (eds) Microbial inoculants, Interdisciplinary biotechnological advances. Springer, Singapore. https://doi.org/10.1007/978-981-97-0633-4_3

Knights HE, Jorrin B, Haskett TL, Poole PS (2021) Deciphering bacterial mechanisms of root colonization. Environmental Microbiology Reports, 13(4):428–444.

Kumar L, Chhogyel N, Gopalakrishnan T, Hasan MK, Jayasinghe SL, Kariyawasam CS, Ratnayake S (2022) Climate change and future of agri-food production. In: Future foods. Academic, pp 49–79

Leal Filho W, Nagy GJ, Setti AFF, Sharifi A, Donkor FK, Batista K, Djekic I (2023) Handling the impacts of climate change on soil biodiversity. Sci Total Environ 869:161671

Li Y, Xu J, Hu J, Zhang T, Wu X, Yang Y (2022) Arbuscular mycorrhizal fungi and glomalin play a crucial role in soil aggregate stability in Pb-contaminated soil. Int J Environ Res Public Health 19(9):5029

Litwin A, Nowak M, Różalska S (2020) Entomopathogenic fungi: unconventional applications. Rev Environ Sci Biotechnol 19(1):23–42

Liu Y, Yang Y, Wang B (2022) Entomopathogenic fungi *Beauveria bassiana* and *Metarhizium anisopliae* play roles of maize (*Zea mays*) growth promoter. Sci Rep 12:15706

Mace GM, Hails RS, Cryle P, Harlow J, Clarke SJ (2015) Towards a risk register for natural capital. J Appl Ecol 52(3):641–653

Manlay RJ, Feller C, Swift MJ (2007) Historical evolution of soil organic matter concepts and their relationships with the fertility and sustainability of cropping systems. Agriculture, Ecosystems & Environment, 119(3-4):217–233.

Martins JLA, Franzin ML, Ferreira DDS, Magina LCR, Martins EF, Mendonça LVP, Venzon M (2024) Metarhizium-inoculated coffee seeds promote plant growth and biocontrol of coffee leaf miner. Microorganisms 12(9):1845

Masson-Delmotte D (2018) IPCC special report on the impacts of global warming of 1.5 C: summary for policy makers. Cambridge University Press

MEA, M. E. A (2005) Ecosystems and human well-being scenarios. Findings of the scenarios working group. The Millennium Ecosystem Assessment

Mirón IJ, Linares C, Díaz J (2023) The influence of climate change on food production and food safety. Environ Res 216:114674

Moonen AC, Bàrberi P (2008) Functional biodiversity: an agroecosystem approach. Agric Ecosyst Environ 127(1–2):7–21

Moran PJ, Patt JM, Cabanillas HE, Adamczyk JL, Jackson MA, Dunlap CA et al (2011) Localized autoinoculation and dissemination of *Isaria fumosorosea* for control of the Asian citrus psyllid in South Texas. Subtrop Plant Sci 63:23–35

Muller DB, Vogel C, Bai Y, Vorholt JA (2016) The plant microbiota: systems-level insights and perspectives. Annu Rev Genet 50(1):211–234

Muniz ER, Paixão FR, Barreto LP, Luz C, Arruda W, Angelo IC, Fernandes ÉK (2020) Efficacy of *Metarhizium anisopliae* conidia in oil-in-water emulsion against the tick *Rhipicephalus microplus* under heat and dry conditions. BioControl 65:339–351

Nannipieri P, Small K, (2006) Nucleic Acids and Proteins in Soil. Soil Biology.Springer Verlag, Berlin, p. 8

Newsham KK, Hopkins DW, Carvalhais LC, Fretwell PT, Rushton SP, O'Donnell AG, Dennis PG (2016) Relationship between soil fungal diversity and temperature in the maritime Antarctic. Nat Clim Chang 6(2):182–186

Paixão FR, Fernandes ÉK, Pedrini N (2019) Thermotolerance of fungal conidia. In: Microbes for sustainable insect pest management: an eco-friendly approach, vol 1. Springer, pp 185–196

Pereira BDJ, Cecílio Filho AB, La Scala Jr N, Figueiredo EBD (2022) Greenhouse gas emissions and carbon footprint of collard greens, spinach and chicory production systems in Southeast of Brazil. Front Plant Sci 13:1015307

Praveen B, Sharma P (2019) A review of literature on climate change and its impacts on agriculture productivity. J Public Aff 19(4):e1960

Rajamanikyam M, Vadlapudi V, Upadhyayula SM (2017) Endophytic fungi as novel resources of natural therapeutics. Braz Arch Biol Technol 60:e17160542

Resquín-Romero G, Garrido-Jurado I, Delso C, Ríos-Moreno A, Quesada-Moraga E (2016) Transient endophytic colonizations of plants improve the outcome of foliar applications of mycoinsecticides against chewing insects. J Invertebr Pathol 136:23–31

Reyad NF (2017) Empact of entomopathogenic fungi on whitefly, *Bemisia tabaci* in tomato crop in Egypt. Int J ChemTech Res 10:372–377

Rinaldi C (2023) From food security to food sustainability: the attitudes of the FAO. Master thesis, Charles University, Faculty of Social Sciences

Rivas-Franco F, Hampton JG, Morán-Diez ME, Narciso J, Rostás M, Wessman P, Glare TR (2019) Effect of coating maize seed with entomopathogenic fungi on plant growth and resistance against *Fusarium graminearum* and *Costelytra giveni*. Biocontrol Sci Tech 29(9):877–900

Russo ML, Pelizza SA, Vianna MF, Allegrucci N, Cabello MN, Toledo AV, Scorsetti AC (2019) Effect of endophytic entomopathogenic fungi on soybean *Glycine max* (L.) Merr. growth and yield. J King Saud Univ Sci 31(4):728–736

Sahana M, Ravetz J, Patel PP, Dadashpoor H, Follmann A (2023) Where is the peri-urban? A systematic review of peri-urban research and approaches for its identification and demarcation worldwide. Remote Sens 15(5):1316

Sahoo S, Singha C, Govind A, Moghimi A (2024) Review of Climate-Resilient Agriculture for Ensuring Food Security: Sustainability Opportunities and Challenges of India. Environmental and Sustainability Indicators, 100544.

Sasan RK, Bidochka MJ (2012) The insect-pathogenic fungus *Metarhizium robertsii* (Clavicipitaceae) is also an endophyte that stimulates plant root development. Am J Bot 99(1):101–107

Savary S, Willocquet L, Pethybridge SJ, Esker P, McRoberts N, Nelson A (2019) The global burden of pathogens and pests on major food crops. Nat Ecol Evol 3(3):430–439

Schlesinger W, Bernhardt E (2013) Biogeochemistry: an analysis of global change. Academic, 672 pp

Seibold S, Gossner MM, Simons NK, Blüthgen N, Müller J, Ambarlı D, Weisser WW (2019) Arthropod decline in grasslands and forests is associated with landscape-level drivers. Nature 574(7780):671–674

Semeraro T, Mastroleo G, Pomes A, Luvisi A, Gissi E, Aretano R (2019) Modelling fuzzy combination of remote sensing vegetation index for durum wheat crop analysis. Comput Electron Agric 156:684–692

Semeraro T, Luvisi A, Lillo AO, Aretano R, Buccolieri R, Marwan N (2020) Recurrence analysis of vegetation indices for highlighting the ecosystem response to drought events: an application to the Amazon forest. Remote Sens 12(6):907

Shah PA, Pell JK (2003) Entomopathogenic fungi as biological control agents. Appl Microbiol Biotechnol 61(5):413–423

Sharma A, Sharma S, Yadav PK (2023) Entomopathogenic fungi and their relevance in sustainable agriculture: a review. Cogent Food Agric 9(1):2180857

Shen FT, Yen JH, Liao CS, Chen WC, Chao YT (2019) Screening of rice endophytic biofertilizers with fungicide tolerance and plant growth-promoting characteristics. Sustain For 11(4):1133

Smith MR, Myers SS (2019) Global health implications of nutrient changes in rice under high atmospheric carbon dioxide. GeoHealth, 3(7), 190–200.

Sinha KK, Choudhary AK, Kumari P (2016) Entomopathogenic fungi. In: Ecofriendly pest management for food security. Academic, pp 475–505

Skendžić S, Zovko M, Pajač Živković I, Lešić V, Lemić D (2021) Effect of climate change on introduced and native agricultural invasive insect pests in Europe. Insects 12(11):985

Sokol NW, Slessarev E, Marschmann GL, Nicolas A, Blazewicz SJ, Brodie EL, Pett-Ridge J (2022) Life and death in the soil microbiome: how ecological processes influence biogeochemistry. Nat Rev Microbiol 20(7):415–430

Song Y, Cui H, Guo W, Sindhu L, Lv S, Li L, Men X (2024) Endophytic fungi improved wheat resistance to *Rhopalosiphum padi* by decreasing its feeding efficiency and population fitness. Ecotoxicol Environ Saf 270:115865

Srivastava P, Chaudhary P, Mishra A, Dasila H, Maithani D (2024) Influence of microbial inoculants on soil health. In: Chaudhary P, Chaudhary A (eds) Microbial inoculants, Interdisciplinary biotechnological advances. Springer, Singapore. https://doi.org/10.1007/978-981-97-0633-4_2

Sui L, WAN T, Lu Y, XU W, ZHANG Z, LI Q (2021) Review of fungal endophytes on plant growth promotion and stress resistance. Chinese Journal of Biological Control, 37(6), 1325.

Sui L, Lu Y, Zhou L, Li N, Li Q, Zhang Z (2023) Endophytic *Beauveria bassiana* promotes plant biomass growth and suppresses pathogen damage by directional recruitment. Front Microbiol 14:1227269

Sun S, Wang Z, Zhang Z, Hoffmann AA, Guo H (2021) The roles of entomopathogenic fungal infection of viruliferous whiteflies in controlling tomato yellow leaf curl virus. Biol Control 156:104552

Swathy K, Parmar MK, Vivekanandhan P (2024) Biocontrol efficacy of entomopathogenic fungi *Beauveria bassiana* conidia against agricultural insect pests. Environ Qual Manag 34(1):e22174

Tilman D, Balzer C, Hill J, Befort BL (2011) Global food demand and the sustainable intensification of agriculture. Proc Natl Acad Sci 108(50):20260–20264

Trivedi P, Leach JE, Tringe SG, Sa T, Singh BK (2020) Plant–microbiome interactions: from community assembly to plant health. Nat Rev Microbiol 18(11):607–621

Tyagi J, Chaudhary P, Jyotsana, Bhagwati U, Bhandari G, Chaudhary A (2022a) Chapter 17: impact of endophytic fungi in biotic stress management. In: Soni R, Suyal D, Goel R (eds) Plant protection: from chemicals to biologicals. De Gruyter, Berlin/Boston, pp 447–462. https://doi.org/10.1515/9783110771558-017

Tyagi J, Chaudhary P, Mishra A, Khatwani M, Dey S, Varma A (2022b) Role of endophytes in abiotic stress tolerance: with special emphasis on *Serendipita indica*. Int J Environ Res 16:62. https://doi.org/10.1007/s41742-022-00439-0

Umapathi R, Ghoreishian SM, Sonwal S, Rani GM, Huh YS (2022) Portable electrochemical sensing methodologies for on-site detection of pesticide residues in fruits and vegetables. Coord Chem Rev 453:214305

Uzman D, Pliester J, Leyer I, Entling MH, Reineke A (2019) Drivers of entomopathogenic fungi presence in organic and conventional vineyard soils. Appl Soil Ecol 133:89–97

Vandenberg JD, Shelton AM, Wilsey WT, Ramos M (1998) Assessment of *Beauveria bassiana* sprays for control of diamondback moth (Lepidoptera: Plutellidae) on crucifers. J Econ Entomol 91(3):624–630

Wang F, Nong X, Hao K, Cai N, Wang G, Liu S, Zhang Z (2020) Identification of the key genes involved in the regulation of symbiotic pathways induced by *Metarhizium anisopliae* in peanut (*Arachis hypogaea*) roots. 3 Biotech 10:1–11

Wilberts L, Rojas-Preciado N, Jacquemyn H, Lievens B (2023) Fungal strain and crop cultivar affect growth of sweet pepper plants after root inoculation with entomopathogenic fungi. Front Plant Sci 14:1196765

Yasin M, Wakil W, Ghazanfar MU, Qayyum MA, Tahir M, Bedford GO (2019) Virulence of entomopathogenic fungi *Beauveria bassiana* and *Metarhizium anisopliae* against red palm weevil, *Rhynchophorus ferrugineus* (Olivier). Entomol Res 49(1):3–12

Zhang X, Meng Y, Huang Y, Zhang D, Fang W (2021) A novel cascade allows *Metarhizium robertsii* to distinguish cuticle and hemocoel microenvironments during infection of insects. PLoS Biol 19(8):e3001360

Zhang MD, Wu SY, Yan JJ, Li J, Ali A, Gao YL (2022) Endophytic *Beauveria bassiana* inhibits the performance of *Phthorimaea operculella* and enhances potato plant growth. Entomol Gen 42(6):911–919

Effect of Climate Change on Beneficial Plant–Microbe Interaction in the Rhizosphere

5

Simran Kaur, Abhishek Das, Susheel Dwivedi, Swati Tripathi, and Neeraj Shrivastava

Abstract

Beneficial plant microorganisms enhance the growth and resilience of plants. These interactions are impacted by climate change, which includes warming, drought, and elevated CO_2. Studies have shown that temperature has mixed effects on these microbes, with positive results when CO_2 levels are high. When plants are stressed by drought, plant growth-promoting microorganisms (PGPM) frequently aid them. Agriculture depends on knowing how plants and microbes interact as the climate changes. Plant stress resistance can be enhanced by reconstructing the microbiome of the rhizosphere. Plant resistance can be directly increased by PGPR inoculation. Even though progress has been made, little is known about how crop–microbe interactions work on a physiological and molecular level in the field. Environmental change, with its expanded recurrence of outrageous occasions, compromises food security by influencing soil-plant-microorganism cooperations. The resilience of plants is largely dependent on microbes. A fundamental methodology is expected to figure out the intricate plant-organism climate collaborations. Tools for investigating these interactions in changing climates are provided by advancements in genomic and sequencing techniques. The significance of interactions between plants and microbes in a changing climate is emphasized in this review. Understanding these interactions and developing resistant crops require additional research.

Authors "Simran Kaur" and "Abhishek Das" have equally contributed to this chapter.

S. Kaur · A. Das · S. Tripathi · N. Shrivastava (✉)
Amity Institute of Microbial Technology, Amity University, Noida, Uttar Pradesh, India
e-mail: nsrivastava1@amity.edu

S. Dwivedi
Kendriya Vidyalaya, Indian Embassy, Kathmandu, Nepal

U. Pankaj et al. (eds.), *Plant-microbiome Interactions for Climate-resilient Agriculture*, https://doi.org/10.1007/978-981-96-3534-4_5

Keywords

PGPR · Chemoattractants · PGPM · Bacteroidetes · Phytohormones · Rhizosphere

1 Introduction

Terrestrial ecosystems depend on interactions between plants and soil organisms for survival and adaptation to changing climate conditions. The biggest challenges of global climate change are sustainable agriculture, human health, and food security (Zhu et al. 2022). Over the past few decades, global climate conditions have been changing due to increased average temperatures and carbon dioxide levels. Variations in these environmental factors, as well as additional changing conditions, are resulting in unpredictable occurrences like increased rainfall and snowfall all over the world, an increase in greenhouse gases and temperature, which causes the soil to dry, and a decrease in moisture content, creating drought situations (Zhu et al. 2022). In the upcoming century, the global climate is predicted to change drastically, which might also affect various parameters with the changing environment (Compant et al. 2010). The agriculture sector is adversely affected by changing climatic conditions because they exert a variety of biotic and abiotic pressures. Climate change-related abiotic stress has a negative impact on crop productivity and growth, as well as the diversity and function of the subsurface microbial population (Ye et al. 2022).

The rhizosphere is a platform for the plant-root interface and is considered the most complex ecosystem of the earth. This narrow zone surrounded by the plant root is a hotspot for various microbial activities (Mendes et al. 2019). The microbiome present in the rhizosphere impacts plant growth and development by making nutrients available for the plants. In return, the rhizospheric microbiome utilizes the nutrient resources released by the plant in the form of rhizodeposits. Thus, soil properties, plant species, and rhizodeposits decide the composition of the microbe community, which finally creates a complex food web in the rhizosphere region of the soil (Mendes et al. 2019). The rhizosphere has a densely populated soil region within this zone of plant–soil interaction, exhibiting an important impact on carbon movement and transformation due to a complex series of interactions between and among species as well as the food web. Understanding the relation between roots and soil microbes improves crop yield and productivity, reducing the dependence on chemical fertilizers and focusing more on sustainable agriculture. The rhizosphere microbiome has the potential to be an important factor in the upcoming green revolution, promoting environmentally friendly and sustainable farming practices (Solomon et al. 2023). Plant metabolism has a major effect on the rhizosphere, by releasing CO_2 and the secretion of various root exudates. By providing microorganisms with energy and functioning as chemical attractants and repellents, root exudates, which include phytohormones, promote interactions between the rhizosphere and other environments (Ahkami et al. 2017).

Plant biodiversity and microbial populations interact in a complex way that supports ecosystem functioning, impacting everything from soil formation and nutrient cycling to plant health and production (Nizamani et al. 2024). All the bacteria, fungi, and viruses that are associated with plants are collectively referred to as the plant microbiome. Within a single plant, various tissues may support completely distinct microbial communities, resulting in complex relationships that might be mutualistic, commensal, or even parasitic (Shi et al. 2023). Various abundant microorganisms surrounding the roots are caused by root exudates that provide nutrients for microbial growth. For example, mycorrhizal fungi and plant growth-promoting rhizobacteria (PGPR) are well known for their ability to support plant development by facilitating the nutrient acquisition and solubilizing the nutrients in soil, even in stressful conditions (Shree et al. 2022). Organic acids, amino acids, fatty acids, carbohydrates, vitamins, plant growth regulators, and other metabolites are among the substances that root exudates have a noticeable impact on the kind and quantity of bacteria around roots. Interactions between microbiota and plants can occur in both above- and below-ground environments, and they may play a role in controlling the plants' ability to withstand changes in climate and other stressors. Microbes such as bacteria (Proteobacteria, Bacteroidetes), actinomycetes, fungus (Ascomycota and Basidiomycota), viruses, and other microorganisms are widely present in phyllospheric regions of the plant, including stems, leaves, flowers, and sometimes seeds also. Due to the dynamic nature of phyllosphere microbiome, it is affected more frequently under changing climatic conditions. Below the ground, many different bacteria, actinomycetes, fungus, and other microorganisms live in soil and interact with different portions of plants; some of these interactions are mutualistically related (PGPR, Rhizobia), whereas others carry out essential ecological tasks. For instance, the microbiome of the citrus rhizospheric zone has a variety of microbiota groups, the most prominent of which are *Pseudomonas*, *Agrobacterium*, *Cupriavidus*, *Rhizobium*, *Sphingomonas*, and several other potentially beneficial microorganisms (Sharma et al. 2022).

2 Plant–Microbe Interaction

The rhizosphere is a common plant-root interface where plant exudates interact with microorganisms and make a microbial reservoir, which develops into the most complex ecosystem. There are various relations between roots or the rhizosphere and microbes like symbiotic, parasite, amensalism and commensalism, etc. The soil microbial populations are influenced by plants and the soil environment (Jacoby et al. 2017). In a single plant, distinct tissues may support completely distinct microbial communities, establishing complex relationships ranging from mutualistic to commensal and even parasitic, among others. In both cultivated and natural populations, it has been demonstrated that the plant microbiome affects the growth, development, and survival of the host (Shi et al. 2023). Microorganisms that help with biological processes, including nutrient intake, growth, and development, are drawn to the rhizosphere by plants (Igiehon and Babalola 2018). The plant–microbe interactions and microbial gene expression are regulated by the recruiting agent, root exudates from the roots like the release of amino acids, cutin monomers, flavonoids, hormones, organic acids, polyphenols, sugars, and nutrients (Compant et al. 2021).

Exudates from different plant species and genotypes can strongly impact the nearby vicinity of plants due to their distinctive characteristics. These exudates from the roots have been linked to several functions, such as regulating nutrient availability and moderating nutrient competition. Exudates from plants have been shown to accelerate mineralization. Plant ion availability was enhanced by the presence of certain acids in the soil, such as carboxylase and phosphatases (Canarini et al. 2019). These microbes regulate the amount, composition, and potential for interaction between the plant and microbiome in the root exudates around the roots (Semchenko et al. 2018). It is observed that species with more widely distributed horizontal root systems than those with deeper, more vertical root systems are in competition with one another. In addition, these plant exudates affected the surrounding microbial populations as well as each other, evaluating root–root interactions including legumes, non-legumes, and interspecies (Montazeaud et al. 2018). The root-root relationship not only focuses on nutrient consumption, but they also have the ability to produce phytotoxins through allelopathy, which reduces competition for resources by inhibiting neighbouring plants' growth or survival, and also influences the root growth of plants in close proximity (Nadarajah and Abdul Rahman 2021).

The microbiomes linked with plants comprise both harmful species, such as pathogenic microbes that hinder plant growth, and advantageous species, such as nitrogen-fixing bacteria, mycorrhizal fungi, other plant growth-promoting rhizobacteria (PGPR), and biocontrol agents (Mohanram and Kumar 2019). Rhizosphere species that fix nitrogen both endophytically and naturally by increasing both aboveground and belowground biomass, such as *Bradyrhizobium*, *Azotobacter*, *Pseudomonas*, *Azospirillum*, *Bacillus*, *Burkholderia*, and *Achromobacter*, have been shown to benefit crops (Igiehon and Babalola 2018). Microorganisms found in the rhizosphere additionally assist in the absorption of trace metals, including zinc (Zn) and iron (Fe) (Nadarajah and Abdul Rahman 2021). Furthermore, the rhizosphere microbiome contributes significantly to the breakdown of organic material, which raises soil fertility and, eventually, increases plant yield. The plant microbiome, therefore, facilitates the growth of plants even in nutrient-poor soils (Ahmed et al. 2018).

3 Climate Change Impact on Rhizosphere Environment

There are various direct and indirect factors influencing the microbes surrounding the environment, which affect the plant. These regions are mainly environmental, rhizospheric, and phyllospheric elements, which impact the presence of microorganisms. Plant species, root architecture, genotype, age, and developmental phases are among the biotic variables that primarily affect the existence of these microbiota in the plant. Abiotic elements that also have an impact include temperature, pH, salt, temperature fluctuations, soil moisture, and organic matter (Sharma et al. 2022).

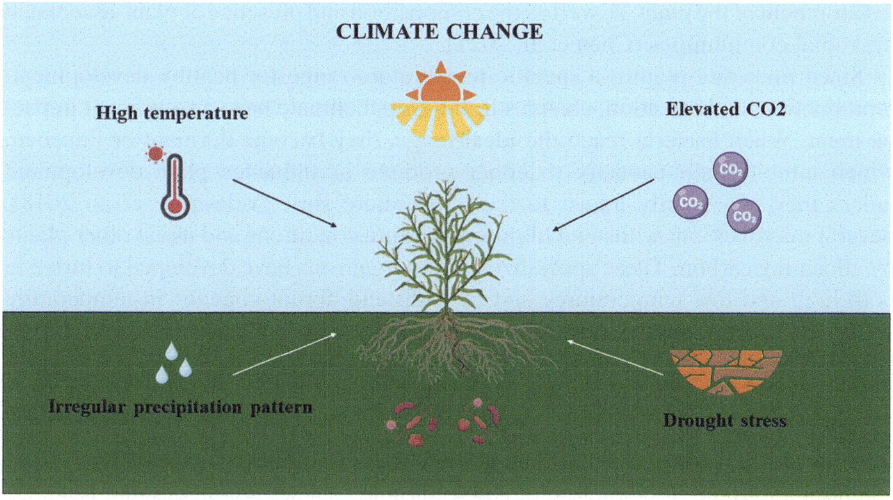

Fig. 5.1 Factors affecting the plant–microbe interaction: the image illustrates how climate change is adversely affecting the intricate relationship between plants and the microbial communities in the soil, particularly in the rhizosphere

Various plant species also have a great impact on the composition of microbiomes present in the rhizosphere. Microbiota from the phyllosphere are affected by anthropogenic and environmental factors. Environmental factors such as ozone, radiation, animal-borne pathogens, release of various volatile compounds, plant phenotype, growth, and health of plants. Urbanization, industrialization, the use of pesticides, herbicides, fungicides, fertilizers, and bio-stimulants are examples of anthropogenic activities that affect the abundance of microbiomes (Gupta and Patil 2021). The soil microbiome is established by soil-associated parameters such as pH, gas gradient, and potentially accessible mineral content. These factors are important in determining the variety of the microbiota (Tecon and Or 2017). Temperature change, differences in UV radiation, and changes in CO_2 concentration can also have an impact on the structure and makeup of the plant microbiome (Islam et al. 2020) (Fig. 5.1).

3.1 Temperature

Temperature is considered to be the primary factor that influences plant development and phenological traits, as well as the distribution and presence of a certain microbial population in the community (Teja et al. 2023). The average temperature has increased over the past few decades due to human involvement and the release of greenhouse gases such as CO_2, water vapour, and methane. It is expected that by the year 2100, the average global temperature is predicted to rise by 1.8–3.6 °C, which might cause drought and water scarcity in many regions of the world (Sharma et al. 2022). Even an increase in average temperature can influence the growth and development of the plant, as well as the composition and presence of plant-associated microbial communities (Chen et al. 2021).

Since microbes require a specific temperature range for healthy development, reproduction, and infection, changes in the global climate have a significant impact on them. When bacteria reach the ideal range, they become dormant or impeded, which inhibits their capacity to either promote or influence plant development unless they can swiftly adjust to the temperature shift (Velásquez et al. 2018). Several microbes can withstand high-temperature conditions and assist other plants by allocating carbon. These specialized microorganisms have developed to thrive in both high and low temperatures and to withstand abrupt changes in temperature without becoming inhibited. Microbes can avoid the harmful effects of severe temperatures with the help of complex networks of secondary metabolites (extremolyte/compatible solutes) such as amino acids, carbohydrates, polyols, and their derivatives (Sharma et al. 2022). It is expected that in each plant–microbe interaction, the consequences of the severe climatic circumstances brought on by climate change will affect both the microbe and the plants. Plant diseases arise from the phenological orientation of both microbes and plants, and the chance of disease occurrence changes whenever either microbe or plant evaluates a change in the climate (Grace et al. 2019). Certain microorganisms, such as endophytic bacteria and arbuscular mycorrhizal fungi, alter their morphology in order to minimize the negative impacts of high temperatures on plants. For example, *B. cereus SA1*, a thermotolerant bacterial strain, increases the biomass and chlorophyll content of soybean plants under high-temperature stress by producing substances such as gibberellin, salicylic acid, and indole-3-acetic acid (Khan et al. 2020) (Fig. 5.2 depicts the effect of temperature and the role of microbes to minimize the effect). The surrounding microbial population might reduce the harmful impacts of temperature stress resulting from global climate change on plants by expanding the temperature range at

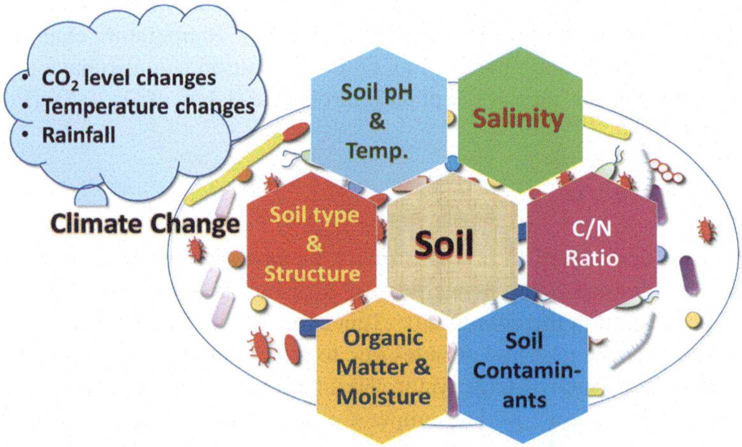

Fig. 5.2 Intricate relationship between climate change, plants, and the microbial communities residing in the soil. The picture depicts how the climatic factors influence the microbiome

which the host plant may thrive. For instance, *Dichanthelium lanuginosum*, a tropical plant, and *Curvularia protuberate*, a bacterium, cannot thrive at higher temperatures when maintained apart. On the other hand, when they are closely grouped together, they develop symbiotically and can withstand high temperatures (Shree et al. 2022). Nonetheless, other environmental factors that affect microbial growth and survival include moisture and UV radiation. The microbial community in the soil, the root endosphere, and the rhizosphere are all impacted by low moisture availability and drought conditions (Cheng et al. 2019).

3.2 Precipitation Pattern

Another observable effect of climate change is the significant variation in precipitation patterns, which affects the amount of moisture in the soil and air, causing drought. Drought stress due to insufficient moisture poses a large risk to food availability since various degrees of exposure levels have parameter amplitude effects on crop productivity (Gupta et al. 2022). Due to deficiency of water, plants undergo various morphological and physiological changes either to stabilize the growth or to prevent after effect (Kumar and Verma 2018). Drought stress causes a reduction in shoot growth during the early phase, but it does not affect root growth, which leads to an increased root/shoot ratio. Severe lack of moisture also wrinkles the cell wall of the plant, which inhibits the growth of new leaves and lowers the plant's fresh weight since the turgor pressure decreases. In response to drought stress, the morphology of the roots also changes, causing the roots to shrink, which modifies the distribution of water and nutrients to different regions of the plant and prevents dehydration, which might cause leaves to lose their capacity to use photosystem II. The availability and absorption of nutrients through roots and their movement to other areas are similarly impacted by a lack of water (Gupta et al. 2022). This drought stress also affects the integrity of the cell wall, generates the reactive oxygen species (ROS), encourages early life senescence, increases the ethylene production, and lowers the chlorophyll concentration. Furthermore, salt concentration rises and starts to accumulate in the rhizosphere due to lack of water and unable to dissolve it, increasing the salinity stress (Kumar and Verma 2018). After being subjected to a lack of water content, plants have started to adapt the condition with genetic adaptation. These adaptations to drought stress also include the use of phytohormones to promote growth in low-water environments, the activation of cellular processes to remove excess salt from plant cells and relocate it elsewhere, production of compatible solutes to lessen the effects of drought stress and preserve water balance, an increase in antioxidant enzymes and chemicals to scavenge excess reactive oxygen species (ROS), etc. (Shree et al. 2022). Lack of water also impacts microbial diversity, cellular activity, and composition of soil (Manzanera 2021). As a result, after being exposed to drought stress, plant-associated bacteria start their defence mechanism and stabilize their growth. These bacteria create compounds that help related plants respond to stress and grow normally by raising the ratio of roots to shoots, producing more biomass, enhancing their ability to absorb water, enhancing the nutrients' bioavailability, boosting production, and enduring drought (Rawal et al. 2022). In exchange, the root exudates provide extra elements that affect these microorganisms' stress response.

3.3 Carbon Dioxide (CO_2) Concentration

The relationship between carbon and the plants, which release root exudates in the soil, is greatly influenced by atmospheric CO_2, which may impact the surrounding environment and help the microbes (Williams et al. 2018). Plants and microbes both have the ability to withstand the stress, which is impacted by elevated CO_2 or high CO_2 levels, via several studies about the correlated relation with increased photosynthetic rate, high biomass production, nutrient utilization, water consumption, and carbohydrate content (Shree et al. 2022). There may be a connection between elevated CO_2, plant growth, and increases in colonization by rhizobacteria that promote plant development, according to early research on plant growth responses to eCO_2 and the presence of certain rhizosphere microorganisms (Williams et al. 2018). Microbes are essential in carbon exchange through various processes like respiration and the breakdown of organic matter, pathogenic or symbiotic relationships with plants, and nutrient availability. In addition to changing the physiology and metabolism of plants, high CO_2 levels have a direct and indirect impact on the microbial population in both rhizospheric and phyllospheric environments. Increased CO_2 levels frequently cause changes in the content of root exudates as well as changes in the availability of nutrients (Sharma et al. 2022).

Various microbes, such as bacteria, fungi, and actinomycetes, show variable responses to elevated CO_2. Fungi and other microbes have a greater ability to store carbon than to use it, and they are also more efficient at assimilating carbon than bacteria. As a result, as plants release more carbon through their roots, the microbial population in the soil system grows and decreases the amount of soil nitrogen that is accessible to the plant because of soil nitrogen immobilization. Recent developments in the study of algae reveal that when CO_2 concentration rises, so do the biomass and lipid levels in microalgae species. Thus, increased CO_2 levels have an impact on plant biological systems and microbial populations, which, in turn, affect future climate conditions. Increased CO_2 levels can affect 'how soil microorganisms acquire nutrients', 'how plants respond to pathogens', and aid in the storage of carbon in the soil (Sharma et al. 2022). Furthermore, in order to practise climate-resilient agriculture, the role of rhizospheric microorganisms in an elevated CO_2 environment must be evaluated. This must be done in conjunction with a rigorous effort to scale up research into the long-term effects of soil microbes in elevated CO_2 conditions and effective carbon sequestration through plant–microbe interactions (Grover et al. 2015) (Fig. 5.2).

3.4 Extreme Weather Events

The abiotic stress that most likely affects soil biota is drought. Drought not only causes osmotic stress but also increases soil heterogeneity, restricts nutrient movement and availability, and raises soil oxygen levels, which frequently result in a significant drop in microbial biomass (Jansson and Hofmockel 2020). The stability of soil particles and the aeration condition of the soil are two physical qualities that are impacted by water content. The pH, gas and solvent diffusion, and nutrient availability are all influenced by the water content of the soil. The development and activity of soil microorganisms are extensively influenced by soil temperature and moisture content (Meisner et al. 2018). Water facilitates the movement of microorganisms in soil and the diffusion of substances between the cells of an organism and its surroundings. Its content controls the pace of mineralization and is also involved in hydrolysis activities (Furtak and Wolińska 2023). Drought can affect the way soil functions and biogeochemical cycles important for nutrients for the plants and especially the microbial diversity, which alters the plant growth (Zhang et al. 2020). Reduced soil moisture not only impacts water movement via soil pores but also impacts the dispersion and diffusion of gases, nutrients, extracellular enzymes, and other substrates (Furtak and Wolińska 2023). Drought can cause an increase in soil osmotic pressure because it can cause solvents to build up in the residual water clusters and create a hypertonic solution, which can cause microbial cells to desiccate and become inactive. Soluble iron precipitation and a significant decrease in soil respiration are further consequences (Lu et al. 2017).

Drought conditions reduce the activity of enzymes that break down different organic molecules. It reduces soil carbon and nitrogen, which in turn reduces the amount of bacteria that are sensitive to variations in soil nitrogen levels, such as Verrucomicrobia, Proteobacteria, and Acidobacteria, and even Cyanobacteria is sensitive to drought (Islam et al. 2020). Gram-negative bacteria, Bacteroidetes, and Gammaproteobacteria showed the biggest decline in relative abundance due to drought conditions. Planctomycetes sp. is one of the Gram-negative bacterial species that is found to be resistant to drought (Chodak et al. 2015; Meisner et al. 2018). It is observed that fungi grow more widely during drought, and studies have indicated that an increase in antibiotic synthesis in soil during droughts is a reaction to compete for decreasing food supplies (Siebert et al. 2019). There are certain fungi that are sensitive to variations in soil moisture, although most research has indicated that fungi are less sensitive to the effects of drought (Hartmann et al. 2017). Ascomycota, especially Sordariomycetes and Dothideomycetes, are among the fungi that are affected by drought conditions (Meisner et al. 2018). The whole soil food chain is impacted by drought. Microbial predators such as phages and nematodes use cilia to pass through holes filled with water. When there is no water in the soil, they become immobile and there is less predatory activity (Furtak and Wolińska 2023).

4 Effects of Climate Change on Beneficial Microbes

Ecology seeks to comprehend the creation and preservation of biodiversity over place and time, with a focus on how climate affects the relationships and structure of communities, as well as their functions (Jain et al. 2020). The long-term change of irregular weather patterns, such as temperature, rainfall, and wind, may thus be characterized as climate change. An unparalleled loss of biodiversity and endangered species on Earth may result from these developments. However, whereas the extinction of plant and animal species has been extensively researched and recorded, little attention is usually paid to how climate change affects microorganisms. In comparison to other suggested environmental drivers, global warming plays a major role in shaping soil microbial diversity by raising the temperature and lowering the moisture content of the surface soil. Naturally, all the environmental factors impacted by climate change, such as variations in precipitation patterns, drought stress, and changes in radiation, global warming has a pervasive influence on the variation in soil microbial diversity (Guo et al. 2018).

Microbes play an important role in the phenomenon of changing climate. They can generate and absorb gases like carbon dioxide (CO_2), methane (CH_4), and nitrous oxide (N_2O), which are responsible for 98% of global warming. Recently, human activity has led to a rise in high amounts of these gases, which are naturally produced by microbes. These changes have given microbes more access to carbon and nitrogen, which they then use to produce these three products. Our options include using microorganisms to recycle waste carbon or nitrogen into stable and beneficial chemicals or understanding and putting into practice methods that inhibit microbial activity to slow the creation of these gases (Tiedje et al. 2022). Fortunately, microbes utilizing these gases for their growth are methane oxidizers (Methanotrophs), denitrifiers (Nitrous oxide reduction), and photo- or chemoautotrophs (Cyanobacteria. algae). The interaction between the microbes and the surroundings is often influenced by human activity, whether they will release the gases or consume it (Grover et al. 2015).

When climate changes and accelerates to high temperatures, it leads to microbial decomposition and the rise of CO_2 emissions. As a result, it will lead to an increase in carbon dioxide in the soil. However, it has been observed that a rise in atmospheric CO_2 has been shown to increase rhizosphere-colonizing bacteria like *Burkholderia* and *Pseudomonas*, as well as fungi, for plant development, while non-rhizospheric species like Bacillus were not encouraged (Kannojia et al. 2019). In controlled conditions, this CO_2 enrichment promotes the growth of rhizobial populations and the rise in N-fixing bacteria; but in natural systems, rhizobial responses are reduced by several resource boundaries. On the other hand, high temperatures cause drought, which inhibits the growth of arbuscular mycorrhizal fungi (Mekala and Polepongu 2019). Second, as a GHG, methane has an approximately 30 times greater impact than CO_2. It is a simple hydrocarbon. Although microorganisms naturally create between 70% and 90% of the methane produced, changes in human activity have contributed to some of the recent increases in microbe production

(Masood et al. 2021). Methanotroph species are essential to maintain global equilibrium, as they have the ability to consume methane as an energy source. Methylococcus and Methylobacter species are two examples of well-known methanotrophs (Vicuña and González 2021). As a result, the interaction between microorganisms (such as bacteria, fungus, unicellular algae, and protozoa) and climate change will be discussed in detail along many human sectors, such as the industrial and health sectors.

5 Plant–Microbe Interaction Under Climate Change

The intricate dance between plants and the microbial communities that live in them is significantly impacted by climate change, which is characterized by rising temperatures, altered patterns of precipitation, and an increase in the frequency of extreme weather events (Wang et al. 2018). Plants face increased stress as the planet warms, affecting their physiology, growth, and development. The surrounding microbial communities' composition and function are altered because of these changes (Solomon et al. 2023). Under stressful conditions, beneficial microorganisms like mycorrhizal fungi and rhizobia may become less prevalent or less effective at promoting plant growth and health. Pathogenic microbes, on the other hand, may thrive in warmer, more humid environments, which can increase the number of disease outbreaks and crop losses. Changes in the chemistry of the soil and the availability of nutrients, both of which are influenced by climate change, further disrupt the delicate balance that exists between beneficial and harmful microbes (Kniazev and Maklakova 2018). Plant interactions with microbes become even more essential for survival and resilience as they struggle to adapt to these new environmental challenges. Developing strategies to mitigate the negative effects of climate change on agriculture and natural ecosystems requires an understanding of these intricate relationships (Vives-Peris et al. 2020).

Mod.ulation of Plant Root Exudates The rhizosphere microbiome, a powerful environment at the plant-soil interface, is significantly impacted by the compound exchange among plants and microorganisms. Vital to this cooperation is the emission of root exudates, a perplexing combination of natural mixtures that fills in as both an energy source and flagging particle for microbial networks (Berendsen et al. 2012). These exudates contain sugars, organic acids, amino acids, and secondary metabolites, and their composition varies greatly depending on the plant species, developmental stage, environmental conditions, and biotic interactions (Huang et al. 2014). Plants apply mind boggling command over exudate creation, changing their synthesis because of different improvements. Upsetting circumstances, like dry season, supplement lack, or microbe assault, frequently trigger changes in exudate profiles (Pantigoso et al. 2022). Additionally, plant hormones like auxin and ethylene are crucial in regulating the patterns of exudation. Hereditary factors additionally add to root exudate variety among plant genotypes (Yuan et al. 2018). Plants can alter the microbiome of their rhizosphere by selectively releasing particular exudates. Chemoattractants are substances that attract beneficial microbes while repelling potential pathogens. The spatial heterogeneity of exudate dissemination inside the rhizosphere makes different microenvironments that help microbial gatherings (Wang et al. 2021). After some time, co-developmental cycles have refined the capacity of plants to control exudates and enlist useful microorganisms (Walker et al. 2003). The utilitarian ramifications of the rhizosphere microbiome, moulded by root exudates, are sweeping. The biocontrol of pathogens, nutrient cycling, solubilizing nutrients, and plant growth, all are supported by residual microbial communities (Berendsen et al. 2012). Rhizosphere microbes may be able to produce siderophores, antibiotics, and growth-promoting substances from specific exudate combinations (Huang et al. 2014). Eventually, the construction and capability of the rhizosphere microbiome, as organized by root exudates, decide its large advantageous limit with respect to establish wellbeing and efficiency (Wang et al. 2021). Describing the exudate intensities associated with microbial enlistment and work will be pivotal for creating techniques to design rhizospheres with upgraded plant execution (Pantigoso et al. 2022). Additionally, sustainable agriculture and ecosystem management will benefit from learning more about the microbiome of the rhizosphere and its role in mitigating biotic and abiotic stresses (Pantigoso et al. 2022).

Changes in Plant Defence Responses To defend themselves against a wide variety of pathogens, plants have developed intricate defence mechanisms. These defences are multifaceted and include immune responses, chemical defences, and physical barriers (Walker et al. 2003). In any case, the powerful idea of plant-microorganism connections frequently prompts changes in plant safeguard reactions. Numerous microbes emit effector proteins to robust plant protections. These effectors can target different parts of the plant invulnerable framework, prompting modified guard reactions (Dodds and Rathjen 2010) Various races of a similar microbe can fluctuate in their destructiveness, prompting different plant protection reactions (Carolan et al. 2017). Plants have a wide variety of resistance genes that make them immune to pathogens. Plant defence responses can be significantly affected by the presence or absence of these genes (Osuna-Cruz et al. 2018). The effectiveness of defence responses can be influenced by plant genetic diversity (Panstruga and Moscou 2020). Plant defences can be compromised by drought, salinity, and nutrient deficiency, making them more susceptible to pathogens (Zhang et al. 2022). They face new challenges because of shifting climatic conditions, which can alter pathogen distribution and prevalence (Juroszek et al. 2020).

Altered Nutrient Acquisition The consumption of nutrients is an essential stage in the development, growth, and survival of plants. It involves transporting vital nutrients from soil to various plant tissues for use in metabolic processes. Plants require a plethora of balanced macronutrients (nitrogen, phosphorus, potassium, calcium, magnesium, and sulphur) and micronutrients (iron, manga-

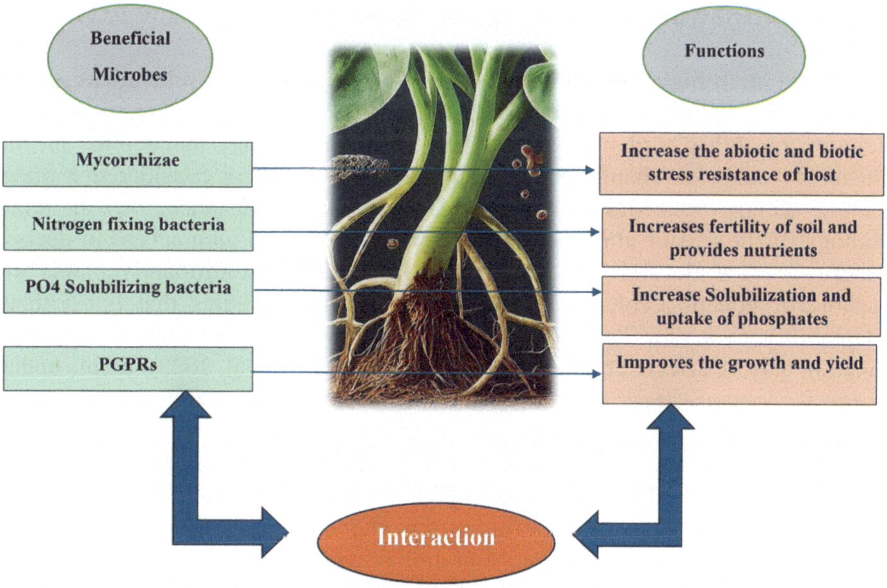

Fig. 5.3 Beneficial microbes and plant health: This image illustrates the beneficial interactions between plants and various microorganisms in the rhizosphere. The complex interactions between these organisms highlight the importance of understanding microbial diversity and harnessing its potential for sustainable agriculture

nese, zinc, copper, boron, molybdenum, chlorine, and nickel) for ideal host development (White et al. 2012) Lack of nutrients restricts plant growth, resulting in stunted growth, smaller leaves, and decreased yield. Leaf yellowing is caused by a lack of chlorophyll caused by nitrogen, magnesium, or iron deficiency. Deficits in elements like potassium, calcium, and magnesium cause tissue death (Lambers et al. 2019b). Supplement inadequacies debilitate photosynthesis, prompting diminished energy creation and development. Lack in phosphorus and potassium can influence bloom and organic product arrangement. Supplement-insufficient plants are more defenceless against assaults (Nieves-Cordones et al. 2020). Extreme supplement levels can be harmful to plants, prompting supplement awkward nature and physiological problems. One nutrient's uptake can be hindered by high concentrations of another. Soil acidification caused by excessive fertilization can affect nutrient availability and microbial activity (Lambers et al. 2019a). Eutrophication can be exacerbated by nutrient runoff contaminating bodies of water. A few supplements seek take-up, prompting irregular characteristics. High potassium levels, for instance, may impede calcium absorption (Rizwan et al. 2018) (Fig. 5.3). A few supplements improve the take-up or use of others. For example, nitrogen and phosphorus frequently cooperate for ideal plant development. For healthy plants, nutrients must be balanced. Unseemly proportions can prompt lack or poison levels (Tan et al. 2019).

6 Mitigation and Adaptation Strategies

Plants and microbial communities are both forced to adapt or relocate as global ecosystems are altered by climate change (Mendes et al. 2016). Phenotypic plasticity, or the ability of plants to adapt their growth, physiology, and defence mechanisms to adapt to changing environmental conditions, is considered a good feature (Kniazev and Maklakova 2018). They might, for instance, alter the morphology of the leaf or alter the architecture of the root to gain access to deeper water sources at elevated CO_2 levels (Wang et al. 2018). Additionally, plants are capable of genetic evolution, acquiring novel characteristics that boost their adaptability to a new environment. Microbial communities' composition and function may be affected by these adaptations (Solomon et al. 2023). Plants under stress, for instance, may release various root exudates, attracting microbes that can provide beneficial services like the acquisition of nutrients or the suppression of pathogens. Plants may, in some cases, follow shifting climatic zones and migrate to more suitable environments (Whithman et al. 2020). However, factors like habitat fragmentation, competition from other species, and limitations on dispersal may impede this migration. Migration or adaptation strategies may also be displayed by microbial communities (Png et al. 2023). Wind, water, and animal vectors are just a few of the many ways in which microbes can spread. They are also capable of rapid evolution, acquiring new genetic characteristics that enable them to adapt to changing environments and thrive. The rate of climate change, the characteristics of the organisms involved, and the conditions of the environment all have an impact on how plant and microbial migration and adaptation interact (Trivedi et al. 2020). Predicting the future of interactions between plants and microbes, developing strategies to preserve biodiversity, and ensuring ecosystem resilience in the face of climate change all require an understanding of these processes (Vives-Peris et al. 2020).

Agronomic Practices to Enhance Rhizosphere Resilience Some agronomic practices like crop rotation, cover cropping, and organic amendments must be applied to mitigate the adverse climatic factors. The rhizosphere, the powerful connection point between plant roots and soil, assumes a pivotal part in plant wellbeing and efficiency. Its strength, or capacity to endure and recuperate from aggravations, is fundamental for practical farming (Bowles et al. 2020). Crop rotation, cover cropping, and organic amendments are examples of agronomic practices that have a significant impact on the health and function of the rhizosphere. The methodical planting of various crop species in succession on the same field is known as crop rotation (Poeplau and Don 2015). Various yields have shifting helplessness to vermin and sicknesses, lessening the development of microorganisms in the dirt. Various underground roots of various harvests upgrade soil accumulation and water penetration (Baysal-Gurel et al. 2018). Preventing nutrient depletion and increasing soil fertility are the major goals of cultivation of successive crops in the same field. The diverse microbial communities that are supported by the various crop root exudates boost the biological

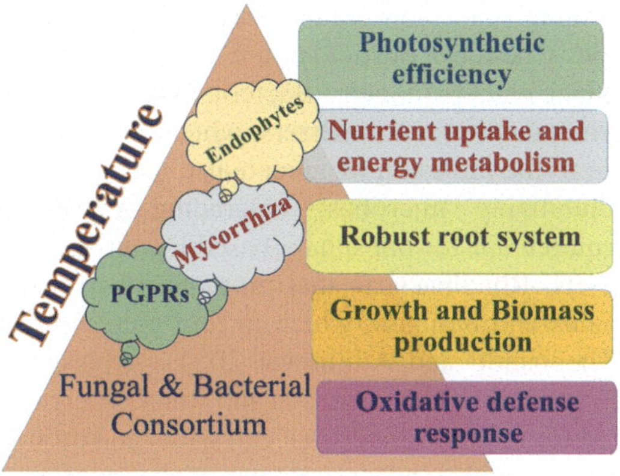

Fig. 5.4 The positive impact of microbes on plant growth and resilience: These microbial partners positively influence various aspects of plant physiology, including photosynthetic efficiency, nutrient uptake, energy metabolism, and the development of a robust root system

activity of the soil (Fairfax and Westbrook 2024). Cover crops are plants that are grown primarily for the purpose of improving the soil and helping the ecosystem, acting as a safeguard for the dirt surface from wind and water disintegration (Poeplau and Don 2015). Their buildups add to soil natural matter, further developing soil design and water-holding limit. Cover harvests can contend with weeds for assets, lessening weed pressure (He et al. 2016). To make nutrients available for subsequent crops, cover crops can either mine nutrients from deeper soil layers or fix nitrogen from the atmosphere, and thus they facilitate the beneficial microorganisms and insects a place for food and survival (Wezel 2017). Natural matter further develops soil conglomeration, air circulation, and water invasion. Organic amendments help to reduce the effects of climate change by storing carbon in the soil (Chi et al. 2022). They release nutrients slowly, increasing plant nutrient availability and lowering the likelihood of nutrient leaching (Amundson and Biardeau 2018). Natural revisions support a different microbial local area, improving supplement cycling and soil fruitfulness (Dierks et al. 2021) (Fig. 5.4).

Inoculation with Beneficial Microbes Global agriculture faces significant challenges from climate change, such as altered precipitation patterns, increased temperatures, and more extreme weather. Crop yields and soil health are negatively impacted by these factors (Rochlani et al. 2022). A promising approach for reducing the effects of climate change and enhancing agricultural sustainability is to inoculate crops with beneficial microbes (Van Der Heijden et al. 2015). Plant growth and development can be significantly aided by beneficial microbes like nitrogen-fixing bacteria, arbuscular mycorrhizal fungi, and plant growth-promoting rhizobacteria (PGPR). PGPR and AMF boost the acquisition of nutrients, particularly nitrogen and phosphorus, lowering the requirement for synthetic fertilizers (Miljaković et al. 2020). Certain microorganisms can further develop plant water use proficiency and stress resilience by regulating chemical creation and cell reinforcement movement, and other soil organisms add to soil conglomeration, further developing water penetration and maintenance (Hasan et al. 2024). By decomposing organic matter and forming stable soil organic matter, soil microbes are essential to carbon sequestration. Chemical pesticides are less necessary because some beneficial microbes reduce plant pathogens. Bacteria that fix nitrogen: These microbes reduce the use of synthetic nitrogen fertilizers, which require a lot of energy to produce and contribute to the emission of greenhouse gases (Lambers and Plaxton 2015). Phosphate-solubilizing microbes: Upgrading phosphorus accessibility diminishes the requirement for phosphate-based manures, frequently mined from non-sustainable assets. Rhizobacteria that promote plant growth (PGPR): PGPR can boost crop yields and biomass, which in turn increases carbon sequestration in plant tissues (Thepbandit and Athinuwat 2024). Arbuscular mycorrhizal fungi (AMF): AMF boosts carbon sequestration and water efficiency while also enhancing soil structure and fertility. Microbes that can withstand drought: Inoculating crops with microbes that can withstand drought can make crops more resilient to water scarcity. Heat-lenient microorganisms: Organisms adjusted to high temperatures can further develop plant heat resistance and efficiency in evolving environments. Salt-lenient microorganisms: In seaside regions, salt-open minded organisms can assist crops with enduring saltiness stress (Abdalla et al. 2023). Microbes for nutrient-poor soils: In nutrient-poor soils, crop performance can be improved by inoculating crops with microbes that can access nutrients from less accessible sources (Chauhan et al. 2023).

Developing Climate-Resilient Plant Genotypes (Breeding for Improved Plant–Microbe Interactions) Global food security is in grave danger from climate change, which is characterized by rising temperatures, erratic precipitation patterns, and extreme weather. As a result, it is becoming increasingly urgent to develop crop cultivars that can withstand a variety of environmental stresses (Clúa et al. 2018). The intricate interaction between plants and the microbes that live in the rhizosphere, the dynamic soil interface that is greatly influenced by root exudates, is crucial to agricultural sustainability (Vandenkoornhuyse et al. 2015). A flourishing rhizosphere microbiome, made of gainful microorganisms, for example, plant development advancing rhizobacteria (PGPR), arbuscular mycorrhizal organisms (AMF), and diazotrophic microscopic organisms, is instrumental in enlarging plant wellbeing, power, and efficiency (Bulgarelli et al. 2023). To encourage versatile agroecosystems, an essential methodology includes rearing for pull models streamlined for productive supplement and water procurement, and subsequently establishing a helpful climate for microbial colonization (Zhang et al. 2023). Furthermore, enhancing plant-microbe symbioses through the manipulation of root exudate profiles to both attract beneficial microbes and discourage pathogens is a promising strategy (Prasad et al. 2023). Plant resilience can be further strengthened through selective breeding for genotypes that can recruit microbial consortia. At the same time, cultivars that are better able to withstand biotic and abiotic stresses can help the microbiome in the rhizosphere indirectly (Pérez-Jaramillo et al. 2016) (Fig. 5.5).

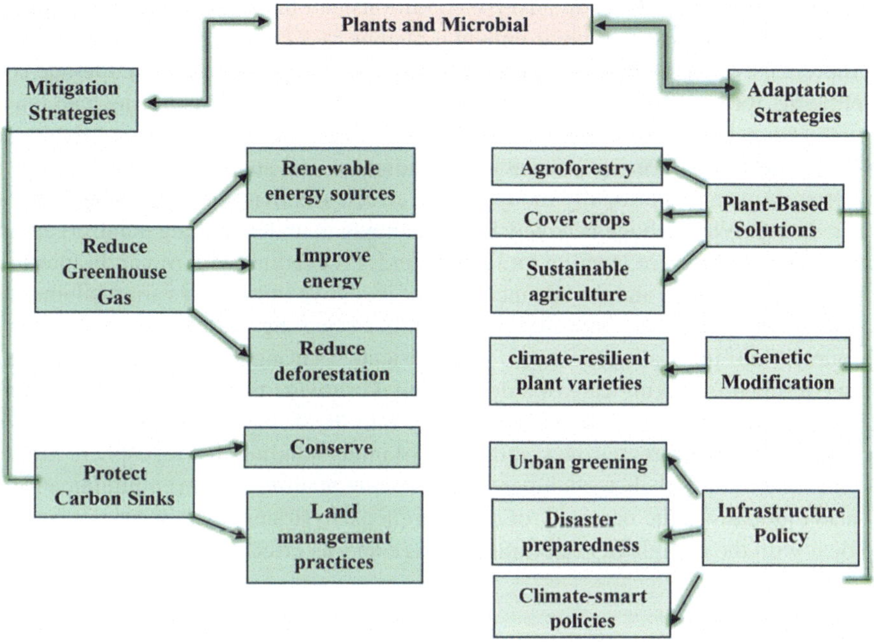

Fig. 5.5 Strategies for mitigating and adapting to climate change through plant–microbe interactions: This image presents a comprehensive framework for addressing climate change through strategies that involve both mitigation and adaptation. Mitigation strategies focus on reducing greenhouse gas emissions, while adaptation strategies aim to prepare for and cope with the inevitable impacts of climate change

7 Conclusion

A crucial component of agricultural productivity and resilience is the rhizosphere, a dynamic ecosystem teeming with microbial life. Notwithstanding, the ghost of environmental change creates a long-shaded area over this multifaceted relationship. As the worldwide environment goes through quick change, described by heightening temperatures, inconsistent precipitation examples, and outrageous climate occasions, the sensitive harmony among plants and their microbial partners is progressively tested. These changes in natural circumstances can upset microbial networks, weaken supplement cycling, and compromise plant wellbeing, eventually affecting farming yields and food security. To explore these difficulties, a far-reaching comprehension of the system's basic plant-organism collaborations is foremost. By explaining the elements that impact microbial local area gathering, capability, and solidness because of environmental change, scientists can recognize procedures to upgrade rhizosphere strength. Reproducing for crop genotypes that have characteristics advancing valuable microbial affiliations, for example, exudation of explicit root compounds or the improvement of specific root structures, is a promising road. In addition, certain strains of microbial inoculants can be used to improve plant performance in adverse conditions and the rhizosphere microbiome. It is basic to perceive that environmental change is a complex test that requires a comprehensive methodology. Incorporating horticultural practices that upgrade soil wellbeing, for example, crop turn, cover editing, and decreased culturing, can make great circumstances for helpful organisms to flourish. Also, investigating the capability of accuracy farming innovations to advance and supplement the board and water system can additionally moderate the effects of environmental change on the rhizosphere. While a huge headway has been made in understanding plant-organism communications, there is still a lot to be found. Proceeding with research, interdisciplinary joint effort, and the advancement of inventive innovations are fundamental to open the maximum capacity of this mind-boggling framework. By putting resources into innovative work, we can fabricate a stronger and practical farming area equipped for taking care of a developing worldwide populace notwithstanding environmental change. This end digs further into the difficulties presented by environmental change, stresses the significance of understanding the rhizosphere microbiome, and frames likely techniques for improving plant-organism communications. It also emphasizes the necessity of a comprehensive strategy and ongoing research to deal with the complexity of climate change and its effects on agriculture.

References

Abdalla M, Bitterlich M, Jansa J, Püschel D, Ahmed MA (2023) The role of arbuscular mycorrhizal symbiosis in improving plant water status under drought. J Exp Bot 74(16):4808–4824

Ahkami AH, White RA III, Handakumbura PP, Jansson C (2017) Rhizosphere engineering: enhancing sustainable plant ecosystem productivity. Rhizosphere 3:233–243

Ahmed S, Rahman S, Hasan M, Paul N, Sajib AA (2018) Microbial degradation of lignocellulosic biomass: discovery of novel natural lignocellulolytic bacteria. Biotechnologia 99(2):137–146

Akhilesh Kumar AK, Verma JP (2018) Does plant-microbe interaction confer stress tolerance in plants: a review? Microbiol Res 207:41–52

Amundson R, Biardeau L (2018) Soil carbon sequestration is an elusive climate mitigation tool. Proc Natl Acad Sci 115(46):11652–11656

Baysal-Gurel F, Liyanapathiranage P, Mullican J (2018) Biofumigation: opportunities and challenges for control of soilborne diseases in nursery production. Plant Health Progress 19(4):332–337

Berendsen RL, Pieterse CM, Bakker PA (2012) The rhizosphere microbiome and plant health. Trends Plant Sci 17(8):478–486

Bowles TM, Mooshammer M, Socolar Y, Calderón F, Cavigelli MA, Culman SW, Deen W, Drury CF, y Garcia AG, Gaudin AC, Harkcom WS (2020) Long-term evidence shows that crop-rotation diversification increases agricultural resilience to adverse growing conditions in North America. One Earth 2(3):284–293

Bulgarelli D, Schlaeppi K, Spaepen S, Van Themaat EVL, Schulze-Lefert P (2023) Structure and functions of the bacterial microbiota of plants. Annu Rev Plant Biol 64(1):807–838

Canarini A, Kaiser C, Merchant A, Richter A, Wanek W (2019) Root exudation of primary metabolites: mechanisms and their roles in plant responses to environmental stimuli. Front Plant Sci 10:157

Carolan K, Helps J, van den Berg F, Bain R, Paveley N, van den Bosch F (2017) Extending the durability of cultivar resistance by limiting epidemic growth rates. Proc R Soc B Biol Sci 284(1863):20170828

Chauhan P, Sharma N, Tapwal A, Kumar A, Verma GS, Meena M, Seth CS, Swapnil P (2023) Soil microbiome: diversity, benefits and interactions with plants. Sustain For 15(19):14643

Chen J, Liu Q, Yu L, Korpelainen H, Niinemets Ü, Li C (2021) Elevated temperature and CO_2 interactively modulate sexual competition and ecophysiological responses of dioecious *Populus cathayana*. For Ecol Manag 481:118747

Cheng YT, Zhang L, He SY (2019) Plant-microbe interactions facing environmental challenge. Cell Host Microbe 26(2):183–192

Chi J, Fan Y, Wang L, Putnis CV, Zhang W (2022) Retention of soil organic matter by occlusion within soil minerals. Rev Environ Sci Biotechnol 21(3):727–746

Chodak M, Gołębiewski M, Morawska-Płoskonka J, Kuduk K, Niklińska M (2015) Soil chemical properties affect the reaction of forest soil bacteria to drought and rewetting stress. Ann Microbiol 65:1627–1637

Clúa J, Roda C, Zanetti ME, Blanco FA (2018) Compatibility between legumes and rhizobia for the establishment of a successful nitrogen-fixing symbiosis. Gene 9(3):125

Compant S, Van Der Heijden MG, Sessitsch A (2010) Climate change effects on beneficial plant–microorganism interactions. FEMS Microbiol Ecol 73(2):197–214

Compant S, Cambon MC, Vacher C, Mitter B, Samad A, Sessitsch A (2021) The plant endosphere world–bacterial life within plants. Environ Microbiol 23(4):1812–1829

Dierks J, Blaser-Hart WJ, Gamper HA, Nyoka IB, Barrios E, Six J (2021) Trees enhance abundance of arbuscular mycorrhizal fungi, soil structure, and nutrient retention in low-input maize cropping systems. Agric Ecosyst Environ 318:107487

Dodds PN, Rathjen JP (2010) Plant immunity: towards an integrated view of plant–pathogen interactions. Nat Rev Genet 11(8):539–548

Fairfax E, Westbrook C (2024) The ecology and evolution of beavers: ecosystem engineers that ameliorate climate change. Annu Rev Ecol Evol Syst 55:323–345

Furtak K, Wolińska A (2023) The impact of extreme weather events as a consequence of climate change on the soil moisture and on the quality of the soil environment and agriculture—a review. Catena 231:107378

Grace MA, Achick TFE, Bonghan BE, Bih ME, Ngo NV, Ajeck MJ, Prudence GTB, Ntungwen FC (2019) An overview of the impact of climate change on pathogens, pest of crops on sustainable food biosecurity. Int J Ecotoxicol Ecobiol 4:114–119

Grover M, Maheswari M, Desai S, Gopinath KA, Venkateswarlu B (2015) Elevated CO_2: plant associated microorganisms and carbon sequestration. Appl Soil Ecol 95:73–85

Guo X, Feng J, Shi Z, Zhou X, Yuan M, Tao X, Hale L, Yuan T, Wang J, Qin Y, Zhou A (2018) Climate warming leads to divergent succession of grassland microbial communities. Nat Clim Chang 8(9):813–818

Gupta A, Bano A, Rai S, Mishra R, Singh M, Sharma S, Pathak N (2022) Mechanistic insights of plant-microbe interaction towards drought and salinity stress in plants for enhancing the agriculture productivity. Plant Stress 4:100073

Gupta R, Patil R (2021) Phyllospheric microbes: diversity, functions, interaction, and applications in agriculture. In: Current trends in microbial biotechnology for sustainable agriculture. Springer, pp 301–323

Hartmann M, Brunner I, Hagedorn F, Bardgett RD, Stierli B, Herzog C, Chen X, Zingg A, Graf-Pannatier E, Rigling A, Frey B (2017) A decade of irrigation transforms the soil microbiome of a semi-arid pine forest. Mol Ecol 26(4):1190–1206

Hasan A, Tabassum B, Hashim M, Khan N (2024) Role of plant growth promoting rhizobacteria (PGPR) as a plant growth enhancer for sustainable agriculture: a review. Bacteria 3(2):59–75

He DC, Zhan JS, Xie LH (2016) Problems, challenges and future of plant disease management: from an ecological point of view. J Integr Agric 15(4):705–715

Huang XF, Chaparro JM, Reardon KF, Zhang R, Shen Q, Vivanco JM (2014) Rhizosphere interactions: root exudates, microbes, and microbial communities. Botany 92(4):267–275

Igiehon NO, Babalola OO (2018) Below-ground-above-ground plant-microbial interactions: focusing on soybean, rhizobacteria and mycorrhizal fungi. Open Microbiol J 12:261

Islam W, Noman A, Naveed H, Huang Z, Chen HY (2020) Role of environmental factors in shaping the soil microbiome. Environ Sci Pollut Res 27:41225–41247

Jacoby R, Peukert M, Succurro A, Koprivova A, Kopriva S (2017) The role of soil microorganisms in plant mineral nutrition—current knowledge and future directions. Front Plant Sci 8:1617

Jain PK, Purkayastha SD, De Mandal S, Passari AK, Govindarajan RK (2020) Effect of climate change on microbial diversity and its functional attributes. In: Recent advancements in microbial diversity. Academic, pp 315–331

Jansson JK, Hofmockel KS (2020) Soil microbiomes and climate change. Nat Rev Microbiol 18(1):35–46

Juroszek P, Racca P, Link S, Farhumand J, Kleinhenz B (2020) Overview on the review articles published during the past 30 years relating to the potential climate change effects on plant pathogens and crop disease risks. Plant Pathol 69(2):179–193

Kannojia P, Sharma PK, Sharma K (2019) Climate change and soil dynamics: effects on soil microbes and fertility of soil. In: Climate change and agricultural ecosystems. Woodhead Publishing, pp 43–64

Khan MA, Asaf S, Khan AL, Jan R, Kang SM, Kim KM, Lee IJ (2020) Thermotolerance effect of plant growth-promoting *Bacillus cereus* SA1 on soybean during heat stress. BMC Microbiol 20:1–14

Kniazev VI, Maklakova EA (2018) Plant-soil feedbacks: the past, the present and future challenges. In: Актуальные проблемы современной науки. pp 112–115

Lambers H, Plaxton WC (2015) Phosphorus: back to the roots. In: Annual plant reviews volume 48: phosphorus metabolism in plants, vol 48. Wiley, pp 1–22

Lambers H, Albornoz FE, Arruda AJ, Barker T, Finnegan PM, Gille C, Gooding H, Png GK, Ranathunge K, Zhong H (2019a) Nutrient-acquisition strategies. In: A jewel in the crown of a global biodiversity hotspot. Kwongan Foundation and the Western Australian Naturalists' Club Inc, Perth, pp 227–248

Lambers H, Oliveira RS, Lambers H, Oliveira RS (2019b) Mineral nutrition. In: Plant physiological ecology. University of Western Australia, pp 301–384

Lu H, Liu S, Wang H, Luan J, Schindlbacher A, Liu Y, Wang Y (2017) Experimental throughfall reduction barely affects soil carbon dynamics in a warm-temperate oak forest, Central China. Sci Rep 7(1):15099

Manzanera M (2021) Dealing with water stress and microbial preservation. Environ Microbiol 23(7):3351–3359

Masood F, Ahmad S, Malik A (2021) Role of methanotrophs in mitigating global warming. In: Microbiomes and the global climate change. Springer, Singapore, pp 43–60

Meisner A, Jacquiod S, Snoek BL, Ten Hooven FC, Van der Putten WH (2018) Drought legacy effects on the composition of soil fungal and prokaryote communities. Front Microbiol 9:294

Mekala S, Polepongu S (2019) Impact of climate change on soil microbial community. In: Plant biotic interactions: state of the art. Springer, Cham, pp 31–41

Mendes R, Garbeva P, Raaijmakers JM (2019) The rhizosphere microbiome: significance of plant beneficial, plant pathogenic, and human pathogenic microorganisms. FEMS Microbiol Rev 37(5):634–663

Miljaković D, Marinković J, Balešević-Tubić S (2020) The significance of Bacillus spp. in disease suppression and growth promotion of field and vegetable crops. Microorganisms 8(7):1037

Mohanram S, Kumar P (2019) Rhizosphere microbiome: revisiting the synergy of plant-microbe interactions. Ann Microbiol 69:307–320

Montazeaud G, Violle C, Fréville H, Luquet D, Ahmadi N, Courtois B, Bouhaba I, Fort F (2018) Crop mixtures: does niche complementarity hold for belowground resources? An experimental test using rice genotypic pairs. Plant Soil 424:187–202

Nadarajah K, Abdul Rahman NSN (2021) Plant–microbe interaction: aboveground to belowground, from the good to the bad. Int J Mol Sci 22(19):10388

Nieves-Cordones M, Rubio F, Santa-María GE (2020) Nutrient use-efficiency in plants: an integrative approach. Front Plant Sci 11:623976

Nizamani MM, Hughes AC, Qureshi S, Zhang Q, Tarafder E, Das D, Acharya K, Wang Y, Zhang ZG (2024) Microbial biodiversity and plant functional trait interactions in multifunctional ecosystems. Appl Soil Ecol 201:105515

Osuna-Cruz CM, Paytuvi-Gallart A, Di Donato A, Sundesha V, Andolfo G, Aiese Cigliano R, Sanseverino W, Ercolano MR (2018) PRGdb 3.0: a comprehensive platform for prediction and analysis of plant disease resistance genes. Nucleic Acids Res 46(D1):D1197–D1201

Panstruga R, Moscou MJ (2020) What is the molecular basis of nonhost resistance? Mol Plant-Microbe Interact 33(11):1253–1264

Pantigoso HA, Newberger D, Vivanco JM (2022) The rhizosphere microbiome: plant–microbial interactions for resource acquisition. J Appl Microbiol 133(5):2864–2876

Pérez-Jaramillo JE, Mendes R, Raaijmakers JM (2016) Impact of plant domestication on rhizosphere microbiome assembly and functions. Plant Mol Biol 90:635–644

Png GK, De Long JR, Fry EL, Heinen R, Heinze J, Morriën E, Sapsford SJ, Teste FP (2023) Plant-soil feedback: the next generation. Plant Soil 485(1):1–5

Poeplau C, Don A (2015) Carbon sequestration in agricultural soils via cultivation of cover crops—a meta-analysis. Agric Ecosyst Environ 200:33–41

Prasad JVNS, Veni VG, Srinivasarao C, Kundu S, Ramakrishna B, Reddy KS, Singh R, Singh SK, Murai AS, Rohilla PP, Makkar GS (2023) Can adoption of climate resilient management practices achieve carbon neutrality in traditional green revolution states of Punjab and Haryana? J Environ Manag 338:117761

Rawal R, Scheerens JC, Fenstemaker SM, Francis DM, Miller SA, Benitez MS (2022) Novel Trichoderma isolates alleviate water deficit stress in susceptible tomato genotypes. Front Plant Sci 13:869090

Rizwan M, Ali S, ur Rehman MZ, Rinklebe J, Tsang DC, Bashir A, Maqbool A, Tack FMG, Ok YS (2018) Cadmium phytoremediation potential of Brassica crop species: a review. Sci Total Environ 631:1175–1191

Rochlani A, Dalwani A, Shaikh N, Shaikh N, Sharma S, Saraf M (2022) Plant growth promoting rhizobacteria as biofertilizers: application in agricultural sustainability. Acta Sci Microbiol (ISSN: 2581–3226) 5(4):12–21

Semchenko M, Lepik A, Abakumova M, Zobel K (2018) Different sets of belowground traits predict the ability of plant species to suppress and tolerate their competitors. Plant Soil 424:157–169

Sharma B, Singh BN, Dwivedi P, Rajawat MVS (2022) Interference of climate change on plant-microbe interaction: present and future prospects. Front Agron 3:725804

Shi X, Zhao Y, Xu M, Ma L, Adams JM, Shi Y (2023) Insight into plant-microbe interactions in rhizosphere for sustainable agriculture in the new crops era. New Crops 1:100004

Shree B, Jayakrishnan U, Bhushan S (2022) Impact of key parameters involved with plant-microbe interaction in context to global climate change. Front Microbiol 13:1008451

Siebert J, Sünnemann M, Auge H, Berger S, Cesarz S, Ciobanu M, Guerrero-Ramírez NR, Eisenhauer N (2019) The effects of drought and nutrient addition on soil organisms vary across taxonomic groups, but are constant across seasons. Sci Rep 9(1):639

Solomon W, Janda T, Molnár Z (2023) Unveiling the significance of rhizosphere: implications for plant growth, stress response, and sustainable agriculture. Plant Physiol Biochem 206:108290

Tan X, Li K, Wang Z, Zhu K, Tan X, Cao J (2019) A review of plant vacuoles: formation, located proteins, and functions. Plan Theory 8(9):327

Tecon R, Or D (2017) Biophysical processes supporting the diversity of microbial life in soil. FEMS Microbiol Rev 41(5):599–623

Teja AR, Leona G, Prasanth J, Yatung T, Singh S, Bhargav V (2023) Role of plant growth–promoting rhizobacteria in sustainable agriculture. In: Advanced microbial technology for sustainable agriculture and environment. Academic, pp 175–197

Thepbandit W, Athinuwat D (2024) Rhizosphere microorganisms supply availability of soil nutrients and induce plant defense. Microorganisms 12(3):558

Tiedje JM, Bruns MA, Casadevall A, Criddle CS, Eloe-Fadrosh E, Karl DM, Nguyen NK, Zhou J (2022) Microbes and climate change: a research prospectus for the future. MBio 13(3):e00800–e00822
Trivedi P, Leach JE, Tringe SG, Sa T, Singh BK (2020) Plant–microbiome interactions: from community assembly to plant health. Nat Rev Microbiol 18(11):607–621
van Der Heijden MG, Martin FM, Selosse MA, Sanders IR (2015) Mycorrhizal ecology and evolution: the past, the present, and the future. New Phytol 205(4):1406–1423
Vandenkoornhuyse P, Quaiser A, Duhamel M, Le Van A, Dufresne A (2015) The importance of the microbiome of the plant holobiont. New Phytol 206(4):1196–1206
Velásquez AC, Castroverde CDM, He SY (2018) Plant–pathogen warfare under changing climate conditions. Curr Biol 28(10):R619–R634
Vicuña R, González B (2021) The microbial world in a changing environment. Rev Chil Hist Nat 94:1–5
Vives-Peris V, De Ollas C, Gómez-Cadenas A, Pérez-Clemente RM (2020) Root exudates: from plant to rhizosphere and beyond. Plant Cell Rep 39(1):3–17
Walker TS, Bais HP, Grotewold E, Vivanco JM (2003) Root exudation and rhizosphere biology. Plant Physiol 132(1):44–51
Wang J, Andersen SU, Ratet P (2018) Molecular and cellular mechanisms of the legume-rhizobia symbiosis. Front Plant Sci 9:1839
Wang NQ, Kong CH, Wang P, Meiners SJ (2021) Root exudate signals in plant–plant interactions. Plant Cell Environ 44(4):1044–1058
Wezel A (ed) (2017) Agroecological practices for sustainable agriculture: principles, applications, and making the transition. World Scientific
White PJ, Broadley MR, Gregory PJ (2012) Managing the nutrition of plants and people. Appl Environ Soil Sci 2012(1):104826
Whitham TG, Allan GJ, Cooper HF, Shuster SM (2020) Intraspecific genetic variation and species interactions contribute to community evolution. Annu Rev Ecol Evol Syst 51(1):587–612
Williams A, Pétriacq P, Beerling DJ, Cotton TA, Ton J (2018) Impacts of atmospheric CO_2 and soil nutritional value on plant responses to rhizosphere colonization by soil bacteria. Front Plant Sci 9:1493
Ye F, Li S, Li X, Lin J (2022) Response of rhizosphere microbiomes to climate change. In: Sustainable crop productivity and quality under climate change. Academic, pp 259–274
Yuan J, Zhao J, Wen T, Zhao M, Li R, Goossens P, Huang Q, Bai Y, Vivanco JM, Kowalchuk GA, Berendsen RL (2018) Root exudates drive the soil-borne legacy of aboveground pathogen infection. Microbiome 6:1–12
Zhang H, Shi L, Lu H, Shao Y, Liu S, Fu S (2020) Drought promotes soil phosphorus transformation and reduces phosphorus bioavailability in a temperate forest. Sci Total Environ 732:139295
Zhang H, Zhu J, Gong Z, Zhu JK (2022) Abiotic stress responses in plants. Nat Rev Genet 23(2):104–119
Zhang Y, Xu J, Li R, Ge Y, Li Y, Li R (2023) Plants' response to abiotic stress: mechanisms and strategies. Int J Mol Sci 24(13):10915
Zhu X, Wang H, Wang X (2022) Interaction of the rhizosphere microbiome and crops under climate change. In: Sustainable crop productivity and quality under climate change. Academic, pp 235–258

Emerging Role of Beneficial Microbes in Plant Stress Management Under Climate Change

6

Yashowardhan Singh, Jonnada Likhita, Sandhya Sinha, Meenakshi Arya, and Priyanshi Garg

Abstract

Climate change presents substantial difficulties to global agriculture, affecting agricultural output and sustainability due to extreme weather events, soil degradation, and heightened disease prevalence. In this context, advantageous microorganisms, such as endophytic bacteria, fungi, and plant growth-promoting rhizobacteria (PGPR), have surfaced as a viable strategy to augment plant resistance against various challenges. These bacteria synergistically interact with plants, altering their physiological and biochemical pathways to enhance tolerance to abiotic challenges such as drought, salt, and temperature variations, as well as biotic pressures like diseases and pests. Microbial inoculants can affect stress-responsive gene expression, augment antioxidant enzyme activity, and promote nutrient absorption, thus enhancing overall plant health and production. This study examines recent progress in comprehending microbial–plant interactions, emphasizing the processes by which these advantageous microorganisms alleviate stress-induced damage. Additionally, it examines novel strategies, like bioformulations and microbial consortia, that might transform sustainable agriculture in response to climate change. Integrating beneficial microorganisms into crop management strategies might enhance agroecosystem resilience and assure food security under climate change.

Y. Singh (✉) · J. Likhita · M. Arya · P. Garg
Department of Plant Pathology, Rani Lakshmi Bai Central Agriculture University, Jhansi, Uttar Pradesh, India

S. Sinha
Department of Entomology, Chhattisgarh Agriculture College, Dhanora, Chhattisgarh, India

U. Pankaj et al. (eds.), *Plant-microbiome Interactions for Climate-resilient Agriculture*, https://doi.org/10.1007/978-981-96-3534-4_6

Keywords

Agriculture · Climate change · Disease · Microorganism · Plant health · Interaction

1 Introduction

Climate is a critical factor influencing agricultural production (Bartlett et al. 2010; Lybbert and Sumner 2012). Climate change poses significant challenges to agriculture, impacting crop yields, water resources, and food security. The increase in global temperatures, altered precipitation patterns, and elevated CO_2 levels are key factors affecting agricultural productivity. These changes influence staple crops and have broader implications for food systems and economic welfare, particularly in vulnerable regions (Srivastav et al. 2021). Climate change influences the prevalence and distribution of pests and pathogens, leading to significant crop losses. This is particularly evident in staple crops like wheat and rice, where increased pest and disease incidence is expected (Wang et al. 2022). The loss of biodiversity due to climate change also affects agricultural ecosystems, reducing resilience and productivity (Ali Mozaffari 2022). Climate change poses significant challenges to agriculture by introducing various stress factors such as drought, salinity, heat, and pathogen pressure, collectively threatening crop productivity and food security. These stressors, often occurring in combination, exacerbate the vulnerability of agricultural systems worldwide.

Climate change poses a significant threat to global agriculture, leading to increased instances of abiotic stresses such as drought, salinity, and extreme temperatures, as well as biotic stresses that alter the distribution and behavior of pests and pathogens (Abewoy 2018; Rivero et al. 2022), biology and life cycles, and increasing the incidence and severity of plant diseases (Gupta et al. 2020; Kumar et al. 2024; Lesk et al. 2016). These environmental stressors adversely affect plant growth, development, and yield, resulting in considerable economic losses and jeopardizing food security worldwide (Zandalinas et al. 2018). Due to climate change, drought and heat waves are becoming more frequent and intense, severely affecting crop yields by impairing plant growth, photosynthesis, and physiological responses (Chaudhry and Sidhu 2021; Rivero et al. 2022). These conditions lead to increased evapotranspiration, reducing water availability for crops and exacerbating water scarcity issues (Ullah et al. 2021). Rising sea levels and altered precipitation patterns contribute to soil salinization, which negatively impacts plant growth and productivity. Salinity stress affects the chemical communication between plants and beneficial microorganisms, which are crucial for plant growth and stress tolerance (Abewoy 2018; Ullah et al. 2021). Heat stress (HS) is a primary consequence of climate change, leading to reduced plant disease resistance and lower crop yields. It affects plant physiology by forcing plants to prioritize defense mechanisms over growth. The frequency and severity of HS events are expected to increase, often occurring alongside biotic stresses, which can further exacerbate their impact on crops (Shelake et al. 2024).

To cope with these challenges, researchers have increasingly focused on exploring the potential of beneficial microbes in enhancing plant resilience and adaptation to such stress conditions (Egamberdieva et al. 2017). Beneficial microbes, including plant growth-promoting rhizobacteria (PGPR), mycorrhizal fungi, and endophytic bacteria, play a vital role in mitigating plant stress through various physiological and biochemical mechanisms (Abd-Allah et al. 2015; Nautiyal et al. 2013). These microorganisms assist plants by improving nutrient uptake, enhancing root architecture, and regulating the production of phytohormones like auxins, cytokinins, and abscisic acid (Egamberdieva et al. 2017). Additionally, they can induce systemic resistance and modulate the expression of stress-responsive genes, thereby priming plants to better tolerate adverse environmental conditions (Grover et al. 2011). For instance, *Bacillus subtilis*, a well-known PGPR, has been reported to induce drought tolerance in plants by enhancing antioxidant enzyme activity and osmolyte accumulation (Puli et al. 2022). Similarly, arbuscular mycorrhizal fungi (AMF) have shown potential in improving plant water relations and alleviating salinity stress by regulating ion balance and enhancing osmotic adjustment (Hashem et al. 2016).

In this context, the present review discusses the emerging role of beneficial microbes in managing plant stress under climate change. This review provides insights into how these microorganisms can be harnessed to develop sustainable and resilient agricultural systems by examining various microbial mechanisms and their interactions with plants.

2 Impact of Climate Change on Pathogen and Host Plant

Climate change is a significant concern for agricultural communities globally, affecting the key components of agriculture: pathogens, hosts, and environmental conditions. These components are crucial for plant infections (Abdou Zayan 2018). The incidence, severity, and geographical distribution of diseases are substantially influenced by climate change, which affects temperature, precipitation, and atmospheric CO_2 levels, thereby altering plant disease dynamics. The distribution of pests and diseases can shift poleward, and the virulence of pathogens can be enhanced, resulting in increased agricultural hazards due to elevated temperatures and CO_2 (Yänez-López et al. 2012).

2.1 Elevated Atmospheric CO_2

The increase in atmospheric CO_2 levels due to climate change is a multifaceted issue with significant implications for global climate systems. The rise in CO_2 concentrations is primarily driven by fossil fuel combustion, global warming, altered precipitation patterns, and potential long-term impacts on Earth's climate stability (Aresta 2010). The increase in atmospheric CO_2 levels has complex implications for agriculture. While elevated CO_2 can enhance photosynthesis and potentially increase crop yields, the overall impact on agriculture is influenced by a combination of factors, including temperature changes, water availability, and nutrient supply. These factors can offset the benefits of CO_2 fertilization, leading to variable outcomes depending on regional conditions (Reddy 2014). In Central and Southeastern Europe, climate change scenarios predict reduced crop-growing durations and yield reductions for certain crops like winter wheat and spring barley. However, when increased CO_2 levels are considered, these negative impacts can be mitigated, resulting in yield increases for some crops (Alexandrov and Eitzinger 2005). High CO_2 levels can enhance the photosynthetic rate, particularly in C_3 plants, which include many staple crops like wheat and rice. This can lead to increased net primary productivity and improved water use efficiency, as plants may require less water for the same amount of photosynthesis (Rosenberg 1981).

Some studies conducted on soybean yields in Iowa indicate that increased CO_2 levels can lead to higher crop yields. The GLYCIM model showed that soybean yields increased with rising CO_2 concentrations, highlighting the potential for CO_2 fertilization to boost agricultural output under certain conditions (Haskett et al. 1997). Elevated CO_2 levels have been shown to aggravate the development of powdery mildew caused by *Sphaerotheca fuliginea* in cucurbit species, as well as lead to significant physiological and biochemical changes in plants, such as increased photosynthesis rates and decreased transpiration, stomatal conductance, salicylic acid, and phenol contents. These changes can influence the plant's defense mechanisms against pathogens, potentially making them more susceptible to diseases like powdery mildew (Khan and Rizvi 2020). In some studies, high carbon dioxide enhances the virulence of pathogens and the susceptibility of wheat (Váry et al. 2015). Elevated CO_2 can alter plant physiology, such as increasing photosynthesis and biomass, which may affect plant–pathogen interactions. For instance, in the case of *Trifolium repens* infected by *Orobanche minor*, elevated CO_2 partially alleviated the negative effects of the parasite on host growth, although it did not affect the parasite's biomass or attachment time (Dale and Press 1998). Changes in plant physiology under elevated CO_2 can also influence the plant's resistance to pathogens. For example, elevated CO_2 and O_3 levels can alter phytohormone-mediated resistance pathways, affecting the interactions between plants, vector insects, and plant viruses (Tenllado and Canto 2020).

2.2 Elevated Atmospheric Temperature

Climate warming could severely impact the population dynamics of diseases, including overwintering and survival, population growth rates, and the number of generations of polycyclic species (Singh et al. 2023). Elevated temperatures significantly impact plant pathogen growth and reproduction, influencing both the distribution and severity of plant diseases. These changes challenge global agriculture by altering pathogen dynamics and potentially increasing infection risks (Chaloner et al. 2020). For example, *Leptosphaeria* species cause stem canker in *Brassica*, which has led to earlier and more intense sporulation due to climate change. This results in a shift in the timing of spore release, increasing the severity of infections and necessitating adjustments in fungicide application strategies (Kaczmarek et al. 2016). Some fungal pathogens increase the infection risk due to high temperatures, particularly in higher latitudes where both crop yields and pathogen pressure are expected to rise. This shift may lead to unfamiliar pathogen assemblages in regions like the United States, Europe, and China, complicating crop protection efforts (Chaloner et al. 2020). Temperature changes can enhance the propagation of plant viruses, increasing their spread within host plants. This is due to the temperature-sensitive nature of virus movement through plasmodesmata, which may lead to new viral outbreaks in agriculture (Amari et al. 2021). Because of interaction between temperature and plant viruses is complex, higher temperatures potentially increase plant susceptibility to viral infections. For example, sugar beet yellows virus shows increased symptom severity at higher temperatures (Kassanis 1957). High temperatures negatively affect pollen viability, particularly during germination, which can impair plant reproduction. This is evident in species like *Pinus edulis*, where elevated temperatures during pollen germination reduce viability, potentially leading to unsuccessful fertilization despite high pollen production (Flores-Rentería et al. 2018). Climate change is expected to shift the geographic distribution of plant pathogens (Kůdela et al. 2018; Lahlali et al. 2024). Heat-loving bacteria may emerge, while cold-tolerant species may decline. This shift is particularly noted in Central Europe, where thermophilic xanthomonads are becoming more prevalent in high temperatures (Kůdela et al. 2018). Rising temperatures may lead to novel strains of diseases that are more adept and virulent (Fisher et al. 2020). The emergence of more aggressive pathogenic strains adapted to higher temperatures poses new threats, as these strains may have a wider host range and increased virulence (Sbeiti et al. 2023). The intensity of Fusarium head blight in wheat is expected to escalate when the prevalence shifts from the less virulent *Fusarium culmorum*, which thrives in cool and moist environments, to the more virulent Fusarium graminearum, which favors warm and humid circumstances (Parikka et al. 2012). Likewise, more virulent and temperature-resistant emerging strains of *Puccinia striiformis* have supplanted earlier strains, resulting in significant outbreaks of wheat rust in the United States, Australia, and Europe (Vidal et al. 2022; Walter et al. 2016). Rising temperatures can expand the distribution of several diseases now constrained by overwintering necessities, such as wheat stem rust produced by *Puccinia graminis* f. sp. *tritici* (Ma et al. 2015).

The molecular mechanisms behind the increased susceptibility of plants to infections at elevated temperatures are inadequately elucidated (Cohen and Leach 2020). High temperatures affect the nuclear localization of NLR proteins, which are crucial for ETI. Additionally, transcription factors and kinases that suppress ETI are induced at higher temperatures, further complicating the plant's immune response. In Arabidopsis, the synthesis of salicylic acid, a hormone essential for plant defense, is inhibited at elevated temperatures (Castroverde and Dina 2021). A recent study indicated that activating jasmonic acid production and signaling genes by *Magnaporthe oryzae* increases rice's vulnerability to rice blast disease under elevated temperatures (Qiu et al. 2022).

Extreme high temperatures (EHTs) negatively affect the fitness and demographic parameters of parasitoids like *Aphelinus asychis*. Increased amplitude and frequency of EHTs prolong development time and reduce adult longevity, fecundity, and intrinsic rate of increase, thereby diminishing their effectiveness as biological control agents (Yu et al. 2022). In the case of mirid predators such as *Cyrtorhinus lividipennis* and *Tytthus chinensis*, elevated temperatures affect their thermal tolerance and predation abilities. *T. chinensis* shows better adaptation to heat stress, maintaining higher predation and reproductive capacities compared to *C. lividipennis*, suggesting its potential as a more effective biological control agent under warming conditions (Bai et al. 2021). Elevated temperatures alter soil bacterial diversity, with significant shifts in the composition of bacterial communities. For instance, there is a net decrease in Proteobacteria and Acidobacteria, while Actinobacteria and Chloroflexi increase, indicating that temperature rise can lead to changes in soil microbial dynamics, potentially affecting soil health and plant–microbe interactions (Mun and Wong Vui Ling 2020). High temperatures induce the expression of heat shock proteins in organisms like *Helicoverpa armigera*, enhancing their resistance to pathogens such as nucleopolyhedrovirus. This response reduces the efficacy of entomopathogens as biological control agents, as the pests become more resilient under heat stress (Alagar Boopathy et al. 2022).

2.3 Drought

Drought stress significantly impacts plant growth and development by altering physiological, biochemical, and molecular processes. It affects various stages of plant growth, from germination to maturity, and can lead to reduced crop yield and quality (Chaudhry and Sidhu 2021). Drought significantly restricts plant growth and entails intricate interactions between reactive oxygen species (ROS) and the hormone abscisic acid (ABA). During drought conditions, ABA accumulation occurs, and the exogenous application of ABA induces hydrogen peroxide (H_2O_2) production in plants. A well-coordinated signaling pathway from root to shoot is essential for maintaining water availability in aerial parts. In these challenging circumstances, roots serve as the primary organs that detect water deficiency and relay signals to above-ground structures via root-to-shoot xylem channels. This signaling mechanism diminishes guard cell turgidity, resulting in stomatal closure and reduced water loss (Carmody et al. 2016; Chaudhry and Sidhu 2021; Jones 2016). During drought stress, the closure of stomata results in a decrease in the rate of photosynthesis in plants, for example, in peas (Fatima et al. 2024). Additionally, the reduced water availability induces a decrease in cell volume, which in turn leads to an increase in the viscosity of cellular contents. This can result in the damage of proteins and enzymes that are responsible for photosynthesis in plants (Impa et al. 2012; Qiao et al. 2024). It also decreases chlorophyll content and photosynthetic rates, affecting overall plant health (Naik et al. 2022). In wheat, drought stress results in reduced plant height, spike length, and grain yield, with significant differences observed between irrigated and rainfed conditions (Razzaq et al. 2024).

2.4 Salinity

Salinity stress is initially detected by the plant's roots, which impedes development in both the short and long term. The immediate effect is from reduced water availability, which produces osmotic stress in plants. Conversely, prolonged exposure results in ion toxicity from salt accumulation in plants (Parihar et al. 2015). Salt stress adversely affects germination, growth, photosynthesis, water balance, oxidative stress, and plant productivity. The germination of seeds is a crucial and basic component of the plant life cycle that influences yield and productivity. The process involves the oxidation of lipids and carbohydrates in seeds, resulting in the degradation of storage proteins to provide energy for plant growth (Chaudhry and Sidhu 2021). Salinity influences germination in several plants; for instance, the exogenous administration of NaCl (50–150 mM) diminished the ultimate germination percentage in *L. sativa*, which is associated with lower activity of phytases and phytate-specific phosphatases in these plants (Nasri et al. 2011).

Excessive soil salinity leads to partial dehydration of cellular cytoplasm, ultimately disrupting plant metabolism (Tsegay and Gebreslassie 2014). The suppression of germination is also ascribed to alterations in the activity of certain enzymes and hormones inside the plants (Miransari and Smith 2014). Singh et al. (2012) said that salinity stress may induce osmotic stress or ion toxicity, which might impair plant germination. An excessive amount of salts in water inhibits plant growth either due to a reduction in the uptake of water leading to water deficit in plants or by entering the transpiring stream and injuring cells causing ion excess in plants (Parihar et al. 2015). Rahneshan et al. (2018) experimented to assess the impact of NaCl (50–150 mM) on the growth of two pistachio (*Pistacia vera*) rootstocks, revealing that moderate to high salt concentrations adversely affect both shoot and root length. The growth inhibition observed under severe stress is supported by diminished carbon assimilation resulting from stomatal limitations or the degradation of metabolic processes in the plants (Fig. 6.1).

3 Beneficial Microbes in Plant Stress Management

Beneficial microorganisms play a crucial role in enhancing plant tolerance to abiotic stresses. Plant growth-promoting rhizobacteria (PGPR), arbuscular mycorrhizal fungi (AMF), and endophytes are key players in mitigating stress effects (Ahmad and Zaib 2020; Al-Surhanee et al. 2021; Khanna et al. 2019; Munir et al. 2022). These microbes improve plant growth and stress tolerance through various mechanisms, including increased hormone and metabolite production, enhanced nutrient uptake, and modulation of root morphology and physiology. They also strengthen plant defense mechanisms by upregulating antioxidant pathways and increasing osmolyte accumulation (Al-Surhanee et al. 2021). Specific microbial genera, such as *Bacillus*, *Pseudomonas*, and *Trichoderma*, have been identified as stress-tolerant species (Khanna et al. 2019). The use of beneficial microbes offers a promising alternative to chemical inputs for sustainable agriculture (Ahmad and Zaib 2020).

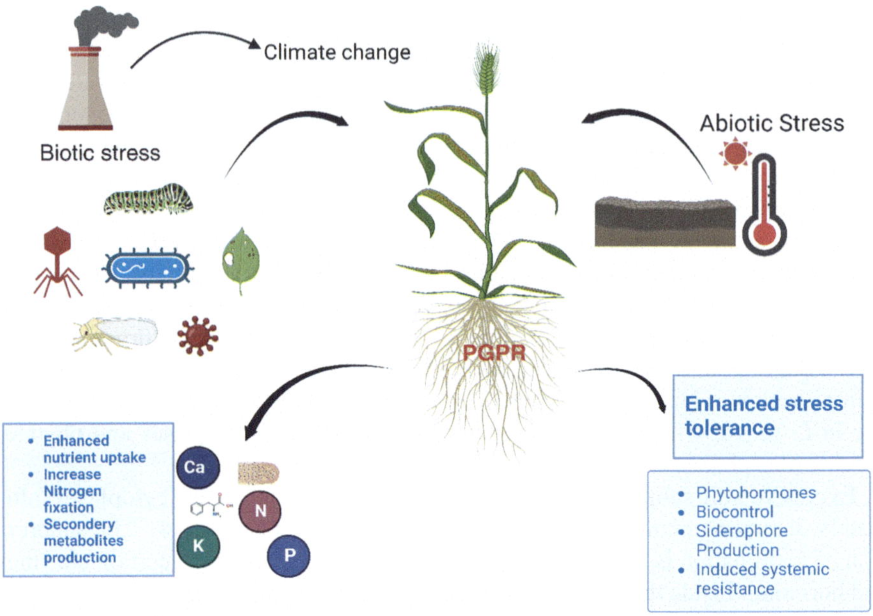

Fig. 6.1 Overview of diagram presenting the impact of climatic change on plant–microbe interaction and activation of biotic and abiotic stress. The presence of plant growth-promoting bacteria (PGPR) in soil activates a positive impact on plant growth

Recent advancements in genetic engineering may further enhance the potential of microbes in abiotic stress mitigation (Munir et al. 2022). Beneficial microbes are essential in managing plant stress by promoting growth and protecting plants from biotic and abiotic stresses. These microbes include bacteria, fungi, and archaea, each with unique mechanisms to help plants. Below are the main types of beneficial microbes involved in plant stress management.

3.1 Endophytes (Bacterial and Fungal)

Endophytes live inside plant tissues without causing harm and can produce bioactive compounds that enhance plant stress tolerance (e.g., drought, salinity, and pathogen resistance). They may also boost plant growth by improving nutrient uptake or modulating hormone levels. Endophytes are microorganisms that colonize plant tissues without causing harm, providing various benefits to their hosts (Kandel et al. 2017; Kumar et al. 2024). The colonization process involves attachment, entry, motility, transmission, and multiplication within the plant (Kumar et al. 2024). Mechanisms such as chemotaxis and quorum sensing facilitate this process (Maela and Serepa-Dlamini 2019). Endophytes promote plant growth through nitrogen fixation, phytohormone production, nutrient acquisition, and stress tolerance (Kandel et al. 2017). The diversity and selection of endophytes are strictly controlled by plants, as colonization is energy-demanding (Basheer 2022). Advanced techniques like metagenomics, metatranscriptomics, and metaproteomics have revolutionized the study of endophytic communities, allowing for in situ analysis and the identification of novel genes (Maela and Serepa-Dlamini 2019). Understanding the colonization process and dynamics is crucial for developing bioinoculants for sustainable agriculture (Basheer 2022).

3.2 Mechanisms of Colonization and Symbiosis

Bacterial endophytes colonize plants through various mechanisms, including chemotaxis, quorum sensing, and the rhizophagy cycle. The colonization process involves attachment, entry, motility, and multiplication within the host plant (Kumar et al. 2020). Endophytes utilize diverse host-interaction systems, such as type IV pili, flagella, secretion systems, and genes for phytohormone synthesis (Frank 2011). These microorganisms promote plant growth through nitrogen fixation, nutrient acquisition, and conferring tolerance to abiotic and biotic stresses (Kandel et al. 2017). Genomic approaches have revealed the genetic potential of endophytes, including genes for motility, colonization, and synergistic interactions with the host (Frank 2011). Post-genomic techniques like metagenomics, metatranscriptomics, and metaproteomics have further advanced our understanding of endophytic communities and their roles in plant growth (Maela and Serepa-Dlamini 2019). Bacterial endophytes are microorganisms that live inside plant tissues without causing harm and often form mutualistic relationships with their hosts. They can help plants overcome stress, promote growth, and provide protection from pathogens. The colonization and symbiosis mechanisms by which bacterial endophytes interact with plants are complex and involve multiple steps, including recognition, adhesion, invasion, and the establishment of symbiosis.

Bacterial endophytes establish a symbiotic relationship with plants through a multistep process that begins with recognizing chemical signals, such as root exudates, that guide them toward the plant surface—a phenomenon known as chemotaxis (Matilla and Krell 2018). Once there, endophytes adhere to the plant surface using specialized structures like pili and flagella (Compant et al. 2010), facilitating their entry through natural openings (stomata, hydathodes, root cracks) or wounds (Hardoim et al. 2008). Some bacteria, like *Bacillus subtilis*, actively penetrate plant tissues by secreting enzymes such as cellulases and pectinases (Reinhold-Hurek and Hurek 2011). Upon successful invasion, endophytes evade plant immune defenses by producing effector proteins and modulating hormone pathways like salicylic acid (SA), jasmonic acid (JA), and ethylene (ET) (Zamioudis and Pieterse 2012). Once they establish a symbiotic niche, they exchange nutrients, such as fixing nitrogen or solubilizing phosphorus, while receiving carbon compounds from the plant (Sessitsch et al. 2012). This symbiosis results in various benefits, including enhanced plant growth through phytohormone production and increased stress tolerance by producing compounds like exopolysaccharides (Glick 2012; Yang et al. 2021) (Table 6.1).

Table 6.1 Roles and mechanisms of endophytes in mitigating abiotic stress in host plants

Abiotic stress	Endophytes	Host	Mode of action	References
Drought	*Burkholderia phytofirmans* PsJN	*Triticum aestivum* L. (Poaceae)	Improved the photosynthetic rate, water use efficiency, and chlorophyll content	Naveed et al. (2013)
	Trichoderma harzianum TH-56	*Oryza sativa* L. (Poaceae)	Inoculation with escalating doses of *T. harzianum* strain Th-56 resulted in the overexpression of aquaporin, dehydrin, and malondialdehyde genes	Pandey et al. (2016)
	Bacillus subtilis B26	*Brachypodium distachyon* (L.) P. Beauv. (Poaceae)	Endophyte-induced upregulation of DREB2B-like, DHN3-like, and LEA-14-A-like genes, along with modification of DNA methylation genes MET1B-like, CMT3-like, and DRM2-like, which enable biochemical alterations to mitigate stress conditions	Gagné-Bourque et al. (2015)
	Trichoderma hamatum DIS 219b	*Theobroma cacao* L.	Bacterial colonization generated alterations in stomatal conductance, net photosynthesis, and green fluorescence emissions due to dryness	Bae et al. (2009)
	Streptomyces padanus	*Kalmia latifolia* L.	Inoculation of *S. padanus* resulted in the buildup and lignification of cell walls in sieve cells, which gave drought resistance in *Kalmia latifolia*	Hasegawa et al. (2005)
	Pantoea alhagi LTYR-11ZT	*Arabidopsis* sp. (Brassicaceae) and wheat (*Triticum* sp., Poaceae)	Strain LTYR-11ZT increased the contents of soluble sugars but decreased proline, MDA, and chlorophyll contents	Chen et al. (2017)
	Ampelomyces sp.	*Solanum lycopersicum* L. (Sola-nacelle)	*Ampelomyces* sp. significantly improved root and shoot systems under drought circumstances. The study posited that the enhanced health of the plant results from synergistic effects	Azizi et al. (2021)

Salt	*Curvularia* sp.	Populus × *tomentosa Carrière*	Production of the antioxidant enzymes superoxide dismutase (SOD) and ascorbate peroxidase (APX). The infected plant exhibited elevated levels of chlorophyll and proline content	Pan et al. (2018)
	P. pseudoalcaligenes	*Arabidopsis thaliana* (L.) Heynh. (Brassicaceae)	*P. pseudoalcaligenes regulates Na+ and K+ ions in response to salt stress, hence maintaining ionic homeostasis*	Win et al. (2018)
	Piriformospora indica	*Brassica rapa* L. (= *B. campestris* subsp. *chinensis* (L.) Makino)	Inoculated plants exhibited elevated activities of antioxidant enzymes and increased expression of genes associated with salt tolerance	Khalid et al. (2018)
	Piriformospora indica	*Oryza sativa* L.	Down-regulation of PiHOG1 confers salt tolerance	Jogawat et al. (2016)
	Bacillus subtilis (BERA 71)	*Cicer arietinum* L. (Fabaceae)	Improvement in plant biomass, photosynthetic pigments, enzymatic and nonenzymatic antioxidant activity, alongside reduced ROS generation and lipid peroxidation	Abd-Allah et al. (2015)
	Penicillium chrysogenum	*Solanum lycopersicum* L.	Inoculation with *P. chrysogenum* demonstrated enhanced salt tolerance at a dose of 300 mM	Morsy et al. (2020)

3.3 Rhizobacteria (Plant Growth-Promoting Rhizobacteria: PGPR)

PGPR colonize the rhizosphere and promote plant growth by producing phytohormones like auxins, cytokinins, and gibberellins. They can also enhance nutrient availability (e.g., nitrogen fixation, phosphorus solubilization) and trigger induced systemic resistance (ISR) to pathogens. Examples include *Pseudomonas*, *Bacillus*, *Azospirillum*, and *Rhizobium*.

3.4 Mechanisms of Root Colonization by Bacterial Endophytes

Bacterial endophytes colonize plant tissues and promote growth through various mechanisms, including nitrogen fixation, phytohormone production, and nutrient acquisition (Kandel et al. 2017). These microbes enhance plant tolerance to biotic and abiotic stresses by inducing stress genes and regulating reactive oxygen species (Nanda et al. 2018). A key trait of "competent" endophytes is the production of 1-aminocyclopropane-1-carboxylate deaminase, which modulates plant ethylene levels and improves colonization (Hardoim et al. 2008). Endophytes employ various strategies to overcome plant defenses, such as producing enzymes to cope with oxidative stress or subtilomycin to affect flagella-induced plant defense. They can also trigger induced systemic resistance through the accumulation of PR1 protein and activation of MAPK signaling and WRKY53 gene expression via the JA/ET signaling pathway (Mengistu 2020). Understanding these mechanisms is crucial for developing biotechnological approaches to improve crop yield under stress conditions (Nanda et al. 2018).

Bacterial endophytes exhibit chemotaxis toward specific root exudates such as sugars, organic acids, and amino acids, which act as chemical signals guiding them to the root surface (Bashan and de-Bashan 2010). For instance, *Azospirillum brasilense*, a well-known plant growth-promoting endophyte, shows chemotaxis toward organic acids and amino acids in maize root exudates, facilitating its colonization of the rhizosphere. Upon reaching the root surface, bacterial endophytes adhere using surface structures like pili, flagella, and exopolysaccharides, establishing a stable interaction with the plant (Compant et al. 2010). *Pseudomonas fluorescens* utilizes its flagella and exopolysaccharides to form biofilms that enhance root colonization. Following attachment, endophytes invade the plant roots through natural openings such as root hairs, lateral root cracks, or damaged areas, spreading to different plant parts via apoplastic or symplastic pathways (Hardoim et al. 2008). For example, *Enterobacter cloacae* enters through root cracks or openings caused by lateral root emergence in various crops like rice and wheat. Once inside, many endophytes form biofilms on root surfaces, which are bacterial communities embedded in a self-produced matrix, enhancing colonization, communication (quorum sensing), and protection against environmental stressors (Raaijmakers et al. 2010). An example is *Pseudomonas putida*, which forms biofilms on rice roots, promoting plant growth and enhancing nutrient uptake.

3.5 Mycorrhizal Fungi

Mycorrhizal fungi are a varied community of soil fungi that establish mutualistic relationships with the roots of the majority of terrestrial plants. This symbiotic association is termed "mycorrhiza," wherein fungi inhabit plant roots, augmenting nutrient and water absorption, while the plant supplies sugars and other organic substances to the fungi. Mycorrhizal fungi can be classified into many varieties, including arbuscular mycorrhizal fungi (AMF) and ectomycorrhizal fungi (EMF), each establishing unique associations with their host plants (Table 6.2).

4 Biocontrol Agents

Biocontrol, short for biological control, refers to the use of living organisms such as beneficial microbes (bacteria, fungi, and viruses), insects, or plants to manage and suppress populations of plant pathogens, pests, or weeds. The aim is to reduce reliance on chemical pesticides and promote sustainable agricultural practices. Biocontrol agents work through various mechanisms, including antagonism, competition, parasitism, and the induction of plant defense responses, to maintain a balanced ecosystem and protect crop health.

4.1 Role of Biocontrol in Plant Stress Management Under Climate Change

Beneficial biocontrol agents such as *Bacillus*, *Pseudomonas*, and *Streptomyces* interact with plant roots to enhance stress tolerance. These microbes can induce ISR by altering plant–pathogen signaling, enhancing root physiology, and producing metabolites like organic acids and enzymes (Adeleke et al. 2024; Kou et al. 2024). Elicitors, including salicylic acid and jasmonic acid, play a pivotal role in ISR by activating pathogenesis-related genes. This leads to increased production of reactive oxygen species, which trigger local cell death and systemic resistance (Meena et al. 2022).

Nutrient management, including the use of biofertilizers, is crucial for alleviating abiotic stress. Nutrients like nitrogen, potassium, and calcium enhance the production of antioxidant enzymes, reducing reactive oxygen species (ROS) and improving plant stress tolerance (Ashokan et al. 2024). Mycorrhizal fungi-enriched biofertilizers, such as those containing arbuscular mycorrhizal fungi (AMF), improve nutrient uptake and enhance stress resilience. AMF increases antioxidant activities and reduces arsenic concentrations in crops, promoting growth under stress conditions (Alam and Dey 2024). Plants possess inherent mechanisms for stress tolerance, including osmoregulation and metabolic adjustments. Understanding these eco-physiological traits can inform the development of biocontrol strategies that enhance plant resilience to abiotic stresses (Rane and Jangid 2024). Combining biocontrol methods with other strategies, such as breeding for stress tolerance and

Table 6.2 Effect of different mycorrhizal fungi on different host plants against abiotic stress

Abiotic stress	Mycorrhizal fungi	Host	Mode of action	References
Salinity	*Claroideoglomus etunicatum*, *Rhizophagus intraradices*	*Cucumis sativus* L.	Augmented growth, elevated antioxidant enzyme activity, increased proline and phenolic concentrations, enhanced mineral element absorption, and reduced sodium ion uptake	Shahvali et al. (2020)
	Glomus mosseae	*Leymus chinensis*	Elevated phosphorus and nitrogen concentrations	Shi-chu et al. (2019)
	Rhizophagus irregularis	*Solanum lycopersicum*	Enhancing comprehensive plant vitality	Hashem et al. (2016)
	Rhizoglomus intraradices, *Funneliformis mosseae*	*Pisum sativum* L.	Substantially improved by increased biomass, chlorophyll concentration, and nutrient uptake	Bijalwan et al. (2021)
	Rhizophagus intraradices	*Euonymus maackii* Rupr	Enhanced nutrition absorption and elevated antioxidant enzyme activity	Klinsukon et al. (2021)
	Glomus mosseae	*Citrullus lanatus* L.	Plants exhibit increased leaf density, greater fruit size, and enhanced root development	Mathur et al. (2018)
High temperature	*Claroideoglomus claroideum*	*Triticum aestivum*	Enhanced nutritional absorption and augmented grain quantities are among the documented impacts	Zhu et al. (2016)
	Rhizophagus intraradices, *Funneliformis mosseae*	*Zea mays* L.	Beneficial plant characteristics, enhanced photosynthetic transpiration rate	Liu and Zhang (2015)
	Glomus tortuosum	*Zea mays* L.	Increased concentrations of shoot nitrogen (N), phosphorus (P), potassium (K), and copper (Cu)	Chu et al. (2016)
	Glomus spp.	*Ipomoea batatas*	Osmo protectants contributed to the modulation of osmotic potential	Pal and Pandey (2016)
	Glomus mosseae	*Triticum aestivum* L.	Beneficial impacts on plant growth metrics and photosynthetic pigments were apparent	Zhang et al. (2019)
	Rhizophagus intraradices	*Triticum durum*	Significant effects included increased grain biomass, nutritional content, and gliadin levels in grains	Wu et al. (2017)
	Rhizophagus irregularis	*Zea mays* L.	Plant roots exhibited diamine oxidase activity, transforming putrescine into aminobutyric acid	Chen et al. (2020)

optimizing cultural practices, can effectively mitigate stress impacts. This integrated approach is essential for sustainable agriculture and food security in the face of climate change (Haq et al. 2024).

4.2 Mechanisms of Microbial-Mediated Drought Stress Management Under Climate Change

Microbial communities adapt to changing environmental conditions by enhancing plant drought tolerance through various mechanisms. Plant growth-promoting rhizobacteria (PGPR) and mycorrhizal fungi modify phytohormonal levels, produce osmolytes, and generate microbial volatile organic compounds (VOCs) that improve water and nutrient uptake. They also induce stress-responsive genes and enhance antioxidant defenses, mitigating oxidative stress. Additionally, these microorganisms can form biofilms that retain soil moisture, further supporting plant resilience under drought conditions. Understanding these interactions is crucial for developing sustainable agricultural practices in the face of climate change (Poudel et al. 2021). The enhanced drought tolerance observed in Arabidopsis is linked to sustained chlorophyll content and increased endogenous abscisic acid (ABA) signaling, along with the stabilization of the diversity and structure of root-associated microbiomes, suggesting that these changes collectively contribute to improved plant health under drought conditions (Yang et al. 2021). In drought-tolerant wheat lines, certain microbes, such as *Curtobacterium flaccumfaciens* and *Arthrobacter* sp., were enriched under drought conditions, promoting plant growth through mechanisms such as producing plant growth-promoting substances and enhancing nutrient cycling. This line-specific enrichment indicates that microbial communities can respond dynamically to environmental stress, ultimately facilitating drought tolerance in crops by improving root and shoot development and overall plant resilience (Hone et al. 2021). *Pseudomonas aeruginosa* HG28-5 and *Bacillus* sp. BSE01 help maintain Na^+/K^+ homeostasis, which is vital for plant survival under salt stress. These bacteria reduce Na^+ accumulation in plant tissues while enhancing K^+ uptake, thereby stabilizing cellular ionic balance (Basu et al. 2023; Dong et al. 2024). Arbuscular mycorrhizal fungi (AMF) significantly improve plant growth under drought conditions, increasing biomass and root morphological traits that enhance the uptake of various nutrients, particularly phosphorus, through an extensive hyphal network and enzyme production (Miransari 2013). They also help plants with abiotic stresses like salinity and drought by improving photosynthesis, gas exchange, and antioxidant defenses (Wahab et al. 2023). These fungi also enhance root hydraulic conductivity and water status under drought stress (Ortas et al. 2020).

PGPR and other beneficial microbes regulate stress-related hormones such as abscisic acid and ethylene, which are crucial for plant responses to drought stress. This hormonal modulation helps in maintaining plant growth and development under adverse conditions (Aroca and Ruiz-Lozano 2009; Barnawal et al. 2019). Rhizobacteria can influence root-to-shoot hormonal signaling, affecting leaf growth and other physiological processes, thereby enhancing drought tolerance (Barnawal et al. 2019). *Azospirillum* species produce IAA, promoting root elongation and increasing root surface area, which enhances water and nutrient uptake, helping the plant tolerate drought stress (Cecagno et al. 2015). Microbial communities induce the accumulation of osmolytes, including sugars, amino acids, and polyamines, which help in osmotic adjustment during drought stress. This metabolic adjustment is crucial for maintaining cell turgor and function under water-deficit conditions (Barnawal et al. 2019; Camaille et al. 2021). The presence of beneficial microbes enhances the activity of antioxidant enzymes in plants, reducing oxidative stress caused by drought. This reduction in oxidative damage is vital for maintaining cellular integrity and function (Aroca and Ruiz-Lozano 2009; Poudel et al. 2021).

Rhizosphere bacteria form biofilms that enhance plant drought stress tolerance by increasing root surface area and root hair density, which promotes better water uptake. For example, *Pseudomonas azotoformans* FAP5 enhances drought tolerance through improved root colonization and stability (Amari et al. 2021). Additionally, microbial volatile organic compounds (mVOCs) emitted by plant growth-promoting rhizobacteria (PGPR) during drought stress act as signals, triggering systemic stress responses in plants and facilitating communication between stressed and neighboring plants (Timmusk et al. 2014). These mVOCs modulate plant physiological responses by improving water use efficiency, nutrient uptake, and influencing the levels of key phytohormones like abscisic acid (ABA), auxins, and ethylene, ultimately enhancing plant drought uptake (Liu and Zhang 2015; Mathur and Roy 2021; Poudel et al. 2021).

4.3 Mechanisms of Microbial-Mediated Salinity Stress Management Under Climate Change

Beneficial microorganisms associated with plants, such as rhizoplane, rhizosphere, and endophytic bacteria, as well as mycorrhizal fungi, significantly contribute to enhancing plant resilience to saline stress (Etesami and Beattie 2017; Etesami and Maheshwari 2018; Gill et al. 2016; Hamilton et al. 2016). PGPR inhabits the rhizosphere, the soil area affected by root exudates, and exhibits greater tolerance to salt stress. These bacteria are commonly located in habitats characterized by recurrent salinity stress or in association with halophytic plants, rendering them well-suited to endure and operate in saline circumstances (Glick et al. 2007). Inoculation with these bacteria induces changes in plants that enhance their tolerance to abiotic stresses, including salinity. This process is known as induced systemic tolerance (IST) (Glick 2014).

PGPR is crucial in enhancing plant tolerance to salinity stress by promoting ion homeostasis, producing antioxidants, and facilitating nutrient uptake. They help maintain water balance and reduce $Na+$ ion concentration in plants, which is critical under saline conditions (Etesami and Maheshwari 2018; Riseh et al. 2021; Sagar et al. 2022). In *Brassica juncea,* PGPR strains like *Bacillus cereus* and *Bacillus megaterium* have been shown to decrease $Na+$ uptake and increase $K+$ and $Ca++$

levels, further supporting ion homeostasis and reducing the detrimental effects of salinity (Daraz et al. 2023).

These bacteria also synthesize phytohormones, including auxin (i.e., IAA), cytokinin, and gibberellins. Phytohormones are recognized for their role in plant growth, development, and responses to stress, particularly salt stress (Shaterian et al. 2005), aiding in stress mitigation by improving root and shoot growth and enhancing photosynthetic efficiency (Riseh et al. 2021). Studies have shown that salinity does not affect hormone synthesis (e.g., auxin) in salt-tolerant plant growth-promoting rhizobacteria (PGPR) (Albacete et al. 2008; Egamberdieva and Kucharova 2009). Previous research demonstrated that salt-tolerant plant growth-promoting rhizobacteria, such as *Azospirillum brasilense*, may synthesize indole-3-acetic acid (IAA) in the presence of 200 mM NaCl (Nabti et al. 2010). Likewise, salt-tolerant plant growth-promoting rhizobacteria (PGPR), such as *Serratia plymuthica* RR2-5-10, *Serratia rhizophila* e-p10, *Pseudomonas chlororaphis* TSAU13, and *Pseudomonas fluorescens* SPB2145, have been documented to synthesize indole-3-acetic acid (IAA) at sodium chloride concentrations reaching 1.5% (Egamberdieva 2011, 2012).

The meta-analysis highlights that endophytic bacteria enhance plant growth under salinity stress through several mechanisms, including improved morphological attributes, increased chlorophyll content, and enhanced antioxidative ability. Specifically, these bacteria promote ion homeostasis by increasing root K+ concentration and decreasing leaf Na+ concentration, which is crucial for mitigating salinity stress (Tufail et al. 2021). Using compatible microbial consortia, including bacterial and fungal symbionts, can synergistically enhance plant stress tolerance. These consortia improve nutrient cycling, stabilize soil aggregates, and enhance osmolyte production, contributing to better plant–water relations and photosynthetic capacity (Talaat and Shawky 2017). Specific microbial inoculants, such as *Trichoderma harzianum* and *Bacillus amyloliquefaciens*, have been shown to enhance salt tolerance in wheat by maintaining Na+/K+ balance and increasing Zn biofortification, thus improving yield and reducing seedling mortality (B. K. Singh et al. 2023).

4.4 Mechanisms of Microbial-Induced Heat Stress Management Under Climate Change

Climate change is leading to an increase in global temperatures, resulting in frequent and prolonged heat stress episodes that significantly impact agricultural productivity and sustainability (Lesk et al. 2016). Heat stress affects plant growth, development, and yield by disrupting physiological processes such as photosynthesis, respiration, and water balance, ultimately leading to oxidative damage and reduced crop performance (Bita and Gerats 2013). To mitigate the adverse effects of heat stress, plants have evolved various physiological and biochemical adaptations. However, these inherent mechanisms are often insufficient under extreme temperature fluctuations, necessitating the exploration of alternative strategies for plant protection (Zandalinas et al. 2018).

Microbial heat stress response mechanisms involve various biochemical, physiological, and molecular adaptations that help protect plants from heat-induced damage. Biochemical adaptation mechanisms involve the synthesis of heat shock proteins (HSPs), the production of phytohormones, and the enhancement of antioxidant activities, which collectively contribute to plant resilience under high-temperature conditions. HSPs play a vital role in maintaining cellular protein integrity and scavenging ROS during heat stress by preventing protein aggregation and assisting in protein refolding. They are synthesized in response to heat stress and other environmental cues, acting as molecular chaperones to stabilize proteins and membranes (Nayak et al. 2023; Siddra 2016). The heat shock response, which triggers HSP synthesis, is a primary defense mechanism against heat stress, helping to maintain cell function and prevent tissue damage (Siddra 2016). In maize, heat stress induces a cytoplasmic heat stress response where heat shock transcription factors activate genes encoding HSPs, contributing to thermotolerance (Li and Howell 2021).

Plant growth-promoting rhizobacteria (PGPR) and endophytes enhance plant heat stress tolerance by producing phytohormones, siderophores, and osmolytes, and also stimulate the production of HSPs, which improve antioxidant activity and nutrient uptake (Ahmad and Zaib 2020; Shaffique et al. 2022). Plant hormones such as auxin, abscisic acid (ABA), and ethylene play significant roles in regulating plant responses to heat stress. These hormones integrate environmental stimuli and endogenous signals to modulate defensive responses. The exogenous application of these hormones can enhance plant thermotolerance, suggesting potential strategies for developing heat-resilient crops (Li et al. 2021). The synthesis of abscisic acid (ABA) is crucial for heat stress tolerance. The gene 9-cis-epoxy carotenoid dioxygenase 1 (OsNCED1) in rice enhances heat tolerance by increasing ABA levels, which in turn boosts antioxidant enzyme activity and reduces reactive oxygen species (ROS) levels (Zhang et al. 2019). Microbial interactions can also modulate phytohormone levels, enhancing plant stress responses. These interactions lead to improved antioxidant activity and the accumulation of osmolytes, which are critical for stress mitigation (Shaffique et al. 2022).

5 Future Challenges

Climate change, mostly induced by human actions, is resulting in substantial alterations in temperature and water accessibility, culminating in severe meteorological phenomena. Agricultural regions are increasingly facing heightened temperatures and less rainfall, leading to significant abiotic stress and consequent reductions in crop production. This situation requires the creation of novel plant kinds that can endure these environmental changes. As climate change intensifies, it brings about a range of abiotic stresses such as heat, drought, and salinity, which are often experienced simultaneously, complicating plant responses and management strategies.

The overall effects of various abiotic stressors are not simply additive; they are influenced by complex relations between the pressures and the plant's responses.

For example, increasing atmospheric CO_2 levels may improve agricultural yields in certain plants; but, when coupled with dryness, this growth advantage may be mitigated by changes in stomatal function and decreased water use efficiency. Recent studies emphasize the necessity of examining the interacting consequences of these stressors, namely increasing CO_2 alongside temperature variations and water availability, as these elements are vital components of climate change scenarios. Considering that climate change models forecast an increase in the frequency and severity of ecological disturbances shortly, plant breeders, researchers, and agricultural professionals must include an understanding of abiotic stress physiology in crop enhancement initiatives. This method will facilitate the creation of robust crop types that can endure projected environmental problems. It is essential to develop climate-resilient crops using both conventional and biotechnology breeding methods. This entails developing genotypes that are resilient to high temperatures, moisture deficiency, and salt. Enhanced water use efficiency, soil moisture conservation techniques, and raised bed farming are effective management options to alleviate the negative effects of climate change.

References

Abd-Allah EF, Hashem A, Alqarawi AA, Bahkali AH, Alwhibi MS (2015) Enhancing growth performance and systemic acquired resistance of medicinal plant *Sesbania sesban* (L.) Merr using arbuscular mycorrhizal fungi under salt stress. Saudi J Biol Sci 22(3):274–283. https://doi.org/10.1016/J.SJBS.2015.03.004

Abdou Zayan S (2018) Impact of climate change on plant diseases and IPM strategies. In: Plant diseases—current threats and management trends. https://doi.org/10.5772/INTECHOPEN.87055

Abewoy D (2018) Review on impacts of climate change on vegetable production and its management practices. Adv Crop Sci Technol 06(01). https://doi.org/10.4172/2329-8863.1000330

Adeleke BS, Ayilara MS, Adeyemi UT, Uthman QO, Chaudhary A, Abou Fayssal S (2024) Role of microbes in plant biotic stress management, pp 145–162. https://doi.org/10.1007/978-981-97-0633-4_7

Ahmad I, Zaib S (2020) Mighty microbes: plant growth promoting microbes in soil health and sustainable agriculture, pp 243–264. https://doi.org/10.1007/978-3-030-44364-1_14

Alagar Boopathy LR, Jacob-Tomas S, Alecki C, Vera M (2022) Mechanisms tailoring the expression of heat shock proteins to proteostasis challenges. J Biol Chem 298(5). https://doi.org/10.1016/J.JBC.2022.101796

Alam MZ, Dey M (2024) The reduction of abiotic stress in food crops through climate-smart mycorrhiza-enriched biofertilizer. AIMS Microbiol 10(3):674–693. https://doi.org/10.3934/MICROBIOL.2024031

Albacete A, Ghanem ME, Martínez-Andújar C, Acosta M, Sánchez-Bravo J, Martínez V, Lutts S, Dodd IC, Pérez-Alfocea F (2008) Hormonal changes in relation to biomass partitioning and shoot growth impairment in salinized tomato (*Solanum lycopersicum* L.) plants. J Exp Bot 59(15):4119–4131. https://doi.org/10.1093/JXB/ERN251

Alexandrov VA, Eitzinger J (2005) The potential effect of climate change and elevated air carbon dioxide on agricultural crop production in Central and Southeastern Europe. J Crop Improv 13(1–2):291–331. https://doi.org/10.1300/J411V13N01_14

Ali Mozaffari G (2022) Climate change and its consequences in agriculture. In: The nature, causes, effects and mitigation of climate change on the environment. https://doi.org/10.5772/INTECHOPEN.101444

Al-Surhanee AA, Soliman MH, Ouf SA (2021) The role of soil microbes in the plant adaptation to stresses: current scenario and future perspective. In: Frontiers in plant-soil interaction, pp 237–258. https://doi.org/10.1016/B978-0-323-90943-3.00006-7

Amari K, Huang C, Heinlein M (2021) Potential impact of global warming on virus propagation in infected plants and agricultural productivity. Front Plant Sci 12. https://doi.org/10.3389/FPLS.2021.649768

Aresta M (2010) Carbon dioxide: utilization options to reduce its accumulation in the atmosphere. In: Carbon dioxide as chemical feedstock, pp 1–13. https://doi.org/10.1002/9783527629916.CH1

Aroca R, Ruiz-Lozano JM (2009) Induction of plant tolerance to semi-arid environments by beneficial soil microorganisms—a review. In: Climate change, intercropping, pest control and beneficial microorganisms, pp 121–135. https://doi.org/10.1007/978-90-481-2716-0_7

Ashokan A, Mini V, Rani B, Anand S (2024) Mitigation of abiotic stresses in plants through nutrient management. Int J Environ Clim Change 14(7):254–267. https://doi.org/10.9734/IJECC/2024/V14I74267

Azizi M, Fard EM, Ghabooli M (2021) *Piriformospora indica* affect drought tolerance by regulation of genes expression and some morphophysiological parameters in tomato (*Solanum lycopersicum* L.). Sci Hortic 287. https://doi.org/10.1016/J.SCIENTA.2021.110260

Bae H, Sicher RC, Kim MS, Kim SH, Strem MD, Melnick RL, Bailey BA (2009) The beneficial endophyte *Trichoderma hamatum* isolate DIS 219b promotes growth and delays the onset of the drought response in Theobroma cacao. J Exp Bot 60(11):3279–3295. https://doi.org/10.1093/JXB/ERP165

Bai Y, Quais MK, Zhou W, Zhu ZR (2021) Consequences of elevated temperature on the biology, predation, and competitiveness of two mirid predators in the rice ecosystem. J Pest Sci 95(2):901–916. https://doi.org/10.1007/S10340-021-01414-Y

Barnawal D, Singh R, Singh RP (2019) Role of plant growth promoting rhizobacteria in drought tolerance. In: PGPR amelioration in sustainable agriculture, pp 107–128. https://doi.org/10.1016/B978-0-12-815879-1.00006-9

Bartlett R, Bharati L, Pant D, Hpsterman H, McCprnick P (2010) Climate change impacts and adaptation in Nepal, p 35. http://www.iwmi.cgiar.org/publications/iwmi-working-papers/iwmi-working-paper-139

Bashan Y, de Bashan LE (2010) How the plant growth-promoting bacterium *azospirillum* promotes plant growth—a critical assessment. Adv Agron 108(C):77–136. https://doi.org/10.1016/S0065-2113(10)08002-8

Basheer J (2022) Colonization, diversity, and distribution of endophytic microbial communities in different parts of plants. In: Biocontrol mechanisms of endophytic microorganisms, pp 1–30. https://doi.org/10.1016/B978-0-323-88478-5.00008-0

Basu S, Kumari S, Subhadarshini P, Rishu AK, Shekhar S, Kumar G (2023) Plant growth promoting *rhizobacterium Bacillus* sp. BSE01 alleviates salt toxicity in chickpeas (*Cicer arietinum* L.) by conserving ionic, osmotic, redox and hormonal homeostasis. Physiol Plant 175(6):e14076. https://doi.org/10.1111/PPL.14076

Bijalwan P, Jeddi K, Saini I, Sharma M, Kaushik P, Hessini K (2021) Mitigation of saline conditions in watermelon with mycorrhiza and silicon application. Saudi J Biol Sci 28(7):3678–3684. https://doi.org/10.1016/J.SJBS.2021.05.019

Bita CE, Gerats T (2013) Plant tolerance to high temperature in a changing environment: scientific fundamentals and production of heat stress-tolerant crops. Front Plant Sci 4(July). https://doi.org/10.3389/FPLS.2013.00273/PDF

Camaille M, Fabre N, Clément C, Barka EA (2021) Advances in wheat physiology in response to drought and the role of plant growth promoting rhizobacteria to trigger drought tolerance. Microorganisms 9(4). https://doi.org/10.3390/MICROORGANISMS9040687

Carmody M, Waszczak C, Idänheimo N, Saarinen T, Kangasjärvi J (2016) ROS signalling in a destabilised world: a molecular understanding of climate change. J Plant Physiol 203:69–83. https://doi.org/10.1016/J.JPLPH.2016.06.008

Castroverde CDM, Dina D (2021) Temperature regulation of plant hormone signaling during stress and development. J Exp Bot 72(21):7436–7458. https://doi.org/10.1093/JXB/ERAB257

Cecagno R, Fritsch TE, Schrank IS (2015) The plant growth-promoting bacteria *Azospirillum amazonense*: genomic versatility and phytohormone pathway. Biomed Res Int 2015. https://doi.org/10.1155/2015/898592

Chaloner TM, Gurr SJ, Bebber DP (2020) Plant pathogen infection risk tracks global crop yields under climate change. Nat Clim Chang 11(8):710–715. https://doi.org/10.1038/S41558-021-01104-8

Chaudhry S, Sidhu GPS (2021) Climate change regulated abiotic stress mechanisms in plants: a comprehensive review. Plant Cell Rep 41(1). https://doi.org/10.1007/S00299-021-02759-5

Chen C, Xin K, Liu H, Cheng J, Shen X, Wang Y, Zhang L (2017) *Pantoea alhagi*, a novel endophytic bacterium with ability to improve growth and drought tolerance in wheat. Sci Rep 7(1):1–14. https://doi.org/10.1038/srep41564

Chen W, Meng P, Feng H, Wang C (2020) Effects of arbuscular mycorrhizal fungi on growth and physiological performance of *Catalpa bungei* C. A. Mey under drought stress. Forests 11(10):1–29. https://doi.org/10.3390/F11101117

Chu XT, Fu JJ, Sun YF, Xu YM, Miao YJ, Xu YF, Hu TM (2016) Effect of arbuscular mycorrhizal fungi inoculation on cold stress-induced oxidative damage in leaves of *Elymus nutans* Griseb. S Afr J Bot 104:21–29. https://doi.org/10.1016/J.SAJB.2015.10.001

Cohen SP, Leach JE (2020) High temperature-induced plant disease susceptibility: more than the sum of its parts. Curr Opin Plant Biol 56:235–241. https://doi.org/10.1016/J.PBI.2020.02.008

Compant S, Clément C, Sessitsch A (2010) Plant growth-promoting bacteria in the rhizo- and endosphere of plants: their role, colonization, mechanisms involved and prospects for utilization. Soil Biol Biochem 42(5):669–678. https://doi.org/10.1016/J.SOILBIO.2009.11.024

Dale H, Press MC (1998) Elevated atmospheric CO_2 influences the interaction between the parasitic angiosperm Orobanche minor and its host *Trifolium repens*. New Phytol 140(1):65–73. https://doi.org/10.1046/J.1469-8137.1998.00247.X

Daraz U, Ahmad I, Li QS, Zhu B, Saeed MF, Li Y, Ma J, Wang XB (2023) Plant growth promoting rhizobacteria induced metal and salt stress tolerance in *Brassica juncea* through ion homeostasis. Ecotoxicol Environ Saf 267:115657. https://doi.org/10.1016/J.ECOENV.2023.115657

Dong H, Wang Y, Di Y, Qiu Y, Ji Z, Zhou T, Shen S, Du N, Zhang T, Dong X, Guo Z, Piao F, Li Y (2024) Plant growth-promoting rhizobacteria *Pseudomonas aeruginosa* HG28-5 improves salt tolerance by regulating Na+/K+ homeostasis and ABA signaling pathway in tomatoes. Microbiol Res 283:127707. https://doi.org/10.1016/J.MICRES.2024.127707

Egamberdieva D (2011) Survival of *Pseudomonas extremorientalis* TSAU20 and *P. chlororaphis* TSAU13 in the rhizosphere of common bean (*Phaseolus vulgaris*) under saline conditions. Plant Soil Environ 57(3):122–127. https://doi.org/10.17221/316/2010-PSE

Egamberdieva D (2012) *Pseudomonas chlororaphis*: a salt-tolerant bacterial inoculant for plant growth stimulation under saline soil conditions. Acta Physiol Plant 34(2):751–756. https://doi.org/10.1007/S11738-011-0875-9

Egamberdieva D, Kucharova Z (2009) Selection for root colonising bacteria stimulating wheat growth in saline soils. Biol Fertil Soils 45(6):563–571. https://doi.org/10.1007/S00374-009-0366-Y

Egamberdieva D, Wirth SJ, Alqarawi AA, Abd-Allah EF, Hashem A (2017) Phytohormones and beneficial microbes: essential components for plants to balance stress and fitness. Front Microbiol 8(October). https://doi.org/10.3389/FMICB.2017.02104

Etesami H, Beattie GA (2017) Plant-microbe interactions in adaptation of agricultural crops to abiotic stress conditions. In: Probiotics and plant health, pp 163–200. https://doi.org/10.1007/978-981-10-3473-2_7

Etesami H, Maheshwari DK (2018) Use of plant growth promoting rhizobacteria (PGPRs) with multiple plant growth promoting traits in stress agriculture: action mechanisms and future prospects. Ecotoxicol Environ Saf 156:225–246. https://doi.org/10.1016/J.ECOENV.2018.03.013

Fatima S, Aslam R, Shahzadi K, Awan H, Nadeem M, Shabir M, Shafiq Y, Hamid M (2024) Effect of drought stress at growth and development of pea (*Pisum sativum* L.). Asian J Res Crop Sci 9(1):76–80. https://doi.org/10.9734/AJRCS/2024/V9I1247

Fisher MC, Gurr SJ, Cuomo CA, Blehert DS, Jin H, Stukenbrock EH, Stajich JE, Kahmann R, Boone C, Denning DW, Gow NAR, Klein BS, Kronstad JW, Sheppard DC, Taylor JW, Wright GD, Heitman J, Casadevall A, Cowen LE (2020) Threats posed by the fungal kingdom to humans, wildlife, and agriculture. MBio 11(3). https://doi.org/10.1128/MBIO.00449-20

Flores-Rentería L, Whipple AV, Benally GJ, Patterson A, Canyon B, Gehring CA (2018) Higher temperature at lower elevation sites fails to promote acclimation or adaptation to heat stress during pollen germination. Front Plant Sci 9. https://doi.org/10.3389/FPLS.2018.00536

Frank AC (2011) The genomes of endophytic bacteria, pp 107–136. https://doi.org/10.1007/978-94-007-1599-8_7

Gagné-Bourque F, Mayer BF, Charron JB, Vali H, Bertrand A, Jabaji S (2015) Accelerated growth rate and increased drought stress resilience of the model grass *Brachypodium distachyon* colonized by *Bacillus subtilis* B26. PLoS One 10(6). https://doi.org/10.1371/JOURNAL.PONE.0130456

Gill SS, Gill R, Trivedi DK, Anjum NA, Sharma KK, Ansari MW, Ansari AA, Johri AK, Prasad R, Pereira E, Varma A, Tuteja N (2016) *Piriformospora indica*: potential and significance in plant stress tolerance. Front Microbiol 7(March). https://doi.org/10.3389/FMICB.2016.00332

Glick BR (2012) Plant growth-promoting bacteria: mechanisms and applications. Scientifica 2012:1–15. https://doi.org/10.6064/2012/963401

Glick BR (2014) Bacteria with ACC deaminase can promote plant growth and help to feed the world. Microbiol Res 169(1):30–39. https://doi.org/10.1016/J.MICRES.2013.09.009

Glick BR, Todorovic B, Czarny J, Cheng Z, Duan J, McConkey B (2007) Promotion of plant growth by bacterial ACC deaminase. Crit Rev Plant Sci 26(5–6):227–242. https://doi.org/10.1080/07352680701572966

Grover M, Ali SZ, Sandhya V, Rasul A, Venkateswarlu B (2011) Role of microorganisms in adaptation of agriculture crops to abiotic stresses. World J Microbiol Biotechnol 27(5):1231–1240. https://doi.org/10.1007/S11274-010-0572-7/FIGURES/1

Gupta A, Rico-Medina A, Caño-Delgado AI (2020) The physiology of plant responses to drought. Science (New York, NY) 368(6488):266–269. https://doi.org/10.1126/SCIENCE.AAZ7614

Hamilton CE, Bever JD, Labbé J, Yang X, Yin H (2016) Mitigating climate change through managing constructed-microbial communities in agriculture. Agric Ecosyst Environ 216:304–308. https://doi.org/10.1016/J.AGEE.2015.10.006

Haq IU, Rahim K, Yahya G, Ijaz B, Maryam S, Paker NP (2024) Eco-smart biocontrol strategies utilizing potent microbes for sustainable management of phytopathogenic diseases. Biotechnol Rep 44. https://doi.org/10.1016/J.BTRE.2024.E00859

Hardoim PR, van Overbeek LS, van Elsas JD (2008) Properties of bacterial endophytes and their proposed role in plant growth. Trends Microbiol 16(10):463–471. https://doi.org/10.1016/J.TIM.2008.07.008

Hasegawa S, Meguro A, Toyoda K, Nishimura T, Kunoh H (2005) Drought tolerance of tissue-cultured seedlings of mountain laurel (*Kalmia latifolia* L.) induced by an endophytic actinomycete II. Acceleration of callose accumulation and lignification. Actinomycetologica 19(1):13–17. https://doi.org/10.3209/SAJ.19.13

Hashem A, Abd Allah EF, Alqarawi AA, Al Huqail AA, Egamberdieva D, Wirth S (2016) Alleviation of cadmium stress in *Solanum lycopersicum* L. by arbuscular mycorrhizal fungi via induction of acquired systemic tolerance. Saudi J Biol Sci 23(2):272–281. https://doi.org/10.1016/J.SJBS.2015.11.002

Haskett JD, Pachepsky YA, Acock B (1997) Increase of CO_2 and climate change effects on Iowa soybean yield, simulated using GLYCIM. Agron J 89(2):167–176. https://doi.org/10.2134/AGRONJ1997.00021962008900020004X

Hone H, Mann R, Yang G, Kaur J, Tannenbaum I, Li T, Spangenberg G, Sawbridge T (2021) Profiling, isolation and characterization of beneficial microbes from the seed microbiomes of drought tolerant wheat. Sci Rep 11(1). https://doi.org/10.1038/S41598-021-91351-8

Impa SM, Nadaradjan S, Jagadish SVK (2012) Drought stress induced reactive oxygen species and anti-oxidants in plants. In: Abiotic stress responses in plants, pp 131–147. https://doi.org/10.1007/978-1-4614-0634-1_7

Jogawat A, Vadassery J, Verma N, Oelmüller R, Dua M, Nevo E, Johri AK (2016) PiHOG1, a stress regulator MAP kinase from the root endophyte fungus *Piriformospora indica*, confers salinity stress tolerance in rice plants. Sci Rep 6(1):1–15. https://doi.org/10.1038/srep36765

Jones AM (2016) A new look at stress: abscisic acid patterns and dynamics at high-resolution. New Phytol 210(1):38–44. https://doi.org/10.1111/NPH.13552

Kaczmarek J, Kedziora A, Brachaczek A, Latunde-Dada AO, Dakowska S, Karg G, Jedryczka M (2016) Effect of climate change on sporulation of the teleomorphs of Leptosphaeria species causing stem canker of brassicas. Aerobiologia 32(1):39–51. https://doi.org/10.1007/S10453-015-9404-4/FIGURES/7

Kandel SL, Joubert PM, Doty SL (2017) Bacterial endophyte colonization and distribution within plants. Microorganisms 5(4). https://doi.org/10.3390/MICROORGANISMS5040077

Kassanis B (1957) Effects of changing temperature on plant virus diseases. Adv Virus Res 4(C):221–241. https://doi.org/10.1016/S0065-3527(08)60600-4

Khalid M, Hassani D, Liao J, Xiong X, Bilal M, Huang D (2018) An endosymbiont *Piriformospora indica* reduces adverse effects of salinity by regulating cation transporter genes, phytohormones, and antioxidants in *Brassica campestris* ssp. Chinensis. Environ Exp Bot 153:89–99. https://doi.org/10.1016/J.ENVEXPBOT.2018.05.007

Khan MR, Rizvi TF (2020) Effect of elevated levels of CO_2 on powdery mildew development in five cucurbit species. Sci Rep 10(1):1–13. https://doi.org/10.1038/s41598-020-61790-w

Khanna K, Kaur R, Bali S, Sharma A, Bakshi P, Saini P, Thukral AK, Ohri P, Mir BA, Choudhary SP, Bhardwaj R (2019) Role of beneficial microorganisms in abiotic stress tolerance in plants. In: Plant tolerance to environmental stress, pp 327–348. https://doi.org/10.1201/9780203705315-20

Klinsukon C, Lumyong S, Kuyper TW, Boonlue S (2021) Colonization by arbuscular mycorrhizal fungi improves salinity tolerance of eucalyptus (*Eucalyptus camaldulensis*) seedlings. Sci Rep 11(1). https://doi.org/10.1038/S41598-021-84002-5

Kou C, Song F, Li D, Xu H, Zhang S, Yang W, Shi W, Gao Z (2024) A necessary considering factor for crop resistance: precise regulation and effective utilization of beneficial microorganisms. New Crops 1:100023. https://doi.org/10.1016/J.NCROPS.2024.100023

Kůdela V, Pokorný R, Lebeda A (2018) Potential impact of climate change on geographic distribution of plant pathogenic bacteria in Central Europe. Plant Prot Sci 45(SPEC. ISS). https://doi.org/10.17221/2832-PPS

Kumar A, Droby S, Singh VK, Singh SK, White JF (2020) Entry, colonization, and distribution of endophytic microorganisms in plants. In: Microbial endophytes: functional biology and applications, pp 1–33. https://doi.org/10.1016/B978-0-12-819654-0.00001-6

Kumar S, Choudhary M, Janakadatta Reddy K, Vishwakarma VK, Kashyap VK, Sahoo S, Mukhopadhyay S (2024) A review on the impact of climate change on plant pathogen interactions. J Adv Microbiol 24(8):11–27. https://doi.org/10.9734/JAMB/2024/V24I8843

Lahlali R, Taoussi M, Laasli SE, Gachara G, Ezzouggari R, Belabess Z, Aberkani K, Assouguem A, Meddich A, El Jarroudi M, Barka EA (2024) Effects of climate change on plant pathogens and host-pathogen interactions. Crop Environ 3(3):159–170. https://doi.org/10.1016/J.CROPE.2024.05.003

Lesk C, Rowhani P, Ramankutty N (2016) Influence of extreme weather disasters on global crop production. Nature 529(7584):84–87. https://doi.org/10.1038/nature16467

Li Z, Howell SH (2021) Heat stress responses and thermotolerance in maize. Int J Mol Sci 22(2):1–19. https://doi.org/10.3390/IJMS22020948

Li N, Euring D, Cha JY, Lin Z, Lu M, Huang LJ, Kim WY (2021) Plant hormone-mediated regulation of heat tolerance in response to global climate change. Front Plant Sci 11:627969. https://doi.org/10.3389/FPLS.2020.627969/BIBTEX

Liu XM, Zhang H (2015) The effects of bacterial volatile emissions on plant abiotic stress tolerance. Front Plant Sci 6(September):158867. https://doi.org/10.3389/FPLS.2015.00774/BIBTEX

Lybbert TJ, Sumner DA (2012) Agricultural technologies for climate change in developing countries: policy options for innovation and technology diffusion. Food Policy 37(1):114–123. https://doi.org/10.1016/J.FOODPOL.2011.11.001

Ma L, Qiao J, Kong X, Zou Y, Xu X, Chen X, Hu X (2015) Effect of low temperature and wheat winter-hardiness on survival of *Puccinia striiformis* f. sp. tritici under controlled conditions. PLoS One 10(6). https://doi.org/10.1371/JOURNAL.PONE.0130691

Maela PM, Serepa-Dlamini MH (2019) Current understanding of bacterial endophytes, their diversity, colonization and their roles in promoting plant growth. Appl Microbiol 5(1):157. https://doi.org/10.35248/2471-9315.19.5.157

Mathur P, Roy S (2021) Insights into the plant responses to drought and decoding the potential of root associated microbiome for inducing drought tolerance. Physiol Plant 172(2):1016–1029. https://doi.org/10.1111/PPL.13338

Mathur S, Sharma MP, Jajoo A (2018) Improved photosynthetic efficacy of maize (*Zea mays*) plants with arbuscular mycorrhizal fungi (AMF) under high-temperature stress. J Photochem Photobiol B 180:149–154. https://doi.org/10.1016/J.JPHOTOBIOL.2018.02.002

Matilla MA, Krell T (2018) Plant growth promotion and biocontrol mediated by plant-associated bacteria. Plant microbiome: stress response, 45–80

Meena M, Yadav G, Sonigra P, Nagda A, Mehta T, Swapnil P, Harish, Marwal A (2022) Role of elicitors to initiate the induction of systemic resistance in plants to biotic stress. Plant Stress 5:100103. https://doi.org/10.1016/J.STRESS.2022.100103

Mengistu AA (2020) Endophytes: colonization, behaviour, and their role in defense mechanism. Int J Microbiol. https://doi.org/10.1155/2020/6927219

Miransari M (2013) Soil microbes and the availability of soil nutrients. Acta Physiol Plant 35(11):3075–3084. https://doi.org/10.1007/S11738-013-1338-2/TABLES/2

Miransari M, Smith DL (2014) Plant hormones and seed germination. Environ Exp Bot 99:110–121. https://doi.org/10.1016/J.ENVEXPBOT.2013.11.005

Morsy M, Cleckler B, Armuelles-Millican H (2020) Fungal endophytes promote tomato growth and enhance drought and salt tolerance. Plants (Basel, Switzerland) 9(7):1–18. https://doi.org/10.3390/PLANTS9070877

Mun CL, Wong Vui Ling CM (2020) Effects of elevated temperature on the tropical soil bacterial diversity. Sains Malays 49(10):2335–2344. https://doi.org/10.17576/JSM-2020-4910-01

Munir N, Hanif M, Abideen Z, Sohail M, El-Keblawy A, Radicetti E, Mancinelli R, Haider G (2022) Mechanisms and strategies of plant microbiome interactions to mitigate abiotic stresses. Agronomy 12(9). https://doi.org/10.3390/AGRONOMY12092069

Nabti E, Sahnoune M, Ghoul M, Fischer D, Hofmann A, Rothballer M, Schmid M, Hartmann A (2010) Restoration of growth of durum wheat (*Triticum durum* var. waha) under saline conditions due to inoculation with the rhizosphere bacterium *Azospirillum brasilense* NH and extracts of the marine alga *Ulva lactuca*. J Plant Growth Regul 29(1):6–22. https://doi.org/10.1007/S00344-009-9107-6/FIGURES/3

Naik PK, Ruban JS, Naidu SA, Fetricia JP (2022) Effect of drought stress and mineral nutrition in plant growth and development. Int J Curr Microbiol App Sci 11(5):109–114. https://doi.org/10.20546/IJCMAS.2022.1105.014

Nanda S, Mohanty B, Joshi RK (2018) Endophyte-mediated host stress tolerance as a means for crop improvement. In: Endophytes and secondary metabolites, Reference series in phytochemistry, pp 1–25. https://doi.org/10.1007/978-3-319-76900-4_28-1

Nasri N, Kaddour R, Rabhi M, Plassard C, Lachaal M (2011) Effect of salinity on germination, phytase activity and phytate content in lettuce seedling. Acta Physiol Plant 33(3):935–942. https://doi.org/10.1007/S11738-010-0625-4

Nautiyal CS, Srivastava S, Chauhan PS, Seem K, Mishra A, Sopory SK (2013) Plant growth-promoting bacteria *Bacillus amyloliquefaciens* NBRISN13 modulates the gene expression profile of leaf and rhizosphere community in rice during salt stress. Plant Physiol Biochem 66:1–9. https://doi.org/10.1016/J.PLAPHY.2013.01.020

Naveed M, Hussain MB, Zahir ZA, Mitter B, Sessitsch A (2013) Drought stress amelioration in wheat through inoculation with *Burkholderia phytofirmans* strain PsJN. Plant Growth Regul 73(2):121–131. https://doi.org/10.1007/S10725-013-9874-8

Nayak L, Lal MK, Tiwari RK, Kumar R, Lal P, Das R, Behera B, Thakur P, Altaf MA, Kumar A (2023) A balancing act: exploring the interplay between HSPs and osmoprotectants in temperature stress responses. S Afr J Bot 162:64–71. https://doi.org/10.1016/J.SAJB.2023.08.069

Ortas I, Rafique M, Çekiç FÖ (2020) Do mycorrhizal fungi enable plants to cope with abiotic stresses by overcoming the detrimental effects of salinity and improving drought tolerance? In: Symbiotic soil microorganisms. Springer, Berlin, pp 391–428. https://doi.org/10.1007/978-3-030-51916-2_23

Pal A, Pandey S (2016) Role of arbuscular mycorrhizal fungi on plant growth and reclamation of barren soil with wheat (*Triticum aestivum* L.) crop. Int J Soil Sci 12(1):25–31. https://doi.org/10.3923/IJSS.2017.25.31

Pan X, Qin Y, Yuan Z (2018) The potential of a halophyte-associated endophytic fungus for sustaining Chinese white poplar growth under salinity. Symbiosis 76(2):109–116. https://doi.org/10.1007/S13199-018-0541-8/METRICS

Pandey V, Ansari MW, Tula S, Yadav S, Sahoo RK, Shukla N, Bains G, Badal S, Chandra S, Gaur AK, Kumar A, Shukla A, Kumar J, Tuteja N (2016) Dose-dependent response of *Trichoderma harzianum* in improving drought tolerance in rice genotypes. Planta 243(5):1251–1264. https://doi.org/10.1007/S00425-016-2482-X

Parihar P, Singh S, Singh R, Singh VP, Prasad SM (2015) Effect of salinity stress on plants and its tolerance strategies: a review. Environ Sci Pollut Res Int 22(6):4056–4075. https://doi.org/10.1007/S11356 014-3739-1

Parikka P, Hakala K, Tiilikkala K (2012) Expected shifts in Fusarium species' composition on cereal grain in Northern Europe due to climatic change. Food Addit Contam Part A Chem Anal Control Expo Risk Assess 29(10):1543–1555. https://doi.org/10.1080/19440049.2012.680613

Poudel M, Mendes R, Costa LAS, Bueno CG, Meng Y, Folimonova SY, Garrett KA, Martins SJ (2021) The role of plant-associated bacteria, fungi, and viruses in drought stress mitigation. Front Microbiol 12. https://doi.org/10.3389/FMICB.2021.743512/PDF

Puli COR, Akila CS, Kandi SR, Naik PNKT, Yammanuri N, Guduru KK, Kunduru S, Bolupalli S, Bramhachari PV, Pamuru RR (2022) Endophytic microbiome-assisted drought tolerance in plants. In: Understanding the microbiome interactions in agriculture and the environment, pp 185–213. https://doi.org/10.1007/978-981-19-3696-8_10

Qiao M, Hong C, Jiao Y, Hou S, Gao H (2024) Impacts of drought on photosynthesis in major food crops and the related mechanisms of plant responses to drought. Plan Theory 13(13). https://doi.org/10.3390/PLANTS13131808

Qiu J, Xie J, Chen Y, Shen Z, Shi H, Naqvi NI, Qian Q, Liang Y, Kou Y (2022) Warm temperature compromises JA-regulated basal resistance to enhance *Magnaporthe oryzae* infection in rice. Mol Plant 15(4):723–739. https://doi.org/10.1016/J.MOLP.2022.02.014

Raaijmakers JM, de Bruijn I, Nybroe O, Ongena M (2010) Natural functions of lipopeptides from *Bacillus* and *Pseudomonas*: more than surfactants and antibiotics. FEMS Microbiol Rev 34(6):1037–1062. https://doi.org/10.1111/J.1574-6976.2010.00221.X

Rahneshan Z, Nasibi F, Moghadam AA (2018) Effects of salinity stress on some growth, physiological, biochemical parameters and nutrients in two pistachio (*Pistacia vera* L.) rootstocks. J Plant Interact 13(1):73–82. https://doi.org/10.1080/17429145.2018.1424355

Rane J, Jangid KK (2024) Impacts of abiotic stresses on eco-physiology of crop in changing climate. Adv Glob Change Res 78:427–445. https://doi.org/10.1007/978-981-99-7935-6_16

Razzaq S, Hamza MA, Khan S, Imam M, Nazir K, Kausar S, Naqvi REF, Shah AA, Abbas W, Rauf A (2024) New insights for promoting the plant growth and germination and applications in agriculture. Sch Int J Biochem 7(02):30–36. https://doi.org/10.36348/SIJB.2024.V07I02.003

Reddy PP (2014) Climate resilient agriculture for ensuring food security. Springer India, pp 1–373. https://doi.org/10.1007/978-81-322-2199-9

Reinhold-Hurek B, Hurek T (2011) Living inside plants: bacterial endophytes. Curr Opin Plant Biol 14(4):435–443. https://doi.org/10.1016/J.PBI.2011.04.004

Riseh RS, Ebrahimi-Zarandi M, Gholizadeh Vazvani M, Skorik YA (2021) Reducing drought stress in plants by encapsulating plant growth-promoting bacteria with polysaccharides. Int J Mol Sci 22(23). https://doi.org/10.3390/IJMS222312979

Rivero RM, Mittler R, Blumwald E, Zandalinas SI (2022) Developing climate-resilient crops: improving plant tolerance to stress combination. Plant J Cell Mol Biol 109(2):373–389. https://doi.org/10.1111/TPJ.15483

Rosenberg NJ (1981) The increasing CO_2 concentration in the atmosphere and its implication on agricultural productivity. Clim Chang 3(3):265–279. https://doi.org/10.1007/BF02423219

Sagar A, Rai S, Ilyas N, Sayyed RZ, Al-Turki AI, El Enshasy HA, Simarmata T (2022) Halotolerant rhizobacteria for salinity-stress mitigation: diversity, mechanisms and molecular approaches. Sustainability (Switzerland) 14(1). https://doi.org/10.3390/SU14010490

Sbeiti AAL, Mazurier M, Ben C, Rickauer M, Gentzbittel L (2023) Temperature increase modifies susceptibility to Verticillium wilt in *Medicago spp* and may contribute to the emergence of more aggressive pathogenic strains. Front Plant Sci 14. https://doi.org/10.3389/FPLS.2023.1109154

Sessitsch A, Hardoim P, Döring J, Weilharter A, Krause A, Woyke T, Mitter B, Hauberg-Lotte L, Friedrich F, Rahalkar M, Hurek T, Sarkar A, Bodrossy L, Van Overbeek L, Brar D, Van Elsas JD, Reinhold-Hurek B (2012) Functional characteristics of an endophyte community colonizing rice roots as revealed by metagenomic analysis. Mol Plant-Microbe Interact 25(1):28–36. https://doi.org/10.1094/MPMI-08-11-0204

Shaffique S, Khan MA, Wani SH, Pande A, Imran M, Kang SM, Rahim W, Khan SA, Bhatta D, Kwon EH, Lee IJ (2022) A review on the role of endophytes and plant growth promoting rhizobacteria in mitigating heat stress in plants. Microorganisms 10(7). https://doi.org/10.3390/MICROORGANISMS10071286

Shahvali R, Shiran B, Ravash R, Fallahi H, Banović Đeri B (2020) Effect of symbiosis with arbuscular mycorrhizal fungi on salt stress tolerance in GF677 (peach × almond) rootstock. Sci Hortic 272. https://doi.org/10.1016/J.SCIENTA.2020.109535

Shaterian J, Waterer D, De Jong H, Tanino KK (2005) Differential stress responses to NaCl salt application in early- and late-maturing diploid potato (*Solanum* sp.) clones. Environ Exp Bot 54(3):202–212. https://doi.org/10.1016/J.ENVEXPBOT.2004.07.005

Shelake RM, Wagh SG, Patil AM, Červený J, Waghunde RR, Kim JY (2024) Heat stress and plant-biotic interactions: advances and perspectives. Plants (Basel, Switzerland) 13(15). https://doi.org/10.3390/PLANTS13152022

Shi-chu L, Yong J, Ma-bo L, Wen-xu Z, Nan X, Hui-hui Z (2019) Improving plant growth and alleviating photosynthetic inhibition from salt stress using AMF in alfalfa seedlings. J Plant Interact 14(1):482–491. https://doi.org/10.1080/17429145.2019.1662101

Siddra I (2016) A review on genetic basis of heat shock proteins. Mol Plant Breed. https://doi.org/10.5376/MPB.2016.07.0004

Singh J, Sastry EVD, Singh V (2012) Effect of salinity on tomato (*Lycopersicon esculentum* Mill.) during seed germination stage. Physiol Mol Biol Plants 18(1):45–50. https://doi.org/10.1007/S12298-011-0097-Z

Singh BK, Delgado-Baquerizo M, Egidi E, Guirado E, Leach JE, Liu H, Trivedi P (2023) Climate change impacts on plant pathogens, food security and paths forward. Nat Rev Microbiol 21(10):640–656. https://doi.org/10.1038/S41579-023-00900-7

Srivastav AL, Dhyani R, Ranjan M, Madhav S, Sillanpää M (2021) Climate-resilient strategies for sustainable management of water resources and agriculture. Environ Sci Pollut Res Int 28(31):41576–41595. https://doi.org/10.1007/S11356-021-14332-4

Talaat NB, Shawky BT (2017) Microbe-mediated induced abiotic stress tolerance responses in plants. In: Plant-microbe interactions in agro-ecological perspectives, vol 2, pp 101–133. https://doi.org/10.1007/978-981-10-6593-4_5

Tenllado F, Canto T (2020) Effects of a changing environment on the defenses of plants to viruses. Curr Opin Virol 42:40–46. https://doi.org/10.1016/J.COVIRO.2020.04.007

Timmusk S, Abd El-Daim IA, Copolovici L, Tanilas T, Kännaste A, Behers L, Nevo E, Seisenbaeva G, Stenström E, Niinemets Ü (2014) Drought-tolerance of wheat improved by rhizosphere bacteria from harsh environments: enhanced biomass production and reduced emissions of stress volatiles. PLoS One 9(5):e96086. https://doi.org/10.1371/JOURNAL.PONE.0096086

Tsegay BA, Gebreslassie B (2014) The effect of salinity (NaCl) on germination and early seedling growth of *Lathyrus sativus* and *Pisum sativum* var. *abyssinicum*. African J Plant Sci 8(5):225–231. https://doi.org/10.5897/AJPS2014.1176

Tufail MA, Bejarano A, Shakoor A, Naeem A, Arif MS, Dar AA, Farooq TH, Pertot I, Puopolo G (2021) Can bacterial endophytes be used as a promising bio-inoculant for the mitigation of salinity stress in crop plants?—a global meta-analysis of the last decade (2011-2020). Microorganisms 9(9). https://doi.org/10.3390/MICROORGANISMS9091861

Ullah A, Bano A, Khan N (2021) Climate change and salinity effects on crops and chemical communication between plants and plant growth-promoting microorganisms under stress. Front Sustain Food Syst 5. https://doi.org/10.3389/FSUFS.2021.618092

Váry Z, Mullins E, Mcelwain JC, Doohan FM (2015) The severity of wheat diseases increases when plants and pathogens are acclimatized to elevated carbon dioxide. Glob Chang Biol 21(7):2661–2669. https://doi.org/10.1111/GCB.12899

Vidal T, Boixel AL, Maghrebi E, Perronne R, du Cheyron P, Enjalbert J, Leconte M, de Vallavieille-Pope C (2022) Success and failure of invasive races of plant pathogens: the case of *Puccinia striiformis* f. sp. tritici in France. Plant Pathol 71(7):1525–1536. https://doi.org/10.1111/PPA.13581

Wahab A, Muhammad M, Munir A, Abdi G, Zaman W, Ayaz A, Khizar C, Reddy SPP (2023) Role of arbuscular mycorrhizal fungi in regulating growth, enhancing productivity, and potentially influencing ecosystems under abiotic and biotic stresses. Plants (Basel, Switzerland) 12(17). https://doi.org/10.3390/PLANTS12173102

Walter S, Ali S, Kemen E, Nazari K, Bahri BA, Enjalbert J, Hansen JG, Brown JKM, Sicheritz-Pontén T, Jones J, de Vallavieille-Pope C, Hovmøller MS, Justesen AF (2016) Molecular markers for tracking the origin and worldwide distribution of invasive strains of *Puccinia striiformis*. Ecol Evol 6(9):2790–2804. https://doi.org/10.1002/ECE3.2069

Wang BX, Hof AR, Ma CS (2022) Impacts of climate change on crop production, pests and pathogens of wheat and rice. Front Agric Sci Eng 9(1):4–18. https://doi.org/10.15302/J-FASE-2021432

Win KT, Fukuyo T, Keiki O, Ohwaki Y (2018) The ACC deaminase expressing endophyte Pseudomonas spp. Enhances NaCl stress tolerance by reducing stress-related ethylene production, resulting in improved growth, photosynthetic performance, and ionic balance in tomato plants. Plant Physiol Biochem 127:599–607. https://doi.org/10.1016/J.PLAPHY.2018.04.038

Wu HH, Zou YN, Rahman MM, Ni QD, Wu QS (2017) Mycorrhizas alter sucrose and proline metabolism in trifoliate orange exposed to drought stress. Sci Rep 7(1):1–10. https://doi.org/10.1038/srep42389

Yänez-López R, Torres-Pacheco I, Guevara-González R, Hernández-Zul MI, Quijano-Carranza JA, Rico-García E (2012) The effect of climate change on plant diseases. Afr J Biotechnol 11(10). https://doi.org/10.5897/AJB10.2442

Yang N, Nesme J, Røder HL, Li X, Zuo Z, Petersen M, Burmølle M, Sørensen SJ (2021) Emergent bacterial community properties induce enhanced drought tolerance in Arabidopsis. NPJ Biofilms Microbiomes 7(1). https://doi.org/10.1038/S41522-021-00253-0

Yu XL, Li JY, Zhou YT, Peng J, Qiu BL (2022) Simulated extreme high temperatures alter the demographic parameters of *Aphelinus asychis* and diminish parasitoid fitness. Biol Control 174:105028. https://doi.org/10.1016/J.BIOCONTROL.2022.105028

Zamioudis C, Pieterse CMJ (2012) Modulation of host immunity by beneficial microbes. Mol Plant-Microbe Interact 25(2):139–150. https://doi.org/10.1094/MPMI-06-11-0179

Zandalinas SI, Mittler R, Balfagón D, Arbona V, Gómez-Cadenas A (2018) Plant adaptations to the combination of drought and high temperatures. Physiol Plant 162(1):2–12. https://doi.org/10.1111/PPL.12540

Zhang X, Zhang H, Lou X, Tang M (2019) Mycorrhizal and non-mycorrhizal *Medicago truncatula* roots exhibit differentially regulated NADPH oxidase and antioxidant response under Pb stress. Environ Exp Bot 164:10–19. https://doi.org/10.1016/J.ENVEXPBOT.2019.04.015

Zhu X, Song F, Liu F (2016) Altered amino acid profile of arbuscular mycorrhizal maize plants under low temperature stress. J Plant Nutr Soil Sci 179(2):186–189. https://doi.org/10.1002/JPLN.201400165

The Biochemistry of Plant-Microbe Interactions in the Rhizosphere

7

Jyotsana Tilgam, Ritu Vishwakrma, Shaloo Verma, Khushboo Upadhayay, and Alok Ranjan

Abstract

The rhizosphere, the soil surrounding plant roots, forms a chemically intricate environment fostering the growth of diverse microbial populations. This microbial community's composition undergoes constant changes influenced by factors such as environmental conditions, soil properties, plant activities, and chemical communication between plants and soil-dwelling microbes linked to the root system. The rhizosphere's microbial population experiences selective pressure from plants, and research conducted at the community level has identified distinct microbial communities associated with different plant species, suggesting a role for plant-produced metabolites in shaping the rhizosphere microbiome. This chapter provides an overview of the intricate interactions between plants and microbes in the rhizosphere, with a focus on factors like the diversity and characteristics of root exudates and the functional groups of microorganisms that contribute to plant growth through various mechanisms. Recent advancements in molecular and microbiological techniques have unveiled the complex interplay among rhizosphere inhabitants at multiple levels.

Keywords

Microbial communities · Plant-microbe interaction · Rhizosphere · Root exudates · Quorum sensing

J. Tilgam (✉) · R. Vishwakrma · S. Verma · K. Upadhayay
ICAR-National Bureau of Agriculturally Important Microorganisms, Maunath Bhanjan, Uttar Pradesh, India

A. Ranjan
Department of Biotechnology, Patna Women's College, Patna, Bihar, India

U. Pankaj et al. (eds.), *Plant-microbiome Interactions for Climate-resilient Agriculture*, https://doi.org/10.1007/978-981-96-3534-4_7

1 Introduction

The rhizosphere represents a dynamic element of the soil ecosystem pivotal in shaping root development, morphology, nutrient uptake, root exudation, and rhizodeposition (Dazzo et al. 2015). Rhizodeposits encompass various materials including root cap and border cells, deceased and lysed root cells, released gases, solutes (root exudates), and mucigel (Jones et al. 2009). These compounds significantly stimulate microbial activity (Kuzyakov et al. 2003; Hinsinger et al. 2009) and enhance enzyme production (Asmar et al. 1994), thereby influencing nutrient availability and soil organic matter (SOM) decomposition (Cheng et al. 1996). The rhizosphere exhibits increased soil enzymatic activity compared to root-free bulk soil, owing to enzymatic production from both roots and microbial activity resulting from root cell breakdown (Jones et al. 2009). Approximately 20% of photosynthesized carbon is released as organic compounds of varying molecular weights (Badri and Vivanco 2009). The quantity and composition of these compounds vary with plant age, influencing enzyme regulation during plant metabolism, and subsequently affecting photosynthetic activity and root growth (Aulakh et al. 2001; Lipiec et al. 2016).

Plant-microbe interactions lead to the formation of microbial habitats within specific plant compartments, starting from the rhizosphere soil surrounding plant roots (Zhang et al. 2017), extending to the endosphere internal tissues of above and below-ground plant organs (Compant et al. 2019), and the phyllosphere above-ground plant parts. Plant signaling molecules and immune responses influence microbial colonization in these compartments, shaping microbial community composition and activity. The rhizosphere harbors dominant prokaryotic phyla including Proteobacteria, Actinobacteria, Bacteroidetes, Firmicutes, and Acidobacteria, while fungal communities are mainly composed of Ascomycota and Basidiomycota, similar to soil environments (Ashajyothi et al. 2022). This niche selection is highly specific to plant species and the compounds secreted, as seen with maize roots altering their microbiome through secondary metabolites like benzoxazinoids (Neal et al. 2012). The composition of root exudates significantly influences the bacterial communities in the rhizosphere, with different rhizobacterial species showing distinct patterns across various plant species. The structure and diversity of the rhizospheric microbiome are influenced by plant species, genotypes, root exudates, and spatiotemporal changes in the rhizosphere's physicochemical conditions. The study of plant-microbial interactions spans multiple disciplines, integrating ecology, microbial biology, plant biology, and molecular biology, supported by informatics, statistics, modeling, and biotechnology (Berg et al. 2020). Advances in genomic techniques, such as marker genes and metagenomics, facilitate the identification of microbial taxa and their functional roles (Song et al. 2020), contributing to a deeper understanding of the plant microbiome (Fierer 2017). By elucidating and harnessing the functional attributes of plant-associated microbial communities, manipulating the plant rhizosphere microbiome holds promise for enhancing plant health and productivity (Xun et al. 2021).

2 Underground Alliances: Root Exudates and Microbial Dynamics

The selection of a plant's root microbiome largely depends on its root exudates, which indicate variations in exudate composition according to the plant's genotype, developmental stage, and environment (Koprivova and Kopriva 2022). The rhizosphere ecology is directly influenced by root exudates, which consist of various high and low molecular weight compounds, including primary and secondary metabolites. These exudates include aromatic acids, sterols, aliphatic acids, fatty acids, sugars, polysaccharides, amino acids, phenolics, proteins, ions, and enzymes (Badri et al. 2009). They are released by plants as a result of root pressure and the dynamic interactions between plants and microbes. Root exudates have a varied chemical makeup and are rich in various nutritive substance which shapes the rhizospheric microbial community (Patel et al. 2020). Furthermore, root exudates facilitate quorum-sensing, mycorrhizal association, endophytic relationships, plant growth-promoting rhizobacterial consortia, nodulation, antibacterial activity, biofilm suppression/formation, and tri-trophic interactions including nematodes (Mondal et al. 2023). In addition to bioremediating contaminated soil, root exudates also manage pests, regulate the carbon cycle, water and nutrient availability, alter microbial populations near the rhizosphere, and influence the physicochemical characteristics of the soil (Balyan and Pandey 2024). Endophytic and rhizosphere bacteria share a close relationship, with many facultative endophytes also capable of colonizing the rhizosphere. Plant's sessile nature has subjected them to a variety of biotic and abiotic stressors, including high temperatures, droughts, flooding, salt, and infections from pathogens or pests that can negatively impact their development and yield. The generation of root exudates is one of the many intricate adaptation processes that plants have developed to deal with these stressful environments. In recent years, many findings have also shown the protective potential of root exudates in enhancing plant tolerance to abiotic stresses and resistance against biotic stresses.

2.1 Role of Secondary Metabolites

In plant-microbe interaction, primary means of communication are metabolites exuded from roots, primarily diverse secondary metabolites. Certain secondary metabolites selectively promote root colonization by bacteria with specialized functions. The exuded metabolites trigger changes in the composition and function of plant-associated microbiome. In recent years, many metabolites have been identified as part of this communication network, modulating specific functions of the root microbiota (Koprivova and Kopriva 2022). Maize root exudation of DIMBOA increases the population of beneficial *Pseudomonas putida* (Neal et al. 2012), while *A. thaliana* actively exudes phytochemicals via ABC transporters (Badri et al. 2012). Phenolic compounds in root exudates also send specific signals to soil bacteria (Badri et al. 2013). Ji et al. (2024) demonstrated that root exudates of *Haloxylon ammodendron* show different effects at varying plant densities, with compounds like O-methylandrocymbine and gibberellin A8 influencing soil nutrient availability and competition. Intraspecific competition affects the bacteria in the rhizosphere, especially impacting Proteobacteria, Acidobacteriota, Cyanobacteria, and Actinobacteriota. Vicinamibacteria and Sphingomonadales are beneficial in this context. Habib et al. (2020) discovered that wheat and *Ageratina adenophora* can attract beneficial bacterial communities through specific compounds released in their root exudates, creating a self-supporting system. While it is well established that root exudates play a key role in this process, these compounds differ in how easily they decompose, their energy and nutrient content, and other chemical properties (Keiluweit et al. 2015). This variation makes it challenging to fully understand how particular secondary metabolites shape microbial community structures and function (Hu et al. 2018).

2.2 Factors Influencing Root Exudates

Root exudation patterns can be affected quantitatively and qualitatively by both biotic and abiotic factors, varying with plant genotype, age, and environmental conditions (Philippot et al. 2013). The root system architecture (RSA) plays a key role in determining exudation patterns, with different root zones secreting specific compounds. The root cap, border cells, and root hairs are all involved in this process, with border cells influencing microbial interactions and heavy metal resistance.

The pattern of root exudation is closely linked to plant age and developmental stages, significantly influenced by root system architecture (RSA). Root exudation increases from seedlings through flowering and decreases at maturity (Aulakh et al. 2001). The chemistry of root exudates also varies with developmental stages, such as the secretion of phenolic compounds and defense-related proteins by *Arabidopsis* roots during flowering, which is associated with age-related resistance (De-la-Pena et al. 2010). Environmental factors such as soil texture, pH, moisture, temperature, and light also impact root exudation. For instance, hypoxia due to high soil moisture can lead to anaerobic respiration and the accumulation of ethanol, lactic acid, and alanine (Rivoal and Hanson 1994). While the deficiencies or excess of minerals or toxic metals trigger the secretion of specific acids and signaling molecules (Wang et al. 2006). Additionally, the presence of neighboring plants and rhizosphere microflora can modify the composition of root exudates (Haldar and Sengupta 2015).

3 Different Interactions and Signaling in the Rhizosphere

3.1 Plant-Microbe Interactions

Plant-microbe interactions in the rhizosphere can be categorized as associative, symbiotic, neutralistic, or parasitic (Haldar and Sengupta 2015). Positive interactions include symbiotic and associative relationships with beneficial microbes such as mycorrhizal fungi, nitrogen-fixing bacteria, and plant growth-promoting rhizobacteria (PGPR). Negative interactions involve parasitic plants, pathogenic bacteria, fungi, oomycetes, nematodes, and herbivores.

During plant-to-microbe communication, numerous root exudates found across most plants function as signals to soil microbes. Plants perceive rhizosphere microbes through specialized pattern recognition receptors PRRs (Venturi and Keel 2016). Signaling mediated by plant-secreted molecules to microorganisms plays a crucial role in fostering beneficial interactions, influencing processes such as nutrient uptake and defense responses (Mathesius et al. 2003; Mhlongo et al. 2018). Research into signaling between plants and rhizosphere microorganisms has predominantly concentrated on close symbiotic relationships, particularly those involving mycorrhizal fungi and rhizobial bacteria (Hassan and Mathesius 2012).

The symbiosis between legumes and Rhizobia is essential for agriculture, with flavonoid signals from plant roots initiating the production of Nod factors or lipochitooligosaccharides (LCOs) by Rhizobia, leading to nodule formation and nitrogen fixation (Spaink 2000). Cai et al. (2009) demonstrated that legumes release antimetabolites such as canavanine that play a role in selectively promoting rhizobia within the complex rhizosphere environment. Arbuscular mycorrhizal fungi (AMF) are associated with over 80% of plant species, where flavonoids and strigolactones act as key signaling molecules for AMF symbiosis (Akiyama et al. 2005). In addition to their roles in rhizosphere mutualisms, flavonoids also play a critical role in plant interactions with soil pathogens, including *Fusarium* species (Steinkellner et al. 2007).

3.2 Microbe-Plant Interactions

Microbe-to-plant interactions involve microorganisms emitting signaling molecules that plants detect, influencing their development, gene expression, and immune and stress responses (Badri and Vivanco 2009; Venturi and Keel 2016). Rhizosphere microorganisms initiate plant responses through microbial elicitors known as microbe-associated molecular patterns (MAMPs), such as lipopolysaccharides, peptidoglycans, flagellin, and chitin (Millet et al. 2010). These MAMPs can induce systemic defense responses, including induced systemic resistance (ISR) and systemic acquired resistance (SAR), primarily activated by beneficial bacteria and pathogens, respectively (Conrath et al. 2015). Additionally, molecules like Nod and Myc factors released by rhizobia and mycorrhizal fungi are critical for initiating symbiotic relationships with plants, as discussed in an earlier paragraph (Oldroyd 2013).

Rhizosphere microbial populations actively communicate with each other as well as with plant roots. Rhizosphere-derived microbial signals can induce specific alterations in the plant transcriptome. These beneficial microbes are categorized into two main groups: (i) biological control agents (BCAs), which indirectly boost plant productivity by managing plant pathogens, and (ii) plant growth-promoting microorganisms (PGPMs), which directly enhance plant growth (Haldar and

Sengupta 2015). Beneficial control agents (BCAs) impact soil-borne pathogens through competition, antagonism, and hyperparasitism. They produce antimicrobial metabolites and lytic enzymes, compete for nutrients and space, produce siderophores, and induce systemic resistance (ISR). For example, *Bacillus amyloliquefaciens* SQR9 protects plants from invasion by *Fusarium oxysporum* by competing for nutrients and space (Albayrak 2019). *Trichoderma* releases biologically active compounds and secondary metabolites that can degrade pathogen cell walls (Kumar and Khurana 2021). Pathogens counteract these defenses with mechanisms like efflux and degradation of antimicrobial compounds (Schouten et al. 2004) and quorum quenching (Rasmussen and Givskov 2006).

Rhizobia interacts synergistically or antagonistically in legume rhizospheres with other diazotrophs and plant growth promoting microbes, enhancing nodulation and nitrogen fixation, plant growth promoting microbes, including *Pseudomonas* and *Bacillus* species, promote plant health and growth through various mechanisms, such as producing antibiotics, competing for nutrients, and inducing plant defenses. *Pseudomonas* species have been shown to enhance plant growth in the absence of pathogens by directly solubilizing insoluble phosphorus sources and/or regulating plant growth regulators, such as indole-acetic acid or 1-amino-1-cyclopropane carboxylate (ACC) deaminase (Naik et al. 2008; Dey et al. 2004). These beneficial microorganisms are used as biofertilizers and phytostimulators to improve crop yields and manage soil-borne diseases.

Endophytes, including bacteria and fungi that live inside plant tissues without harming the host, play a significant role in plant biology. Endophytic bacteria vary widely in their host specificity; some are closely associated with specific plant species or cultivars, while others, like *Burkholderia cepacia* Bu72, can benefit multiple plant species (Badri et al. 2009). Endophytic and rhizosphere bacteria share a close relationship, with many facultative endophytes also capable of colonizing the rhizosphere. Endophytic bacteria enhance plant growth both directly and indirectly. Directly, they promote growth through nitrogen fixation, production of growth regulators (auxins, cytokinins, gibberellins), suppression of stress ethylene, and alterations in sugar sensing. For example, some endophytes can metabolize trehalose, improving drought, salt, and cold tolerance (Taghavi et al. 2009). Indirectly, they benefit plants by competing with pathogens for resources, producing antibiotics and hydrolytic enzymes, inhibiting pathogen toxins, and inducing plant defense mechanisms (Badri et al. 2009).

3.2.1 Role of Quorum Sensing Molecules

Inter-microbial communication within microbial communities, encompassing intraspecies and interspecies interactions, primarily relies on synthesizing, releasing, and detecting autoinducers (AIs), a process known as Quorum Sensing (QS). These autoinducers regulate the transcription of numerous QS-controlled genes involved in biofilm formation, chemotaxis, and virulence factors, enabling microorganisms to monitor cell density and coordinate collective behavioral changes (Venturi and Keel 2016). Gram-negative bacteria, such as *Pseudomonas* spp., *Burkholderia* spp., and *Serratia* spp., commonly employ N-acyl homoserine lactones (AHL) as QS signaling molecules. In contrast, Gram-positive rhizobacteria utilize autoinducer peptides (AIPs) for intra- and interspecies communication (Quadriya et al. 2018). QS also plays a pivotal role in interkingdom communication between bacteria and plant-associated fungi, although specific mechanisms remain largely unexplored (Deveau et al. 2018). AHL-type QS molecules, akin to their role in microbe-microbe signaling, also influence plant gene expression and functions related to development, stress response, and immunity.

Non-rhizospheric bacteria offer better advantage over rhizospheric bacteria in terms of root colonization through metabolization of various plant root exudates (Ledger et al. 2012). Mechanism of quorum sensing (QS) offer them competitive advantage and better survival rates in particular niche. This encourages the quest to engineer these neutral bacteria to contribute towards the plant growth. *Cupriavidus pinatubonensis* JMP134 is one such non-PGPR bacterium which is amenable to genetic manipulations. Zúñiga et al. (2018), constructed an QS dependent IAA biosynthesis device in strain *C. pinatubonensis* JMP134. A genetic circuit was created using Plux promoter and *luxI*, *luxR* genes from bacterium *V. fscheri* to *Cupriavidus pinatubonensis*. A positive feedback QS circuit could produce IAA and responding to N-(3-Oxohexanoyl)-L-homoserine lactone, 3-oxo-C6- HSL. *iaaM gene* (2-tryptophan monooxygenase), and *iaaH* gene (indole-3-acetamide hydrolase) from *P. savastanoi* 48 were assembled downstream the Plux-luxI sequence in the plasmid pSEVA-Plux-luxI. In-vivo and in-vitro test showed the self-induction of the device at higher concentrations of bacterial cells leading to significant enhancement in lateral root expansion, primary root length, fresh weight and rosette area of *Arabidopsis thaliana* plants. This study emphasizes the need for non-rhizospheric bacteria that can be better utilized as a vehicle for agriculture applications.

Antibiotics can serve as quorum sensing (QS) signals for communication between bacteria, both within and between species. A recent study identified antibiotic signals in *Lysobacter capsica*, a bacterial strain from the tobacco rhizosphere, which showed biocontrol potential due to its production of antifungal and antibiotic signals (Brescia et al. 2020). Additionally, fungi produce QS signals such as γ-heptalactone and farnesol that interact with bacterial species (Hartmann and Schikora 2012). These QS signals from both fungi and bacteria contribute to microbial competition in the rhizosphere for host infection (Jamil et al. 2022).

3.2.2 Role of Volatile Organic Compounds (VOCs)

Volatile organic compounds (VOCs) are crucial signaling molecules involved in long-distance communication within microbial communities and between microbes and plants (Jamil et al. 2022). Various bacterial and fungal species produce these low molecular weight, lipophilic compounds through species-specific metabolic pathways (Kanchiswamy et al. 2015). Bacterial VOCs include alkanes, ketones, alkenes, terpenoids, and sulfurs. VOCs function as antimicrobial quorum sensing signals, influencing microbial behaviors such as virulence, stress resistance, and biofilm formation. For example, low concentrations of nitric oxide (NO) affect biofilm formation in several bacterial species (Barraud et al. 2009). VOCs also play a role in regulating plant growth and immunity against stresses. Further research is needed to explore additional signaling mechanisms associated with microbial VOCs.

3.3 Host-Pathogen Interaction

Four primary groups of plant pathogens exist, but only two—fungi and nematodes—are significant in the soil (Agrios 2005). Like plant–mutualist interactions, pathogens use chemical signals in the initial stages of host recognition and infection. Most plants produce antimicrobial secondary metabolites, either as part of their natural growth and development process or as a defense mechanism against pathogen attacks. These antimicrobial compounds protect a broad spectrum of pathogens. Pre-existing antifungal compounds, known as phytoanticipins, are present in healthy plants and serve as chemical barriers against fungal pathogens. In contrast, phytoalexins are antimicrobial compounds that are produced in response to pathogen attacks and are not typically found in healthy plants. Both groups of compounds are highly effective against a wide range of fungal pathogens (Badri et al. 2009).

Most of the crop damage is attributed to plant-parasitic nematodes, particularly root-knot nematodes (RKN) and cyst nematodes. It is widely believed that nematodes sense their environment through chemosensory perception. Typically, RKNs must locate and penetrate a root, migrate into the vascular cylinder, and establish a permanent feeding site. These processes involve extensive signaling between the nematode and the host, with well-documented research on the proteins secreted by nematodes during the migratory phase. However, identifying the initial signaling molecules (non-protein compounds) released by the host to attract nematodes remains in its early stages. Earlier studies have highlighted the significant role of plant hormones in the interaction between nematodes and their host plants especially, auxin and cytokinin (Gheysen 1998). Horiuchi et al. (2005) reported that Medicago roots release a volatile compound, dimethyl sulfide, which attracts nematodes (*C. elegans*) and enables them to act as vectors for rhizobia, facilitating efficient nitrogen fixation and enhancing nodulation, even when the legume and rhizobium are not nearby.

4 Bioengineering Rhizospheric Soils for Disease Suppression and Stress Resilience

The health and productivity of soil are fundamental for successful food crop production. However, biotic and abiotic stresses can lead to significant yield losses or even total crop failure when the delicate balance between a plant and its environment is disrupted (Kopecká et al. 2023; Mwangi et al. 2023). Plants communicate with their surroundings through chemical signals released as root exudates, which help them adapt to environmental changes. During stress, plants increase root exudation to attract beneficial microbes, a process known as a "cry for help" (Bakker et al. 2018). These microbes can aid plants in overcoming stress, restoring ecological balance, and preventing disease. If these microbial interventions and the plant's defenses fail, the plant may succumb to disease or stress.

The beneficial microbes like Plant Growth-Promoting microbes, can significantly enhance plant growth and disease suppression. Beneficial microbes perform alleviation of stresses and antagonism against pathogenic microorganisms through mechanisms like releasing gaseous substances such as hydrogen cyanide (HCN), inducing systemic resistance (ISR) and systemic acquired resistance (SAR), and producing ACC deaminase enzyme to reduce ethylene (ET) concentration in plants. Engineering the rhizosphere to optimize plant-microbe interactions and enhance disease resilience is a promising field (Tilgam et al. 2022). Recent research supports that microbial consortia often provide greater benefits than single-species inoculations. For instance, Kumawat et al. (2022) found that combining *Bradyrhizobium* sp. LSBR-3 with phosphate-solubilizing *Pseudomonas oryzihabitans* LSE-3 significantly improved soybean yield by 11% compared to single inoculations. Similarly, a consortium of *Trichoderma asperellum* and *Bacillus amyloliquefaciens* demonstrated enhanced wheat growth and biocontrol potential (Karuppiah et al. 2019). Effective microbial communities, especially those with diverse genera, are stable and efficient in controlling pathogens like *Rhizoctonia solanacearum* in tomatoes (Hu et al. 2016). Microbes also help plants endure abiotic stress by inducing stress mitigation mechanisms and supporting essential nutrient and hormone production. For example, Ouhaddou et al. (2023) showed that biostimulants could help lettuce tolerate water stress by boosting secondary metabolites and antioxidant activity. Sharma et al. (2019) found that bio-inoculants such as *Rhizobium*, PSB, AM fungi, and *Azotobacter* significantly increased chickpea yield and soil health, demonstrating that reduced phosphorus use can maintain productivity while improving soil enzyme activities.

Tailoring microbial consortia to specific crops and environmental conditions allows for precision agriculture and boosts overall productivity (Kumar et al. 2023). Synthetic microbial communities can be designed using computational techniques, such as machine learning, to respond to specific plant needs, like nutrient deficiencies or pathogen attacks (Compant et al. 2019; Ali and Glick 2024). These engineered communities work synergistically with plant defense systems to improve plant health and resilience. Success in creating these communities depends on selecting the right plant, microbe, and formulation at the appropriate time. Combining synthetic microbial communities with proven chemicals can further improve microbial survival and competitiveness in the rhizosphere. However, balancing the complexity of synthetic communities with the feasibility of large-scale production and application remains a challenge.

5 Conclusion

Rhizosphere chemical dialogues are crucial for communication between plant roots and microbes in the soil-root interface. Root-secreted exudates and microbial compounds serve as signaling agents that drive these interactions. While some factors initiating these dialogues have been identified, many remain unknown. The rhizosphere is a key area of study for ecologists, molecular biologists, and plant scientists. Advances in "omics" technologies, like proteomics, secretomics, and metabolomics, are helping to unravel these complex interactions, enhancing our understanding and potential for agricultural applications. This chapter explores the complex interactions between plants and microbes in the rhizosphere, highlighting the importance of root exudates in shaping beneficial microbial communities. By advancing our knowledge of these interactions we can develop more reliable and sustainable agricultural practices. Future research should focus on these specific interactions to improve food production systems and support global food security.

References

Agrios GN (2005) Plant pathology, 5th edn. Elsevier, New York

Akiyama K, Matsuzaki K, Hayashi H (2005) Plant sesquiterpenes induce hyphal branching in arbuscular mycorrhizal fungi. Nature 435:824–827

Albayrak ÇB (2019) Bacillus species as biocontrol agents for fungal plant pathogens. In: Bacilli and agrobiotechnology: phytostimulation and biocontrol: volume 2. Springer, Cham, pp 239–265

Ali S, Glick BR (2024) Root exudate metabolites alter food crops microbiomes, impacting plant biocontrol and growth. Crops 4(1):43–54

Ashajyothi M, Charishma K, Patel A, Paul S, Venkatesh YN, Prakash I, Tilgam J (2022) Rhizosphere microbiome: significance in sustainable crop protection. In: Rhizosphere microbes: biotic stress management. Springer Nature Singapore, Singapore, pp 283–309

Asmar F, Eiland F, Nielsen NE (1994) Effect of extracellular-enzyme activities on solubilization rate of soil organic nitrogen. Biol Fertil Soils 17(1):32–38

Aulakh MS, Wassmann R, Bueno C, Kreuzwieser J, Rennenberg H (2001) Characterization of root exudates at different growth stages of ten rice (*Oryza sativa* L.) cultivars. Plant Biol 3(2):139–148

Badri DV, Vivanco JM (2009) Regulation and function of root exudates. Plant Cell Environ 32(6):666–681

Badri DV, Weir TL, van der Lelie D, Vivanco JM (2009) Rhizosphere chemical dialogues: plant–microbe interactions. Curr Opin Biotechnol 20(6):642–650

Badri DV, Chaparro JM, Manter DK, Martinoia E, Vivanco JM (2012) Influence of ATP-binding cassette transporters in root exudation of phytoalexins, signals, and in disease resistance. Front Plant Sci 3:149

Badri DV, Chaparro JM, Zhang R, Shen Q, Vivanco JM (2013) Application of natural blends of phytochemicals derived from the root exudates of Arabidopsis to the soil reveal that phenolic-related compounds predominantly modulate the soil microbiome. J Biol Chem 288(7):4502–4512

Bakker PAHM, Pieterse CMJ, de Jonge R, Berendsen RL (2018) The soil-borne legacy. Cell 172(6):1178–1180

Balyan G, Pandey AK (2024) Root exudates, the warrior of plant life: revolution below the ground. S Afr J Bot 164:280–287

Barraud N, Schleheck D, Klebensberger J, Webb JS, Hassett DJ, Rice SA, Kjelleberg S (2009) Nitric oxide signaling in *Pseudomonas aeruginosa* biofilms mediates phosphodiesterase activity, decreased cyclic di-GMP levels, and enhanced dispersal. J Bacteriol 191(23):7333–7342
Berg G, Rybakova D, Fischer D, Cernava T, Vergès MCC, Charles T et al (2020) Microbiome definition re-visited: old concepts and new challenges. Microbiome 8(1):1–22
Brescia F, Marchetti-Deschmann M, Musetti R, Perazzolli M, Pertot I, Puopolo G (2020) The rhizosphere signature on the cell motility, biofilm formation and secondary metabolite production of a plant-associated Lysobacter strain. Microbiol Res 234:126424
Cai T, Cai W, Zhang J, Zheng H, Tsou AM, Xiao L, Zhong Z, Zhu J (2009) Host legume-exuded antimetabolites optimize the symbiotic rhizosphere. Mol microbiol 73(3):507–517
Cheng W, Zhang Q, Coleman DC, Carroll CR, Hoffman CA (1996) Is available carbon limiting microbial respiration in the rhizosphere? Soil Biol Biochem 28(10):1283–1288
Compant S, Samad A, Faist H, Sessitsch A (2019) A review on the plant microbiome: ecology, functions, and emerging trends in microbial application. J Adv Res 19:29–37
Conrath U, Beckers GJ, Langenbach CJ, Jaskiewicz MR (2015) Priming for enhanced defense. Annu Rev Phytopathol 53(1):97–119
Dazzo FB, Yanni YG, Jones A, Elsadany AY (2015) CMEIAS bioimage informatics that define the landscape ecology of immature microbial biofilms developed on plant rhizoplane surfaces. AIMS Bioeng 2(4):469–486
De-la-Peña C, Badri DV, Lei Z et al (2010) Root secretion of defense-related proteins is development-dependent and correlated with flowering time. J Biol Chem 285:30654–30665
Deveau A, Bonito G, Uehling J, Paoletti M, Becker M, Bindschedler S et al (2018) Bacterial–fungal interactions: ecology, mechanisms and challenges. FEMS Microbiol Rev 42(3):335–352
Dey R, Pal KK, Bhatt DM, Chauhan SM (2004) Growth promotion and yield enhancement of peanut (*Arachis hypogaea* L.) by application of plant growth-promoting rhizobacteria. Microbiol Res 159(4):371–394
Fierer N (2017) Embracing the unknown: disentangling the complexities of the soil microbiome. Nat Rev Microbiol 15(10):579–590
Gheysen G (1998) Chemical signals in the plant–nematode interaction: a complex system? In: Romeo JT, Downum KR, Verpoorte R (eds) Phytochemical signals and plant–microbe interactions, vol 32. Plenum Press, New York, pp 95–117
Habib U, Humaira Y, Saqib M et al (2020) Multitrait *pseudomonas* spp. isolated from monocropped wheat (*Triticum aestivum*) suppress Fusarium root and crown rot. Phytopathology 110(5):582–592
Haldar S, Sengupta S (2015) Plant-microbe cross-talk in the rhizosphere: insight and biotechnological potential. Open Microbiol J 9:1–7
Hartmann A, Schikora A (2012) Quorum sensing of bacteria and trans-kingdom interactions of N-acyl homoserine lactones with eukaryotes. J Chem Ecol 38(6):704–713
Hassan S, Mathesius U (2012) The role of flavonoids in root–rhizosphere signalling: opportunities and challenges for improving plant–microbe interactions. J Exp Bot 63(9):3429–3444
Hinsinger P, Bengough AG, Vetterlein D, Young IM (2009) Rhizosphere: biophysics, biogeochemistry and ecological relevance. Plant Soil 321(1–2):117–152
Horiuchi J, Prithiviraj B, Bais HP, Kimball BA, Vivanco JM (2005) Soil nematodes mediate positive interactions between legume plants and rhizobium bacteria. Planta 222(5):848–857
Hu J, Wei Z, Friman V-P, Gu S-H, Wang X-F, Eisenhauer N, Yang T-J, Ma J, Shen Q-R, Xu Y-C et al (2016) Probiotic diversity enhances rhizosphere microbiome function and plant disease suppression. MBio 7:e01790-16
Hu L, Robert CA, Cadot S, Zhang XI, Ye M, Li B et al (2018) Root exudate metabolites drive plant-soil feedbacks on growth and defense by shaping the rhizosphere microbiota. Nat Commun 9(1):2738
Jamil F, Mukhtar H, Fouillaud M, Dufossé L (2022) Rhizosphere signaling: insights into plant–rhizomicrobiome interactions for sustainable agronomy. Microorganisms 10(5):899

Ji S, Wu D, Li W et al (2024) Mediating role of root-exuded secondary metabolites in intra-specific interactions with *Haloxylon ammodendron*. Plant Soil. https://doi.org/10.1007/s11104-024-06783-8

Jones DL, Nguyen C, Finlay RD (2009) Carbon flow in the rhizosphere: carbon trading at the soil–root interface. Plant Soil 321(1–2):5–33

Kanchiswamy CN, Malnoy M, Maffei M (2015) Bioprospecting bacterial and fungal volatiles for sustainable agriculture. Trends Plant Sci 20(4):206–211

Karuppiah V, Sun J, Li T, Vallikkannu M, Chen J (2019) Co-cultivation of *Trichoderma asperellum* GDFS1009 and *Bacillus amyloliquefaciens* 1841 causes differential gene expression and improvement in the wheat growth and biocontrol activity. Front Microbiol 10:1068

Keiluweit M, Bougoure JJ, Nico PS et al (2015) Mineral protection of soil carbon counteracted by root exudates. Nat Clim Chang 5:588–595

Kopecká R, Kameniarová M, Černý M, Brzobohatý B, Novák J (2023) Abiotic stress in crop production. Int J Mol Sci 24(7):6603

Koprivova A, Kopriva S (2022) Plant secondary metabolites altering root microbiome composition and function. Curr Opin Plant Biol 67:102227

Kumar N, Khurana SP (2021) Trichoderma-plant-pathogen interactions for benefit of agriculture and environment. In: Biocontrol agents and secondary metabolites. Woodhead Publishing, pp 41–63

Kumar A, Das A, Singh D, Das MK, Srivastava GP, Singh JP et al (2023) Soil health restoration in degraded lands: a microbiological perspective. Land Degrad Dev 34(17):5155–5170

Kumawat KC, Singh I, Nagpal S, Sharma P, Gupta RK, Sirari A (2022) Co-inoculation of indigenous *Pseudomonas oryzihabitans* and *Bradyrhizobium* sp. modulates the growth, symbiotic efficacy, nutrient acquisition, and grain yield of soybean. Pedosphere 32(3):438–451

Kuzyakov Y, Leinweber P, Sapronov D, Eckhardt KU (2003) Qualitative assessment of rhizodeposits in non-sterile soil by analytical pyrolysis. J Plant Nutr Soil Sci 166(6):719–723

Ledger ME, Harris RM, Armitage PD, Milner AM (2012) Climate change impacts on community resilience: evidence from a drought disturbance experiment. Adv Ecol Res 46:211–258

Lipiec J, Siczek A, Sochan A, Bieganowski A (2016) Effect of sand grain shape on root and shoot growth of wheat seedlings. Geoderma 265:1–5

Mathesius U, Mulders S, Gao M, Teplitski M, Caetano-Anollés G, Rolfe BG, Bauer WD (2003) Extensive and specific responses of a eukaryote to bacterial quorum-sensing signals. Proc Natl Acad Sci 100(3):1444–1449

Mhlongo MI, Piater LA, Madala NE, Labuschagne N, Dubery IA (2018) The chemistry of plant–microbe interactions in the rhizosphere and the potential for metabolomics to reveal signaling related to defense priming and induced systemic resistance. Front Plant Sci 9:112

Millet YA, Danna CH, Clay NK, Songnuan W, Simon MD, Werck-Reichhart D, Ausubel FM (2010) Innate immune responses activated in Arabidopsis roots by microbe-associated molecular patterns. Plant Cell 22(3):973–990

Mondal S, Pramanik K, Pal P, Mitra S, Ghosh SK, Mondal T, Soren T, Maiti TK (2023) Multifaceted roles of root exudates in light of plant-microbe interaction. In: Unravelling plant-microbe synergy. Academic, pp 49–76

Mwangi RW, Mustafa M, Charles K, Wagara IW, Kappel N (2023) Selected emerging and reemerging plant pathogens affecting the food basket: a threat to food security. J Agric Food Res 14:100827

Naik PR, Sahoo N, Goswami D, Ayyadurai N, Sakthivel N (2008) Genetic and functional diversity among fluorescent pseudomonads isolated from the rhizosphere of banana. Microb Ecol 56(3):492–504

Neal AL, Ahmad S, Gordon-Weeks R, Ton J (2012) Benzoxazinoids in root exudates of maize attract *Pseudomonas putida* to the rhizosphere. PLoS One 7(4):e35498

Oldroyd GE (2013) Speak, friend, and enter: signalling systems that promote beneficial symbiotic associations in plants. Nat Rev Microbiol 11(4):252–263

Ouhaddou R, Ech-Chatir L, Anli M, Ben-Laouane R, Boutasknit A, Meddich A (2023) Secondary metabolites, osmolytes and antioxidant activity as the main attributes enhanced by biostimulants for growth and resilience of lettuce to drought stress. Gesunde Pflanzen 75(5):1737–1175

Patel JS, Yadav SK, Bajpai R, Teli B, Rashid M (2020) PGPR secondary metabolites: an active syrup for improvement of plant health. In: Molecular aspects of plant beneficial microbes in agriculture. Academic, pp 195–208

Philippot L, Raaijmakers JM, Lemanceau P, van der Putten WH (2013) Going back to the roots: the microbial ecology of the rhizosphere. Nat Rev Microbiol 11(11):789–799

Quadriya H, Adeeb Mujtaba Ali S, Parameshwar J, Manasa M, Yahya Khan M, Hameeda B (2018) Microbes living together: exploiting the art for making biosurfactants and biofilms. In: Implication of quorum sensing system in biofilm formation and virulence. Springer, Singapore, pp 161–177

Rasmussen TB, Givskov M (2006) Quorum sensing inhibitors: a bargain of effects. Microbiology 152(4):895–904

Rivoal J, Hanson AD (1994) Metabolic control of anaerobic glycolysis overexpression of lactate dehydrogenase in transgenic tomato roots supports the Davies-Roberts hypothesis and points to a critical role for lactate secretion. Plant Physiol 106(4):1179–1185

Schouten A, van den Berg G, Edel-Hermann V et al (2004) Defense responses of *Fusarium oxysporum* to 2,4-diacetylphloroglucinol, a broad-spectrum antibiotic produced by *Pseudomonas fluorescens*. Mol Plant-Microbe Interact 17(11):1201–1211

Sharma V, Sharma S, Sharma S, Kumar V (2019) Synergistic effect of bio-inoculants on yield, nodulation and nutrient uptake of chickpea (*Cicer arietinum* L.) under rainfed conditions. J Plant Nutr 42(4):374–383

Song C, Zhu F, Carrión VJ, Cordovez V (2020) Beyond plant microbiome composition: exploiting microbial functions and plant traits via integrated approaches. Front Bioeng Biotechnol 8:896

Spaink HP (2000) Root nodulation and infection factors produced by rhizobial bacteria. Ann Rev Microbiol 54(1):257–288

Steinkellner S, Lendzemo V, Langer I, Schweiger P, Khaosaad T, Toussaint JP, Vierheilig H (2007) Flavonoids and strigolactones in root exudates as signals in symbiotic and pathogenic plant-fungus interactions. Molecules 12(7):1290–1306

Taghavi S, Garafola C, Monchy S, Newman L, Hoffman A, Weyens N, Barac T, Vangronsveld J, van der Lelie D (2009) Genome survey and characterization of endophytic bacteria exhibiting a beneficial effect on growth and development of poplar. Appl Environ Microbiol 75(3):748–757

Tilgam J, Sreeshma N, Priyadarshini P, Bhavyasree RK, Choudhury S, Bharati A, Ashajyothi M (2022) Rhizosphere engineering for systemic resistance/tolerance to biotic and abiotic stress. In: Re-visiting the rhizosphere eco-system for agricultural sustainability. Springer Nature Singapore, Singapore, pp 271–300

Venturi V, Keel C (2016) Signaling in the rhizosphere. Trends Plant Sci 21(3):187–198

Wang C, Meek DJ, Panchal P et al (2006) Isolation of poly-3-hydroxybutyrate metabolism genes from complex microbial communities by phenotypic complementation of bacterial mutants. Appl Environ Microbiol 72(1):384–391

Xun W, Shao J, Shen Q, Zhang R (2021) Rhizosphere microbiome: functional compensatory assembly for plant fitness. Comput Struct Biotechnol J 19:5487–5493

Zhang R, Vivanco JM, Shen Q (2017) The unseen rhizosphere root–soil–microbe interactions for crop production. Curr Opin Microbiol 37:8–14

Zúñiga A, Fuente FDL, Federici F, Lionne C, Bônnet J, de Lorenzo V, González B (2018) An engineered device for indoleacetic acid production under quorum sensing signals enables *Cupriavidus pinatubonensis* JMP134 to stimulate plant growth. ACS Synth Biol 7(6):1519–1527

Plant Root Exudation: Mechanism and Role in Plant-Microbe Communication Under Stress

8

Bhaskar Jyoti Parasar, Indrani Sharma, Sampurna Kashyap, Ramen Barman, Sneha Devi Marme, Priyanka Das, and Niraj Agarwala

Abstract

Climate change-induced biotic and abiotic stressors negatively affect the growth and development of plants, posing a serious threat to agricultural productivity. Root exudates comprises of organic compounds that can directly or indirectly assist in stress tolerance by recruiting or suppressing rhizomicrobiota. Diverse range of metabolites such as amino acids, aliphatic acids, proteins, sugars, flavonoids, or fatty acids secreted by roots in the rhizosphere play a significant role as nutrients or signals and can alter microbial community under stress conditions. Chemical composition, exudation pattern, and interaction of root exudate with microbe determines positive and negative crosstalk in this synergy and aids the plant with enhanced resource use efficiency. This chapter highlights the substantial gap in our understanding of this crosstalk, its significance, and potential application, particularly in addressing stress regimes that can open new avenues to sustainable and climate-smart agriculture.

Keywords

Root exudate · Microbiome · Root system architecture · Climate-smart agriculture · Stress factors

B. J. Parasar · I. Sharma · S. Kashyap · R. Barman · S. D. Marme · P. Das · N. Agarwala (✉)
Department of Botany, Gauhati University, Guwahati, Assam, India
e-mail: niraj_botany@gauhati.ac.in

U. Pankaj et al. (eds.), *Plant-microbiome Interactions for Climate-resilient Agriculture*, https://doi.org/10.1007/978-981-96-3534-4_8

Abbreviations

AMF	Arbuscular mycorrhizal fungi
CSA	Climate-smart agriculture
CWR	Crop wild relative
EPS	Exopolysaccharides
RSA	Root system architecture
SL	Strigolactone
VOC	Volatile organic compound

1 Introduction

In the changing climate scenario, an increase in frequency and intensity of extreme weather events such as alterations in precipitation patterns or rise in temperature, impact not only agricultural productivity but also the overall biodiversity. These changes in the environment may have an impeding effect on plants' growth and viability due to alterations in their physiology, morphology, and metabolic activities (Zhao 2023). However, plants have developed various sophisticated mechanisms for adapting to changing environmental conditions including the modulation of the rhizosphere microbiome. Plants employ key adaptive strategies such as the release of root exudates facilitating plant-microbe interactions to mitigate any stress (Williams et al. 2024). Root exudates comprise a variety of metabolites, secreted either actively or passively into the rhizosphere, creating a unique habitat of distinct microbial communities (Spooren et al. 2024). The rhizomicrobiome perceives the plant-derived stressed signals in the form of root exudates generating an exudate-microbe communication pathway to assist in the plant's defence mechanisms. Stress-induced alteration of soil microorganisms can lead to a decline in various ecosystem functions, including plant community diversity, nutrient recycling, and retention (Vimal et al. 2024). Changes in the rhizospheric microbial diversity enable plants to thrive during and after the stress conditions. Besides this, the root system architecture (RSA) plays a significant role in determining changes in the rhizospheric microbial diversity. RSA is significant in the acquisition of diverse microbial genera around the roots by creating differences in exudate compounds across various root zones (Galindo-Castañeda et al. 2024). Root exudation-mediated co-evolution of microbes and plants engaged in the rhizospheric interactions has been well documented in many previous literatures (Fields and Friman 2022; Mohapatra et al. 2024). This chapter highlights the role of root exudates in shaping the rhizomicrobiota under environmental constraints, thereby reflecting its stress-relieving potential. Furthermore, the chapter also discusses how the exploitation of root exudate traits from various crops and their wild relatives can form a potential solution for climate-smart agriculture (CSA). In addition, harnessing the potential of root exudate and their interaction with microbes may provide substantial knowledge to develop stress-tolerant plants in future.

2 Root Exudate: Below-Ground Wireless Crosstalk

Below-ground microbial biodiversity serves as a sustainable resource for maintaining ecosystem functions. Plant-soil feedback is one of the crucial factors for plant health, reproduction, community structure, and plant succession (Kong et al. 2019). These feedbacks are directly linked with soil microbial community dynamics that are recognized as an important player in this context. Plants can release up to 20% of their photosynthetically fixed carbon (C) and 15% of nitrogen (N) in the form of simple molecules like sugars, secondary metabolites, and organic acids (Prescott et al. 2020). The rhizodeposition of these compounds has pronounced effects on selective microbial populations that can respond accordingly and, finally colonize the host plant rhizosphere. Root-secreted organic acids have more deterministic effects than sugar in shaping the soil microbial community (Richardson et al. 2021). The greater influence of organic acids on soil microbiome can be explained by their higher amount of C contents, SOM (soil organic matter) solubilization, and changes in soil pH (Sindhu et al. 2022). The exudation patterns of plants vary significantly depending on their genotypes, age (Chen et al. 2023), environmental constraints such as temperature, solar radiation, seasonal variation, soil nutrient conditions (Jiang et al. 2022; Gao et al. 2024), etc. encountered by plants. Thereby, plant defines a biased rhizosphere microbial community by modulating their root exudate profiles (McLaughlin et al. 2023). Root exudates are usually rich in C compounds and act as nutrient and energy sources for rhizospheric microorganisms. The concentration of root-derived C compound can directly increase the extracellular enzymatic activity of their associated root microbiome that is involved in the breakdown of various organic compounds (Ma et al. 2022). Exudates also help to acquire and mobilize insoluble minerals by releasing complex molecules such as carboxylates, phenols, and siderophores (Sardans et al. 2023). Besides facilitating microbial growth by providing C and N substrate, exudates can also regulate plant-microbe interactions near the root vicinity by releasing some signalling molecules, attractants, repellents, or stimulators (Balasubramanian et al. 2020). For instance, a major secondary metabolite in maize root exudates, benzoxazinoids, not only acts as a stimulant for plant defence response (Hu et al. 2018), but also a chemoattractant for the plant growth-promoting bacterium *Pseudomonas putida* KT2440 (López-Farfán et al. 2019). Furthermore, root exudates are also a key mediator for plant-to-plant interactions. The chemical compounds such as strigolactones (SL), ethylene, jasmonic acid, and allantoin, present in the root exudates act as signalling molecules among the coexisting plant community (Wang et al. 2021a).

Root exudation process is regulated as a passive secretion process, facilitated via various pathways: the movement across the root plasma membrane by diffusion, ionic channels, and vesicle transport (Vives-Peris et al. 2020). The secretion pathway of the exudate molecules highly depends upon their chemical nature. Most of the plant's primary metabolites are released into the soil by the process of diffusion via the less differential region of the root tips (Canarini et al. 2019). Mainly, the low molecular weight compounds (like sugars, amino acids, and phenolic compounds) are released through diffusion, as the concentration gradient is developed in the root cytoplasm and rhizosphere (Massalha et al. 2017). While carbohydrate and carboxylate substances like malate and oxalate are usually transported through protein-meditated ion channels. Studies revealed that various types of plasma membrane-associated transporters are involved in the channel-mediated exudation process, namely SLAC (Slow Anion Channels), QUAC (Quick Anion Channel), and ALMT (Aluminium activated malate transporter). However, the secretion of high molecular weight substances takes place through the process of exocytosis by the involvement of vesicles (Vives-Peris et al. 2020). Vacuoles act as a temporary storage intracellular organelle for most of the compounds secreted via root exudates. On the other hand, the active transportation of root-secreted metabolites is mediated by plasma membrane-located transporters, namely ABC (ATP-Binding Cassette) and MATE (Multidrug and toxic compound extrusion) (Kang et al. 2011). In addition to these, a diverse array of other membrane transporters plays a significant role in the root exudates secretion process. For example, ABCG-type transporters are responsible for flavonoid exudation, SLs are exuded by PDR1 transporters (Ziegler et al. 2017), ALMT1 transporters are responsible for malic acid exudation (Rodrigues et al. 2019), MATE transporter takes part in citrate exudation (Wang et al. 2017), AtPDR9 helps in the exudation of phenolic compounds (Lefèvre et al. 2018), and ATPases regulate the protons transport that maintains the rhizospheric pH. This comprehensive and highly regulated root exudation process not only facilitates the release of various metabolites into the root-soil vicinity but also plays a pivotal role in belowground plant-microbe interactions.

Plant can regulate their microbial dynamics by altering their exudation profile, thereby attracting the beneficial microorganisms and deterring the harmful ones, effectively distinguishing between allies and adversaries (Olanrewaju and Babalola 2022). This crosstalk shapes the microbiome traits, defining its mutualistic, commensal and pathogenic relationship. The interactions between host plant and their associative microbiota play a crucial role in mitigating various stress conditions (Liu et al. 2020). However, the resilience of rhizospheric microbiota to stress conditions is species-specific and depends on their compatibility with the host plant's exudates under those conditions. Root exudates serve as belowground chemical signalling molecules, influencing rhizomicrobiome and host plant adaptability to stressful environmental circumstances (Li et al. 2023b). Moreover, certain microorganisms, such as Rhizobacteria, possess the ability to alter the expression profile of host stress-responsive genes and metabolites to enhance the host plant resilience (Dukare et al. 2022). From recent studies, plausible mechanism involving the modulation of root exudation patterns and changes in the rhizomicrobiome interactions under stressful conditions can derived (Upadhyay et al. 2022; Orozco-Mosqueda et al. 2022), however, comprehensive knowledge of below-ground chemical signalling under various stress events will further elucidate the molecules to ecosystem scale picture suggesting their potential applications for sustainable agriculture.

3 Abiotic Stress-Induced Root Exudation Shapes Rhizomicrobial Dynamics

Abiotic stress such as drought, heat, cold, nutrient and salt stress affects plant growth and productivity by altering the physical, chemical, or biological characteristics of the rhizosphere through root exudation and optimize the plant's resource-use efficiency (Chai and Schachtman 2022). The 'cry for help' defensive response against stressors baits root exudate as a chemoattractant for beneficial microorganisms to overcome the stress (Parasar et al. 2024) (Table 8.1). For instance, Huang et al. (2022), found various root exudate components comprising different sugars, organic acids, and amino acids acting as chemoattractant for *Bacillus velezensis* SQR9 in the cucumber rhizosphere. A direct correlation between root exudate and rhizospheric microbes during abiotic stress was reported by Tiziani et al. (2022), revealing exudate-mediated alteration in the microbial metabolic pathways linked to stress reactions (Fig. 8.1).

Table 8.1 Abiotic and biotic factors influencing rhizospheric microbial community through changes in the root exudate compositions

Environmental stress factors	Plant species	Primary root exudate composition	Altered rhizospheric microbial community	References
Drought	Soybean	Malic acid	*Bacillus subtilis*	Feng et al. (2021)
	Tomato	SL	*Rhizophagus irregularis*	Leventis et al. (2021)
	Tomato and spinach	Flavonoid and SL	AMF	Wahab et al. (2023)
	Sorghum	Glycerol-3-phosphate	Actinobacteria	Xu and Coleman-Derr (2019)
	Maize	Soluble sugars, amino acids, and organic acids	Gram-positive bacteria	Yuan et al. (2024)
	Peanuts	Oleic, citric, and lactic acid and flavanones such as naringenin	*Bradyrhizobium* and *Azospirillum brasilense*	Solanki et al. (2024)
	Maize	L-valine	Gemmatimonadota	Yuan et al. (2024)
Heat	Maize and soybeans	Malate	*Bacillus subtilis* and *Bacillus cereus*	Tiziani et al. (2022)
	Cucumber	Benzoic, palmitic, and phthalic acids	*Glomus mosseae*	Unnikrishnan et al. (2023)
	Maize	Soluble sugars, amino acids, and organic acids	Gram-positive bacteria	Yuan et al. (2024)

(continued)

Table 8.1 (continued)

Environmental stress factors	Plant species	Primary root exudate composition	Altered rhizospheric microbial community	References
Cold	Tomato	Malondialdehyde	*Glomus mosseae*	Liu et al. (2016)
	Grapes	Proline, starch, and phenolics	*Burkholderia phytofirmans*	Kushwaha et al. (2020)
Nutrient deficiency	False wheatgrass	Glutaric, butyric acid, and 4-methyl catechol	*Nitrospira*, *Mortierella*, and *Haliangium*	Lin et al. (2022)
Fe deficiency	Red clover	Phenolic compounds	*Pseudomonas* sp.	Dhankhar et al. (2022)
N stress	Leguminous and non-leguminous plants	Flavonoid	N_2-fixing bacteria	Knežević et al. (2022)
	Alfalfa	Proline-rich root exudation	*Sinorhizobium meliloti*	Sabbioni and Forlani (2022)
	Sea-lavender	Methyl butyric acid and palmitic acid	*Bacillus flexus* KLBPM 491	Kumar et al. (2024)
	Cotton	Proline-rich root exudation	*Pseudomonas putida*	Tiwari et al. (2023)
Low pH	Lupin	Carboxylate	*Burkholderia* spp.	Tilgam et al. (2022)
Aphids, *Ralstonia solanacearum* SL1931	Pepper	L-malic acid	*Bacillus subtilis*	Badri et al. (2009) and Park and Ryu (2021)
Pseudomonas syringae	Arabidopsis	Malic acid	*Bacillus subtilis*	Lakshmanan et al. (2013) and Dastogeer et al. (2020)
Pythium ultimum	Barley	Phenolic compounds	*Pseudomonas fluorescens*	Jousset et al. (2011)
Fusarium culmorum	Carex	Monoterpene (Z)-limonene-oxide	*Collimonas*, *Janthinobacterium*, and *Paenibacillus*	Schulz-Bohm et al. (2018)
Fusarium oxysporum f. sp. *cucumerinum*	Cucumber	Citric acid and fumaric acid	*Bacillus amyloliquefaciens*	Dastogeer et al. (2020)
Fusarium oxysporum	Tomato	Terpenes, formic acid, dimethyl trisulfide, benzothiazole, and benzonitrile	*Bacillus* sp.	Gulati et al. (2020)

(continued)

Table 8.1 (continued)

Environmental stress factors	Plant species	Primary root exudate composition	Altered rhizospheric microbial community	References
Delia radicum	Oilseed rape	Indolyl glucosinolates, sulfur, and trehalose	*Stenotrophomonas, Pseudomonas, Paenibacillus,* and *Bacillus*	Ourry et al. (2018) andTaye et al. (2022)
Diabrotica virgifera virgifera	Maize	Sesquiterpene, (E)-b-caryophyllene (Ebc)	Entomopathogenic nematode	Rasmann et al. (2005) and Degenhardt et al. (2009)
Meloidogyne incognita	Tomato	SL	AMF	Xu et al. (2019)

Here, *AMF* Arbuscular mycorrhizal fungi, *SL* strigolactone, *Fe* iron, *N* Nitrogen

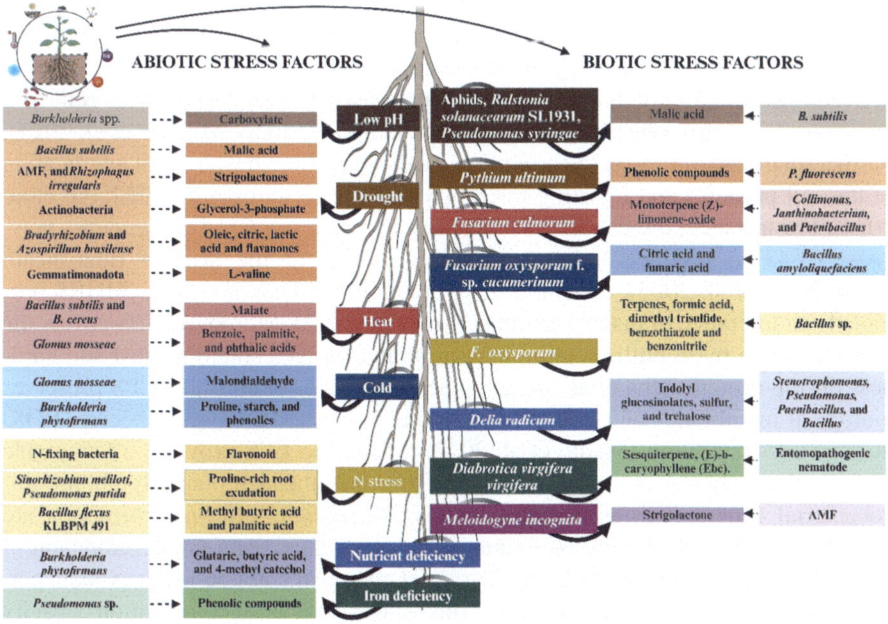

Fig. 8.1 Stress-induced changes in root exudate composition and microbes recruited to cope with different stress conditions. Here, AMF Arbuscular mycorrhizal fungi; the dotted arrow represents the microbe recruited in response to the specific exudate

Root exudates based on their molecular weight can be classified as high molecular weight exudates including mucilage, complex polysaccharides, some proteins and lipids, while low molecular weight exudate comprises primary metabolites such as sugars, amino acids, phenolics, and secondary metabolites like flavonoids, coumarin, and sorgoleone (Balasubramanian et al. 2020). The host plant, its developmental phases, physio-chemical makeup of the soil, and the microbes present in the surrounding environment affect the composition and specificity of root exudates released upon various abiotic stresses (Upadhyay et al. 2022).

3.1 High Molecular Weight Exudates

Plant exudes a large amount of high molecular weight compounds, primarily mucilages with a viscous mixture of exopolysaccharides (EPS), proteins, and lipids during abiotic stress (Mandal et al. 2022). A proteomic study revealed the presence of proteins in the mucilage of maize roots, sensitive to various stressors. This includes water and salt stress proteins, temperature-sensitive histone proteins, heat shock proteins and chaperonins that help the plant in stress tolerance (Nazari 2021). An aid to water absorbance promotes nutrient retention for the host plant and elicits positive chemotactic responses of beneficial microbes that assist the plant in withstanding harsh conditions (Bhat et al. 2023). Nazari et al. (2023) reported mucilage comprising polysaccharides such as hexoses, pentoses, and uronic acids as a significant source of C for attracting the rhizosphere microbes. Reciprocally, microbe-derived extracellular polymeric substances also have a pivotal role in stress-relieving mechanisms by rhizosheath development (Aslam et al. 2022; Mueller et al. 2024). Further, plants secrete EPS to form the rhizosheath, which have role in aggregation of the soil particles and boosting microbial activities (Guhra et al. 2022; Ali et al. 2024). For instance, the exudation of sucrose in wheat triggers *Paenibacillus polymyxa* CF43 to produce levan, a key determinant of the size of rhizosheath (Dobbelaere et al. 2003). Similarly, EPS-producing *Pseudomonas* sp. MHR6 promoted salt tolerance in maize seedlings (Liu et al. 2022). Moreover, EPS covers the root forming a hydrophilic biofilm around it, and forms a shield against soil hardness during drought (Jha et al. 2022). Other than the tolerant traits, EPS preserves biological membranes and help in detoxifying the reactive oxygen species (ROS) during stress condition (Mahto et al. 2022). Moreover, EPS provides a favourable niche in the early stages of microbial colonization in the host plant. Symbiotic rhizobacterial genera such as *Bradyrhizobium*, *Rhizobium*, and *Mesorhizobium* are recruited by EPS, promoting N_2 fixation and hence, improving the development of legumes (Atieno and Lesueur 2019). Different chemical compounds in exudate perform specific functions, influencing the rhizospheric bacterial colonization capacity and its plant stress alleviating properties (Beauregard 2015). The precise mechanisms concerning the role of EPS and mucilage in recruiting microbes are yet undeciphered and require further research insights into this complex interaction.

3.2 Low Molecular Weight Root Exudates

Plant root recruits beneficial soil microorganisms through exudation of low molecular weight compounds having chemotactic properties and attracts stress-mitigating microbes towards the rhizosphere. Specificity in exudate composition secreted by roots attracts certain microbes which use them as C sources assisting their sustenance in stressful environments (Bhagat et al. 2023). Low molecular weight exudate compounds can further be segregated into primary and secondary metabolites based on their functions (Canarini et al. 2019). These components of exudate take part directly or indirectly in stress alleviation mechanisms.

3.2.1 Primary Metabolites

Abiotic stress impacts the rhizospheric microbial community structure influencing the overall organic composition of the rhizosphere (Bogati and Walczak 2022). For instance, drought and heat stress induce higher exudation of soluble sugars, amino acids, and organic acids. This attributes to a shift in the structure of the microbial community through the expression of functional genes, with higher dominance of gram-positive bacteria while having insignificant changes in the fungal composition (Yuan et al. 2024). Mostly, the plant-microbe interaction shows feedback responses where microbes also secrete exogenous compounds to alleviate abiotic stresses (Ali et al. 2024). For example, *Achromobacter xylosoxidans* Cm4, *Variovorax paradoxus* 5C-2, and *Pseudomonas oryzihabitans* Ep4 through secretion of ACC deaminase improved drought tolerance in *Solanum tuberosum* (Gupta et al. 2022). Li et al. (2023a) found significant changes in microbial populations upon drought stress significantly correlating to the root secretions, highlighting the enrichment of Actinobacteria in response to organic acid and amino acid contents in root exudates. Glycerol-3-phosphate exudation directs Actinobacteria to cope with drought stress and supports its thriving in low-fertility situations (Xu and Coleman-Derr 2019). The abundance of gram-positive bacteria such as Actinobacteria, *Verticillium virens*, and *Sclerotinia* are positively correlated to particular amino sugars, sugar alcohols, polysaccharides, and glycoside hydrolases (Chen et al. 2022). For example, upon drought stress root exudes malic acid, which serves as a chemoattractant for *Bacillus subtilis* in soybean plants (Feng et al. 2021). Major organic acids exuded by plants include acetic acid, oxalic acid, tartaric acid, propionic acid, lactic acid, succinic acid, butyric acid, citric acid, valeric acid, and malonic acid (Afridi et al. 2024). The organic acid content increases with stress intensity, as evident in maize under drought and mediates the plant-microbe interaction for its mitigation (Rafique 2023). For example, oleic, citric, lactic acid, and flavanones such as naringenin recruit plant growth-promoting rhizobacteria (PGPR), *Bradyrhizobium*, and *Azospirillum brasilense* in *Arachis hypogaea* under drought conditions (Solanki et al. 2024). Similarly, under salt stress, *Limonum sinense* secretes methyl butyric acid and palmitic acid, which positively affect *Bacillus flexus* KLBPM 491 population by promoting its proliferation and helping the host to endure stress conditions (Kumar et al. 2023). Again, Lin et al. (2022) reported that *Leymus chinensis* exudes glutaric, butyric acid, and 4-methyl catechol attracting *Nitrospira*, *Mortierella*, and *Haliangium* species under soil C, N, and phosphorous (P) deficiency.

Osmolytes including soluble sugars, trehalose, proline, and amino acids serve as biochemical markers of salt stress and help scavenge free radicals while stabilizing subcellular structures. Proline-rich root exudation promotes colonization of *Sinorhizobium meliloti* in the roots of alfalfa (Sabbioni and Forlani 2022) and promotes the proliferation of *Ochrobactrum* sp., *Microbacterium* sp., and *Enterobacter* sp. in *Sorghum* roots (Govindasamy et al. 2020). Similarly, proline, starch, and phenolics trigger *Burkholderia phytofirmans* in *Vitis vinifera* to sustain cold stress (Kushwaha et al. 2020). It has been found that proline exudation gets upregulated by beneficial microorganisms such as *Burkholderia*, *Arthrobacter*, and *Bacillus* (Parasar et al. 2024). Again, exudate comprising proline mitigates salt stress in cotton plants by attracting *Pseudomonas putida* (Tiwari et al. 2023). In maize, proline drives stress alleviating *Pseudomonas aeruginosa* and *Proteus penneri*, while uronic acid helps in relieving stress by favouring Arbuscular mycorrhizal fungi (AMF) colonization (Quiroga et al. 2019; Poudel et al. 2021). Exudation comprising L-valine secreted upon drought stress attracts Gemmatimonadota, which enhances the roots' ability to absorb nitrate and ammonium (Yuan et al. 2024). Low pH conditions in soil along with nutrient deficiency result in the exudation of carboxylate, recruiting stress relieving *Burkholderia* spp. in the Lupin plant (Tilgam et al. 2022). Similarly, malate abundant in the exudates from the roots of maize and soybeans under heat stress is associated with the recruitment of advantageous microbes like *Bacillus subtilis* and *Bacillus cereus* (Tiziani et al. 2022). Phenolic compounds stimulate siderophore production in *Trifolium pratense*. This activates *Pseudomonas* sp. aiding in solubilization of iron (Fe) and boosting its absorption by plants in Fe deficiency (Dhankhar et al. 2022). Likewise, exudates comprising phenolic acid and quinate compounds secreted from *Cupressus* roots led to the growth of *Bacillus subtilis* and *Pseudomonas stutzeri* (Oppenheimer-Shaanan et al. 2022). Hence, abiotic stress-induced root exudates involving the primary metabolites serve as sensors and integrators of the plant-microbe interaction responding to the effect of stress-induced alteration in the root vicinity.

3.2.2 Secondary Metabolites

Secondary metabolites exuded by plants during abiotic stress include several compounds such as triterpenes, sorgoleone, SLs, flavonoids, and coumarin. Coumarin-containing root exudate is triggered upon soil Fe deficiency and acts as a mediator in host-microbe interaction that increases the availability of Fe in the host (Stringlis et al. 2018). Similarly, *Arabidopsis* releases coumarin eliciting PGPR responses in stress conditions (Stringlis et al. 2019). Again, the exudation of scopoletin by *Arabidopsis* in Fe-deficient alkaline stress conditions, converts Fe^{3+} to Fe^{2+} and makes it available to the plant during stress (Ning et al. 2023). N deficiency affects a variety of secondary metabolites in root exudates mediating microbial activity to increase the amount of N required by plants. Both leguminous and non-leguminous plants exhibit an increase in flavonoid exudation under N stress, which facilitates root interactions with N-fixing bacteria by activating the rhizobial nodulation genes (Knežević et al. 2022). Moreover, exudate containing flavonoids, naringenin, and daidzein promotes the root colonization of non-legume plants by the diazotrophic bacterium *Azorhizobium caulinodans*, providing the host plants with a stable N supply in exchange for C (Kumar et al. 2024). Similarly, exudate rich in flavones such as luteolin and apigenin, recruits Oxalobacteraceae population in maize (Pantigoso et al. 2022).

Flavonoid and SL in exudates assist in plants' adaptation to drought stress in tomato and spinach, playing a crucial role by promoting the formation of AMF hyphae early in the root colonization process (Wahab et al. 2023). Plants experiencing drought decrease SL content in the exudate, which acts as a signal for adherence of AMF spores to their roots and the association results in increase of SL content in the exudate (Bedini et al. 2018; Mitra et al. 2021). For instance, exudate containing SL increases in response to *Rhizophagus irregularis* colonization in tomato under drought stress. SL serve as host recognition signals for AMF to respond against the 'call-for-help' signals and trigger positive feedback of *Rhizophagus irregularis* colonization (Leventis et al. 2021). Triterpenes such as thalianin, thalianyl fatty acid esters, and arabidin exudation serve as C sources promoting bacterial proliferation (Chen and Liu 2024). Furthermore, many N inhibitors in exudate such as sakuranetin and sorgoleone in *Sorghum bicolor* (Wang et al. 2021b), 1,9-decanediol and syringic acid in rice, and brachialactone in *Brachiaria humidicola* (Lu et al. 2022) regulate the enzyme activities of nitrifying bacteria. *Cucumis sativus* exudes benzoic, palmitic, and phthalic acids to tolerate heat stress through the interaction of mycorrhiza such as *Glomus mosseae* (Unnikrishnan et al. 2023). *Glomus mosseae* recruited by malondialdehyde secreted by the roots assist tomato plants in enduring cold stress (Liu et al. 2016). A better understanding of the mechanisms and combined effects of different abiotic stimuli during stress in complex root exudate-rhizomicrobial interplay can provide a holistic insight into the ecosystem functioning.

4 Root Exudate Mediated Microbial Interplays in Response to Biotic Stress

Root-secreted chemical compounds attribute to complex interactions in the vicinity of plant root system (Aziz et al. 2021). These species-specific compounds change their composition in response to fluctuating environmental circumstances, resulting in alterations in the rhizospheric microbiome (Badri and Vivanco 2008; Sharma et al. 2023). Biotic factors, comprising plant-parasites/pathogens or symbiotic microbial communities, associated with the host plant rhizosphere play an important role in shaping the rhizomicrobiome (Fig. 8.1). These factors including saprophytic and pathogenic microbes influence the RSA. For example, Zolobowska and Van Gijsegem (2006), reported the change in root morphology in Petunia plants after infection with *Ralstonia solanacearum*. Similarly, alterations in the root development have been observed in *Arabidopsis thaliana* after being colonized by *Bacillus megaterium* (Allard-Massicotte et al. 2016). It suggests that the microbial dynamics of the host plant rhizosphere is correlated with changes in the exudation compounds modulated through the RSA influenced by biotic stressors (Table 8.1). This section highlights how the biotic factors influence the plant and alter its belowground exudates, thereby affecting the composition of the rhizospheric microbial community.

4.1 Significance of Biotic Stress Factors Upon Root Exudates and Soil Microbiome

Plants encountering environmental stimuli either in the physical (pathogen attack or herbivory) or chemical form, incorporate a blend of phytohormones and volatile organic compounds (VOCs) in their root exudates. Unlike non-stimulated plants, this alteration eventually restructures the rhizomicrobial community (Dastogeer et al. 2020). For example, the external application of defense-related compounds like nitric oxide, methyl jasmonate, and salicylic acid, induces the host plant to synthesize a wide range of secondary metabolites such as rosmarinic acid, caffeic acid, and glucosinolates (Alvarado et al. 2019; Upadhyay et al. 2024). Moreover, under biotic stress, there is an increase in the phenolic compounds in root exudates which can directly suppress the growth of pathogenic microbes or indirectly alter the expression of various antifungal genes among beneficial microorganisms (Tak and Kumar 2020). For example, Jousset et al. (2011), in a split root experiment, infected one-half of the barley roots with the pathogenic fungus *Pythium ultimum* while the other half was inoculated with *Pseudomonas fluorescens*. Exudate comprising abundant phenolic compounds from the side containing *Pseudomonas fluorescens* facilitates the recruitment of beneficial PGPR and also leads to the expression of antibiotic-related genes present in *Pseudomonas fluorescens*.

Intriguingly, Paddock et al. (2021), observed changes in maize plant's exudation profile in response to western corn rootworm larvae. Infected maize plants produced phenolics as exudates, significantly affecting the composition of the bacterial diversity. Park and Ryu (2021), reported the abundance of *Bacillus subtilis* GB03 in the root rhizospheric zone of *Capsicum annuum* during aphids infestation and co-infection with *Ralstonia solanacearum* SL1931. Earlier experimental finding (Badri et al. 2009) indicates that secretion of L-malic acid might play a significant role towards the recruitment of *Bacillus subtilis* FB17 to the root rhizospheric regions. Additionally, alterations in rhizomicrobial diversity were observed in the rhizosphere of tomato plants infected with *Ralstonia solanacearum*. This detailed study indicates that these changes were linked to differences in the composition of root exudates compared to non-infected plants (Gu et al. 2016). Similarly, during the infection of *Pseudomonas syringae*, *Arabidopsis thaliana* roots exude malic acid recruiting *Bacillus subtilis*. This infection alters the expression of malate transporter genes in the roots, which results in the alteration of exudation compounds, leading to increased accumulation of *Bacillus subtilis* in the rhizosphere (Lakshmanan et al. 2013; Dastogeer et al. 2020). Modulation in the bacterial communities in the vicinity of oilseed rape roots was also observed during soil herbivorous insect, *Delia radicum* (cabbage root fly) infestation. This shift in exudation patterns includes an increase in indolyl glucosinolates, sulphur, and trehalose, alongside a decrease in sugar and amino acid levels. Consequently, it alters the microbial community composition around the roots and in the rhizosphere of oilseed rape, enhancing the abundance of bacterial groups such as *Stenotrophomonas*, *Pseudomonas*, *Paenibacillus*, and *Bacillus* (Ourry et al. 2018; Taye et al. 2022). Furthermore, secretion of monoterpene (Z)-limonene-oxide in the course of *Fusarium culmorum* infection in the

Carex plant leads to the accumulation of *Collimonas, Janthinobacterium*, and *Paenibacillus* towards the rhizospheric zone (Schulz-Bohm et al. 2018). *Fusarium oxysporum* f. sp. *cucumerinum* infection also results in the accumulation of citric acid and fumaric acid in the rhizospheric region of cucumber plants. This alteration in the exudate composition aids in *Bacillus amyloliquefaciens* proliferation and biofilm formation (Dastogeer et al. 2020). Additionally, tomato plant roots infected with *Fusarium oxysporum*, lead to the secretion of terpenes, formic acid, dimethyl trisulfide, benzothiazole, and benzonitrile that aids in the recruitment of *Bacillus* sp. in the rhizosphere (Gulati et al. 2020). Furthermore, Akiyama and Hayashi (2006) documented the importance of SL, in root exudates for promoting AMF colonization to the host plant, which aids in structuring the root-associated microbiota. For example, tomato plants infested with parasitic nematode *Meloidogyne incognita*, releases SL which enhances the root-AMF association, improving the host plant resilience via altering host root associated microbiome (Xu et al. 2019). Additionally, the application of SL analogue racGR24, in JA-deficient tomato mutants, improved plant defence against nematode infection. Similarly, maize root infested with *Diabrotica virgifera virgifera* (western corn rootworm) larvae triggers the secretion of a discrete blend of exudates comprising sesquiterpene, (E)-b-caryophyllene (Ebc). This exudates help the plant to recruit soil-borne entomopathogenic nematode, which acts as a natural predator to the beetle larvae, thereby providing biocontrol activity (Rasmann et al. 2005; Degenhardt et al. 2009). Ebc secretion may not always lead to biocontrol activity in the root-soil vicinity. However, Fantaye et al. (2015), in a study reported that, Ebc secretion in the rhizosphere promotes hyphal branching in fungal pathogen, namely, *Fusarium graminearum* and *Colletotrichum graminicola*. Thus, secretion of Ebc as exudates might either enhance host plant resistance or susceptibility depending on the type of biotic stress. Various studies suggested that the overexpression of the Ebc gene could significantly impact soil-residing microbes (Rolfe et al. 2019). These studies collectively demonstrate that plants respond to diverse environmental challenges by altering their root exudates, which in turn influence the recruitment of beneficial microbes in the root rhizospheric regions, thereby contributing towards plant health and defence.

In addition to the regulation of root exudates in response to various environmental stimuli, plants secrete a wide array of antimicrobial compounds that play a significant role in shaping the unique rhizosphere microbial diversity. Plant roots generally secrete a mixture of chemical compounds with antimicrobial properties. These phytochemicals vary in concentration and composition among different plant species, playing a significant role in determining host-specific rhizomicrobial diversity (Pascale et al. 2020). For example, pisatin, a pea phytoalexin with antimicrobial properties, is a crucial compound reported in legumes (Tiku 2020). Notably, the amino acid canavanine may acts as a stimulator for one group of microbes while suppressing many other soil-dwelling microbes (Shah and Gupta 2023). Similarly, metabolites comprising phenolics effectively aid in attracting specific soil-borne microbes, beneficially influencing native soil microbial diversity (Lone et al. 2023). Additionally, compounds belonging to the same class can differently impact the soil environment by altering microbial diversity and their interactions with the host plant and other nearby microbial communities. For example, a compound phytoalexin secreted from the roots of *Arabidopsis* confers resistance to *Phytophthora capsici* (Dye and Bostock 2021); but not to the pathogen *Phytophthora cinnamomi* (Van Den Berg et al. 2021). In contrast, tryptophan-derived secondary metabolites such as glucosinolates and indole-derivative camalexin, which are reported as root exudates from the *Arabidopsis,* exhibit antifungal and antibacterial properties (Mulat et al. 2020). Likewise, diterpene rhizathalene A is reported as a constitutively secreted compound from the roots of *Arabidopsis thaliana* that has a crucial role in conferring resistance to insect herbivory (Erb and Reymond 2019). This indicates that plants not only secrete specific compounds upon pathogen infection, creating an infection court in the root-soil proximity but also establish a preventive microbial buffer zone that protects against further infection (Baetz and Martinoia 2014).

4.2 Remodulation of Rhizomicrobiome Through Exudation of Neighbouring Crop Plants

Several studies have documented the impact of invasive plant species on economically important crop plants. The invasiveness of these wild plants intensifies markedly as they establish themselves in new environments, avoiding native insects and pathogens while adapting their physiology and biochemistry to improve resilience (Sharma et al. 2023). In addition, weeds modify their exudation profile to exhibit allelopathy and attract beneficial microbes (Trivedi et al. 2022). For example, *Parthenium hysterophorus*, commonly found as an invasive plant in India and other countries, eliminates native plants through the secretion of allelopathic compounds like parthenin and phenolic acids (De La Fuente et al. 2000). Contrastingly, exudation of glucosinolates was reported from the root of *Arabidopsis* plants growing in a community having large number of conspecifics, however, the secretion of glucosinolates was found to be decreased in *Arabidopsis* plants grown with fewer conspecific (Badenes-Perez et al. 2013). Bressan et al. (2009), in a study, reported the abundance of *Mesorhizobium amorphae*, *Syncephalis depressa*, *Rhizobium* sp., *Bose* sp., and *Agrobacterium* sp., in the rhizospheric region of *Arabidopsis thaliana* transgenic line CYP79A1 that exogenously produces glucosinolates, as compared to the wild type plants. Moreover, intercropping of *Arachis hypogea* L. with *Manihot esculenta* Crantz stimulates the production of ethylene hormone in *Arachis hypogaea*. The release of ethylene aids in the recruitment of Actinobacteria species *Catenulispora* towards the host plant roots thereby enhancing plant fitness in the intercropping system (Chen et al. 2020). These findings emphasize how neighbouring plants reshape the rhizomicrobiota by modifying their root exudates, thereby enhancing their adaptation to diverse environmental conditions.

5 Exploitation of Root Exudates Towards CSA

Sustainably achieving food security has been endangered by various socio-environmental factors, including recent climate change scenarios, demanding high crop production worldwide. A diverse array of strategies has been employed to meet this challenge since the emergence of the first Green Revolution, particularly focusing on plant aboveground traits (Preece and Peñuelas 2020). However, modifying root exudates can improve plant functional traits like resistance to biotic and abiotic stressors. Furthermore, understanding the cascade of feedback loops in the form of root exudates under different environmental circumstances and exploring its associated traits seems a promising strategy for attaining CSA.

Exudation of chemicals through the roots occurs in an array of interactions with the soil biota, especially with rhizomicrobiota (Tian et al. 2020). Root exudates can exhibit antimicrobial or nematicidal activities depending on chemical composition, influencing plant defence against detrimental microbes (Lamichhane et al. 2024). Studies focusing on defined pathosystems including single host and soil-borne pathogen have been carried out in context with root exudation, however, such studies provide semiquantitative data in response to changes in root exudate composition. Meanwhile, in nature pathogenic complexes are mainly responsible for various soil-borne diseases rather than a single pathogen (Back et al. 2002; You et al. 2020; Barbetti et al. 2021; Ahmadi et al. 2022). Thus, unveiling the alteration in root exudate composition under the effect of soil-borne pathogenic complexes might provide better ways to combat plant diseases. Moreover, a meta-analysis incorporating experimental data from contrasting studies and natural field observations will help determine specific root exudate composition for enhancing overall plant health under diseased conditions. Additionally, the exuded chemicals utilized by beneficial as well as pathogenic microbes to proliferate in the rhizosphere remain unknown to date, since a particular microbe can be beneficial as well as harmful at the same time for different plant species (Lamichhane et al. 2024). Furthermore, both pathogenic and useful microbes may be attracted to root exudates with the same group of molecules. Thus, an in-depth study decoding root exudate metabolites can reveal whether beneficial and pathogenic microbes utilize the same types of exudates or not. Moreover, understanding the variation in exuded molecules might help in designing effective root exudate composition to resist pathogenic infection under diseased scenarios.

Modern agriculture focusing on a small group of crops provides a shallow potential for crop improvement under climate change-induced environmental constraints. However, the crop wild relatives (CWRs) are the storehouse of information and seem a potential candidate for exploiting various plant traits (Brozynska et al. 2016; Dempewolf et al. 2017). The CWRs pose desirable traits associated with plant stress tolerance (Roy et al. 2011), mycorrhizal association (Martín-Robles et al. 2018), and pest resistance (Chen et al. 2015). Further, linking the gaps in knowledge regarding the root exudation of crops and their wild relatives could be a focal area for achieving sustainable agriculture (Kashyap et al. 2024). From various studies, it has been established that modification in root exudates can enhance pest resistance in crops. However, the CWRs are more resistant to pest attacks than the present crops. For example, teosinte along with wild European maize varieties release Ebc as root exudate to attract entomopathogenic nematode from the surrounding soil to feed on the larvae of *Diabrotica virgifera virgifera*. On the other hand, such chemical signalling molecules are not secreted by most North American maize varieties and thus can't withstand the havoc of *D. virgifera virgifera* (Rasmann et al. 2005; Köllner et al. 2008). The study indicates that exploiting the root exudation traits associated with wild maize varieties and incorporating them into present varieties using gene editing or breeding techniques would reduce the attack of pests and the use of pesticides (Preece and Peñuelas 2020). Similarly, wild crop species might have traits associated with better nutrient acquisition from soil. For instance, the CWRs might produce root exudates in higher concentrations having different metabolite compositions with higher efficiency towards soil nutrients such as P. In a study, Iannucci et al. (2017), revealed that the wild emmer wheat and modern durum wheat subspecies secret root exudate having different constituents, indicating how domestication has altered the root exudation composition. Similarly, the difference in exudation composition and pattern in different CWRs and crops, influence plant symbiotic association with mycorrhiza, thereby modulating the plant-mycorrhiza interaction. For example, the mycorrhizal colonization in wild relatives of breadfruit was found to be higher than in domesticated ones (Xing et al. 2012).

Likewise, root exudation plays a pivotal role in maintaining the symbiotic association between legumes and N-fixing bacteria. Leguminous plants release flavonoid compounds as exudate to attract rhizobia. Evidence from chickpea, faba beans, and pea has established that as compared to their wild progenitors the domesticated varieties loosely associate with the rhizomicrobes, indicating that wild and modern varieties differ in their exudation pattern (Mutch and Young 2004; Kim et al. 2014; Pérez-Jaramillo et al. 2016; Bourion et al. 2018). Similar to this, understanding the differences in the exudate composition and pattern in crops and their wild relatives under abiotic stressors might help in designing effective agronomic strategies. Thus, the knowledge regarding rhizomicrobiome can be used as a mirror to unravel the changes in root exudation between wild relatives and domesticated crops. Since exudates serve as major substrates for rhizosphere microbes, creating different metabolic niches for microbes to utilize (Maitra et al. 2024), thus recovering lost traits from CWRs might help to achieve sustainable agriculture. Furthermore, understanding the differences in root exudation due to domestication and predicting how new traits function beyond controlled conditions seeks attention in future research.

6 Conclusion

Root exudates are species-specific and shape the rhizospheric microbiota of plants. Mostly carbonaceous in nature, exudate serves as the organic C source for the survival and thriving of microbes. Depending upon various abiotic and biotic stress conditions, the composition of root exudates released by a plant varies, recruiting certain microbes corresponding to the exuded compounds in the rhizospheric region, which ultimately help the plant in withstanding adverse conditions. It is crucial to understand this specific response of the microbes to the compounds in the root exudates, for deciphering the molecular mechanism behind this crosstalk. Root-soil-microbe interaction results in maximum energy utilization which helps in the optimization of localized nutrient supply. Understanding of this exudate-based regulation of the rhizomicrobiome might pave the way for sustainable crop management. The recent advances in molecular biology techniques can assist us in studying different root exudation compositions in various plants under different environmental conditions. Additionally, the traits corresponding to the root exudation of the CWRs might be exploited for their effective uses in agriculture. However, challenges occur due to the irreplicability of complex interaction processes in laboratory conditions as *in vivo*, where different stressors influence the plants in a combined manner. Moreover, the exudate composition also alters upon adsorption on the soil particles and in the process of further microbial breakdown. Besides, the root exudation pattern is also affected by the soil microbe-derived exudates. Therefore, a holistic understanding of this complex intricate relationship between the stressed plants and associated microbes is necessary to develop stress-tolerant plants and effective application of rhizospheric microbiota to improve sustainable agriculture practices ensuring food security and ecosystem management.

Acknowledgments The authors are thankful to DST, Govt. of India, for providing PURSE support to Gauhati University and DST-FIST Support to the Department of Botany, Gauhati University.

Declaration of Competing Interest The authors have no relevant financial or non-financial interests to disclose.

Author Contributions NA was responsible for writing guidance and manuscript preparation. BJP, RB, SK, IS, SDM, PD were responsible for writing the first draft of the review article. NA revised the article. All authors contributed to the article and approved the submitted version.

Funding Core Research Grant (CRG/2023/002423) provided by Anusandhan National Research Foundation (ANRF), Start-Up grant provided by University Grant Commission, Govt. of India (grant number F.30-386/2017) and Scheme for promoting research among young faculty (GU/Acad./YFPGC/50/2018/1738-79/05) provided to NA by Gauhati University.

References

Afridi MS, Kumar A, Javed MA, Dubey A, De Medeiros FHV, Santoyo G (2024) Harnessing root exudates for plant microbiome engineering and stress resistance in plants. Microbiol Res 279:127564. https://doi.org/10.1016/j.micres.2023.127564

Ahmadi M, Mirakhorli N, Erginbas-Orakci G, Ansari O, Braun H-J, Paulitz T, Dababat AA (2022) Interactions among cereal cyst nematode *Heterodera filipjevi*, dryland crown rot *Fusarium culmorum*, and drought on grain yield components and disease severity in bread wheat. Can J Plant Pathol 44(3):415–431. https://doi.org/10.1080/07060661.2021.2013947

Akiyama K, Hayashi H (2006) Strigolactones: chemical signals for fungal symbionts and parasitic weeds in plant roots. Ann Bot 97(6):925–931. https://doi.org/10.1093/aob/mcl063

Ali N, Abbas SAAA, Sharif L, Shafiq M, Kamran Z, Masah, Haseeb M, Shahid MA (2024) Microbial extracellular polymeric substance and impacts on soil aggregation. In: Bacterial secondary metabolites. Elsevier, pp 221–237

Allard-Massicotte R, Tessier L, Lécuyer F, Lakshmanan V, Lucier J-F, Garneau D, Caudwell L, Vlamakis H, Bais HP, Beauregard PB (2016) *Bacillus subtilis* early colonization of *Arabidopsis thaliana* roots involves multiple chemotaxis receptors. MBio 7(6):e01664-16. https://doi.org/10.1128/mBio.01664-16

Alvarado AM, Aguirre-Becerra H, Vázquez-Hernández MAC, Magaña-Lopez E, Parola-Contreras I, Caicedo-Lopez LH, Contreras-Medina LM, Garcia-Trejo JF, Guevara-Gonzalez RG, Feregrino-Perez AA (2019) Influence of elicitors and eustressors on the production of plant secondary metabolites. In: Akhtar MS, Swamy MK, Sinniah UR (eds) Natural bio-active compounds. Springer Singapore, Singapore, pp 333–388

Aslam MM, Karanja JK, Dodd IC, Waseem M, Weifeng X (2022) Rhizosheath: an adaptive root trait to improve plant tolerance to phosphorus and water deficits? Plant Cell Environ 45(10):2861–2874. https://doi.org/10.1111/pce.14395

Atieno M, Lesueur D (2019) Opportunities for improved legume inoculants: enhanced stress tolerance of rhizobia and benefits to agroecosystems. Symbiosis 77(3):191–205. https://doi.org/10.1007/s13199-018-0585-9

Aziz U, Rehmani MS, Wang L, Luo X, Xian B, Wei S, Wang G, Shu K (2021) Toward a molecular understanding of rhizosphere, phyllosphere, and spermosphere interactions in plant growth and stress response. Crit Rev Plant Sci 40(6):479–500. https://doi.org/10.1080/07352689.2022.2031728

Back MA, Haydock PPJ, Jenkinson P (2002) Disease complexes involving plant parasitic nematodes and soilborne pathogens. Plant Pathol 51(6):683–697. https://doi.org/10.1046/j.1365-3059.2002.00785.x

Badenes-Perez FR, Reichelt M, Gershenzon J, Heckel DG (2013) Interaction of glucosinolate content of *Arabidopsis thaliana* mutant lines and feeding and oviposition by generalist and specialist lepidopterans. Phytochemistry 86:36–43. https://doi.org/10.1016/j.phytochem.2012.11.006

Badri DV, Vivanco JM (2008) Regulation and function of root exudates. Plant Cell Environ. https://doi.org/10.1111/j.1365-3040.2008.01926.x

Badri DV, Weir TL, Van Der Lelie D, Vivanco JM (2009) Rhizosphere chemical dialogues: plant–microbe interactions. Curr Opin Biotechnol 20(6):642–650. https://doi.org/10.1016/j.copbio.2009.09.014

Baetz U, Martinoia E (2014) Root exudates: the hidden part of plant defense. Trends Plant Sci 19(2):90–98. https://doi.org/10.1016/j.tplants.2013.11.006

Balasubramanian V, Sur A, Nayak KK, Singh RK (2020) Plant root exudates as determinant of rhizomicrobiome. In: Sharma SK, Singh UB, Sahu PK, Singh HV, Sharma PK (eds) Rhizosphere microbes. Springer Singapore, Singapore, pp 105–126

Barbetti MJ, Khan TN, Pritchard I, Lamichhane JR, Aubertot J-N, Corrales DC, You MP (2021) Challenges with managing disease complexes during application of different measures against foliar diseases of field pea. Plant Dis 105(3):616–627. https://doi.org/10.1094/PDIS-07-20-1470-RE

Beauregard PB (2015) Not just sweet talkers. In: Advances in botanical research. Elsevier, pp 1–20

Bedini A, Mercy L, Schneider C, Franken P, Lucic-Mercy E (2018) Unraveling the initial plant hormone signaling, metabolic mechanisms and plant defense triggering the endomycorrhizal symbiosis behavior. Front Plant Sci 9:1800. https://doi.org/10.3389/fpls.2018.01800

Bhagat S, Shete P, Jain A (2023) Plant health: feedback effect of root exudates and rhizobiome interactions. In: Rhizobiome. Elsevier, pp 345–375

Bhat MA, Mishra AK, Jan S, Bhat MA, Kamal MA, Rahman S, Shah AA, Jan AT (2023) Plant growth promoting rhizobacteria in plant health: a perspective study of the underground interaction. Plants 12(3):629. https://doi.org/10.3390/plants12030629

Bogati K, Walczak M (2022) The impact of drought stress on soil microbial community, enzyme activities and plants. Agronomy 12(1):189. https://doi.org/10.3390/agronomy12010189

Bourion V, Heulin-Gotty K, Aubert V, Tisseyre P, Chabert-Martinello M, Pervent M, Delaitre C, Vile D, Siol M, Duc G, Brunel B, Burstin J, Lepetit M (2018) Co-inoculation of a pea core-collection with diverse rhizobial strains shows competitiveness for nodulation and efficiency of nitrogen fixation are distinct traits in the interaction. Front Plant Sci 8:2249. https://doi.org/10.3389/fpls.2017.02249

Bressan M, Roncato M-A, Bellvert F, Comte G, Haichar FEZ, Achouak W, Berge O (2009) Exogenous glucosinolate produced by *Arabidopsis thaliana* has an impact on microbes in the rhizosphere and plant roots. ISME J 3(11):1243–1257. https://doi.org/10.1038/ismej.2009.68

Brozynska M, Furtado A, Henry RJ (2016) Genomics of crop wild relatives: expanding the gene pool for crop improvement. Plant Biotechnol J 14(4):1070–1085. https://doi.org/10.1111/pbi.12454

Canarini A, Kaiser C, Merchant A, Richter A, Wanek W (2019) Root exudation of primary metabolites: mechanisms and their roles in plant responses to environmental stimuli. Front Plant Sci 10:157. https://doi.org/10.3389/fpls.2019.00157

Chai YN, Schachtman DP (2022) Root exudates impact plant performance under abiotic stress. Trends Plant Sci 27(1):80–91. https://doi.org/10.1016/j.tplants.2021.08.003

Chen L, Liu Y (2024) The function of root exudates in the root colonization by beneficial soil rhizobacteria. Biology 13(2):95. https://doi.org/10.3390/biology13020095

Chen YH, Gols R, Benrey B (2015) Crop domestication and its impact on naturally selected trophic interactions. Annu Rev Entomol 60(1):35–58. https://doi.org/10.1146/annurev-ento-010814-020601

Chen Y, Bonkowski M, Shen Y, Griffiths BS, Jiang Y, Wang X, Sun B (2020) Root ethylene mediates rhizosphere microbial community reconstruction when chemically detecting cyanide produced by neighbouring plants. Microbiome 8(1):4. https://doi.org/10.1186/s40168-019-0775-6

Chen Y, Yao Z, Sun Y, Wang E, Tian C, Sun Y, Liu J, Sun C, Tian L (2022) Current studies of the effects of drought stress on root exudates and rhizosphere microbiomes of crop plant species. IJMS 23(4):2374. https://doi.org/10.3390/ijms23042374

Chen M, Yao X, Cheng H, Fan A, Lin R, Wang X, Yang Y, Chen G (2023) Changes in Chinese fir plantations root exudation strategies seasonally and as tree age. For Ecol Manag 545:121239. https://doi.org/10.1016/j.foreco.2023.121239

Dastogeer KMG, Tumpa FH, Sultana A, Akter MA, Chakraborty A (2020) Plant microbiome—an account of the factors that shape community composition and diversity. Curr Plant Biol 23:100161. https://doi.org/10.1016/j.cpb.2020.100161

De La Fuente JR, Uriburu ML, Burton G, Sosa VE (2000) Sesquiterpene lactone variability in *Parthenium hysterophorus* L. Phytochemistry 55(7):769–772. https://doi.org/10.1016/S0031-9422(00)00288-0

Degenhardt J, Hiltpold I, Köllner TG, Frey M, Gierl A, Gershenzon J, Hibbard BE, Ellersieck MR, Turlings TCJ (2009) Restoring a maize root signal that attracts insect-killing nematodes to control a major pest. Proc Natl Acad Sci USA 106(32):13213–13218. https://doi.org/10.1073/pnas.0906365106

Dempewolf H, Baute G, Anderson J, Kilian B, Smith C, Guarino L (2017) Past and future use of wild relatives in crop breeding. Crop Sci 57(3):1070–1082. https://doi.org/10.2135/cropsci2016.10.0885

Dhankhar R, Gupta S, Gulati P (2022) Insights on plant–microbe interactions in soil in relation to iron dynamics. Vegetos 36(3):750–767. https://doi.org/10.1007/s42535-022-00467-3

Dobbelaere S, Vanderleyden J, Okon Y (2003) Plant growth-promoting effects of diazotrophs in the rhizosphere. Crit Rev Plant Sci 22(2):107–149. https://doi.org/10.1080/713610853

Dukare A, Mhatre P, Maheshwari HS, Bagul S, Manjunatha BS, Khade Y, Kamble U (2022) Delineation of mechanistic approaches of rhizosphere microorganisms facilitated plant health and resilience under challenging conditions. 3 Biotech 12(3):57. https://doi.org/10.1007/s13205-022-03115-4

Dye SM, Bostock RM (2021) Eicosapolyenoic fatty acids induce defense responses and resistance to *Phytophthora capsici* in tomato and pepper. Physiol Mol Plant Pathol 114:101642. https://doi.org/10.1016/j.pmpp.2021.101642

Erb M, Reymond P (2019) Molecular interactions between plants and insect herbivores. Annu Rev Plant Biol 70(1):527–557. https://doi.org/10.1146/annurev-arplant-050718-095910

Fantaye CA, Köpke D, Gershenzon J, Degenhardt J (2015) Restoring (E)-β-caryophyllene production in a non-producing maize line compromises its resistance against the fungus *Colletotrichum graminicola*. J Chem Ecol 41(3):213–223. https://doi.org/10.1007/s10886-015-0556-z

Feng H, Fu R, Hou X, Lv Y, Zhang N, Liu Y, Xu Z, Miao Y, Krell T, Shen Q, Zhang R (2021) Chemotaxis of beneficial rhizobacteria to root exudates: the first step towards root–microbe rhizosphere interactions. IJMS 22(13):6655. https://doi.org/10.3390/ijms22136655

Fields B, Friman V-P (2022) Microbial eco-evolutionary dynamics in the plant rhizosphere. Curr Opin Microbiol 68:102153. https://doi.org/10.1016/j.mib.2022.102153

Galindo-Castañeda T, Hartmann M, Lynch JP (2024) Location: root architecture structures rhizosphere microbial associations. J Exp Bot 75(2):594–604. https://doi.org/10.1093/jxb/erad421

Gao Y, Wang H, Yang F, Dai X, Meng S, Hu M, Kou L, Fu X (2024) Relationships between root exudation and root morphological and architectural traits vary with growing season. Tree Physiol 44(1):tpad118. https://doi.org/10.1093/treephys/tpad118

Govindasamy V, George P, Kumar M, Aher L, Raina SK, Rane J, Annapurna K, Minhas PS (2020) Multi-trait PGP rhizobacterial endophytes alleviate drought stress in a senescent genotype of sorghum [*Sorghum bicolor* (L.) Moench]. 3 Biotech 10(1):13. https://doi.org/10.1007/s13205-019-2001-4

Gu Y, Wei Z, Wang X, Friman V-P, Huang J, Wang X, Mei X, Xu Y, Shen Q, Jousset A (2016) Pathogen invasion indirectly changes the composition of soil microbiome via shifts in root exudation profile. Biol Fertil Soils 52(7):997–1005. https://doi.org/10.1007/s00374-016-1136-2

Guhra T, Stolze K, Totsche KU (2022) Pathways of biogenically excreted organic matter into soil aggregates. Soil Biol Biochem 164:108483. https://doi.org/10.1016/j.soilbio.2021.108483

Gulati S, Ballhausen M-B, Kulkarni P, Grosch R, Garbeva P (2020) A non-invasive soil-based setup to study tomato root volatiles released by healthy and infected roots. Sci Rep 10(1):12704. https://doi.org/10.1038/s41598-020-69468-z

Gupta A, Bano A, Rai S, Mishra R, Singh M, Sharma S, Pathak N (2022) Mechanistic insights of plant-microbe interaction towards drought and salinity stress in plants for enhancing the agriculture productivity. Plant Stress 4:100073. https://doi.org/10.1016/j.stress.2022.100073

Hu L, Robert CAM, Cadot S, Zhang X, Ye M, Li B, Manzo D, Chervet N, Steinger T, Van Der Heijden MGA, Schlaeppi K, Erb M (2018) Root exudate metabolites drive plant-soil feedbacks on growth and defense by shaping the rhizosphere microbiota. Nat Commun 9(1):2738. https://doi.org/10.1038/s41467-018-05122-7

Huang R, Feng H, Xu Z, Zhang N, Liu Y, Shao J, Shen Q, Zhang R (2022) Identification of adhesins in plant beneficial rhizobacteria *Bacillus velezensis* SQR9 and their effect on root colonization. MPMI 35(1):64–72. https://doi.org/10.1094/MPMI-09-21-0234-R

Iannucci A, Fragasso M, Beleggia R, Nigro F, Papa R (2017) Evolution of the crop rhizosphere: impact of domestication on root exudates in tetraploid wheat (*Triticum turgidum* L.). Front Plant Sci 8:2124. https://doi.org/10.3389/fpls.2017.02124

Jha S, Singh AA, Thakur N (2022) Role of exopolysaccharide and biofilms in microorganisms for alleviating salt stress. In: Mitigation of plant abiotic stress by microorganisms. Elsevier, pp 205–230

Jiang Z, Thakur MP, Liu R, Zhou G, Zhou L, Fu Y, Zhang P, He Y, Shao J, Gao J, Li N, Wang X, Jia S, Chen Y, Zhang C, Zhou X (2022) Soil P availability and mycorrhizal type determine root exudation in sub-tropical forests. Soil Biol Biochem 171:108722. https://doi.org/10.1016/j.soilbio.2022.108722

Jousset A, Rochat L, Lanoue A, Bonkowski M, Keel C, Scheu S (2011) Plants respond to pathogen infection by enhancing the antifungal gene expression of root-associated bacteria. MPMI 24(3):352–358. https://doi.org/10.1094/MPMI-09-10-0208

Kang J, Park J, Choi H, Burla B, Kretzschmar T, Lee Y, Martinoia E (2011) Plant ABC transporters. Arabidopsis Book 9:e0153. https://doi.org/10.1199/tab.0153

Kashyap S, Agarwala N, Sunkar R (2024) Understanding plant stress memory traits can provide a way for sustainable agriculture. Plant Sci 340:111954. https://doi.org/10.1016/j.plantsci.2023.111954

Kim DH, Kaashyap M, Rathore A, Das RR, Parupalli S, Upadhyaya HD, Gopalakrishnan S, Gaur PM, Singh S, Kaur J, Yasin M, Varshney RK (2014) Phylogenetic diversity of Mesorhizobium in chickpea. J Biosci 39(3):513–517. https://doi.org/10.1007/s12038-014-9429-9

Knežević M, Buntić A, Delić D, Stajković-Srbinović O (2022) Root nodule bacteria-rhizobia: exploring the beneficial effects on non-legume plant growth. In: Maheshwari DK, Dobhal R, Dheeman S (eds) Nitrogen fixing bacteria: sustainable growth of non-legumes. Springer Nature Singapore, Singapore, pp 129–168

Köllner TG, Held M, Lenk C, Hiltpold I, Turlings TCJ, Gershenzon J, Degenhardt J (2008) A maize (*E*)-β-caryophyllene synthase implicated in indirect defense responses against herbivores is not expressed in most American maize varieties. Plant Cell 20(2):482–494. https://doi.org/10.1105/tpc.107.051672

Kong HG, Song GC, Ryu C (2019) Inheritance of seed and rhizosphere microbial communities through plant–soil feedback and soil memory. Environ Microbiol Rep 11(4):479–486. https://doi.org/10.1111/1758-2229.12760

Kumar N, Haldar S, Saikia R (2023) Root exudation as a strategy for plants to deal with salt stress: an updated review. Environ Exp Bot 216:105518. https://doi.org/10.1016/j.envexpbot.2023.105518

Kumar GA, Kumar S, Bhardwaj R, Swapnil P, Meena M, Seth CS, Yadav A (2024) Recent advancements in multifaceted roles of flavonoids in plant–rhizomicrobiome interactions. Front Plant Sci 14:1297706. https://doi.org/10.3389/fpls.2023.1297706

Kushwaha P, Kashyap PL, Kuppusamy P (2020) Microbes for cold stress resistance in plants: mechanism, opportunities, and challenges. In: Goel R, Soni R, Suyal DC (eds) Microbiological advancements for higher altitude agro-ecosystems & sustainability. Springer Singapore, Singapore, pp 269–292

Lakshmanan V, Castaneda R, Rudrappa T, Bais HP (2013) Root transcriptome analysis of *Arabidopsis thaliana* exposed to beneficial *Bacillus subtilis* FB17 rhizobacteria revealed genes for bacterial recruitment and plant defense independent of malate efflux. Planta 238(4):657–668. https://doi.org/10.1007/s00425-013-1920-2

Lamichhane JR, Barbetti MJ, Chilvers MI, Pandey AK, Steinberg C (2024) Exploiting root exudates to manage soil-borne disease complexes in a changing climate. Trends Microbiol 32(1):27–37. https://doi.org/10.1016/j.tim.2023.07.011

Lefèvre F, Fourmeau J, Pottier M, Baijot A, Cornet T, Abadía J, Álvarez-Fernández A, Boutry M (2018) The *Nicotiana tabacum* ABC transporter NtPDR3 secretes O-methylated coumarins in response to iron deficiency. J Exp Bot 69(18):4419–4431. https://doi.org/10.1093/jxb/ery221

Leventis G, Tsiknia M, Feka M, Ladikou EV, Papadakis IE, Chatzipavlidis I, Papadopoulou K, Ehaliotis C (2021) Arbuscular mycorrhizal fungi enhance growth of tomato under normal and drought conditions, via different water regulation mechanisms. Rhizosphere 19:100394. https://doi.org/10.1016/j.rhisph.2021.100394

Li G, Wang K, Qin Q, Li Q, Mo F, Nangia V, Liu Y (2023a) Integrated microbiome and metabolomic analysis reveal responses of rhizosphere bacterial communities and root exudate composition to drought and genotype in rice (*Oryza sativa* L.). Rice 16(1):19. https://doi.org/10.1186/s12284-023-00636-1

Li J-H, Muhammad Aslam M, Gao Y-Y, Dai L, Hao G-F, Wei Z, Chen M-X, Dini-Andreote F (2023b) Microbiome-mediated signal transduction within the plant holobiont. Trends Microbiol 31(6):616–628. https://doi.org/10.1016/j.tim.2022.12.005

Lin Y, Mei L, Wei Q, Li B, Zhang P, Sun S, Cui G (2022) *Leymus chinensis* resists degraded soil stress by modulating root exudate components to attract beneficial microorganisms. Front Microbiol 13:951838. https://doi.org/10.3389/fmicb.2022.951838

Liu A, Chen S, Wang M, Liu D, Chang R, Wang Z, Lin X, Bai B, Ahammed GJ (2016) Arbuscular mycorrhizal fungus alleviates chilling stress by boosting redox poise and antioxidant poten-

tial of tomato seedlings. J Plant Growth Regul 35(1):109–120. https://doi.org/10.1007/s00344-015-9511-z
Liu W, Sikora E, Park S-W (2020) Plant growth-promoting rhizobacterium, *Paenibacillus polymyxa* CR1, upregulates dehydration-responsive genes, RD29A and RD29B, during priming drought tolerance in arabidopsis. Plant Physiol Biochem 156:146–154. https://doi.org/10.1016/j.plaphy.2020.08.049
Liu X, Chai J, Zhang Y, Zhang C, Lei Y, Li Q, Yao T (2022) Halotolerant rhizobacteria mitigate the effects of salinity stress on maize growth by secreting exopolysaccharides. Environ Exp Bot 204:105098. https://doi.org/10.1016/j.envexpbot.2022.105098
Lone R, Baba SH, Khan S, Al-Sadi AM, Kamili AN (2023) Phenolics: key players in interaction between plants and their environment. In: Lone R, Khan S, Mohammed Al-Sadi A (eds) Plant phenolics in abiotic stress management. Springer Nature Singapore, Singapore, pp 23–46
López-Farfán D, Reyes-Darias JA, Matilla MA, Krell T (2019) Concentration dependent effect of plant root exudates on the chemosensory systems of *Pseudomonas putida* KT2440. Front Microbiol 10:78. https://doi.org/10.3389/fmicb.2019.00078
Lu Y, Zhang X, Ma M, Zu W, Kronzucker HJ, Shi W (2022) Syringic acid from rice as a biological nitrification and urease inhibitor and its synergism with 1,9-decanediol. Biol Fertil Soils 58(3):277–289. https://doi.org/10.1007/s00374-021-01584-y
Ma W, Tang S, Dengzeng Z, Zhang D, Zhang T, Ma X (2022) Root exudates contribute to belowground ecosystem hotspots: a review. Front Microbiol 13:937940. https://doi.org/10.3389/fmicb.2022.937940
Mahto KU, Vandana, Priyadarshanee M, Samantaray DP, Das S (2022) Bacterial biofilm and extracellular polymeric substances in the treatment of environmental pollutants: beyond the protective role in survivability. J Clean Prod 379:134759. https://doi.org/10.1016/j.jclepro.2022.134759
Maitra P, Hrynkiewicz K, Szuba A, Jagodziński AM, Al-Rashid J, Mandal D, Mucha J (2024) Metabolic niches in the rhizosphere microbiome: dependence on soil horizons, root traits and climate variables in forest ecosystems. Front Plant Sci 15:1344205. https://doi.org/10.3389/fpls.2024.1344205
Mandal M, Chatterjee S, Majumdar S (2022) Outside the cell surface: encoding the role of exopolysaccharide producing rhizobacteria to boost the drought tolerance in plants. In: Roy S, Mathur P, Chakraborty AP, Saha SP (eds) Plant stress: challenges and management in the new decade. Springer, Cham, pp 295–310
Martín-Robles N, Lehmann A, Seco E, Aroca R, Rillig MC, Milla R (2018) Impacts of domestication on the arbuscular mycorrhizal symbiosis of 27 crop species. New Phytol 218(1):322–334. https://doi.org/10.1111/nph.14962
Massalha H, Korenblum E, Tholl D, Aharoni A (2017) Small molecules below-ground: the role of specialized metabolites in the rhizosphere. Plant J 90(4):788–807. https://doi.org/10.1111/tpj.13543
McLaughlin S, Zhalnina K, Kosina S, Northen TR, Sasse J (2023) The core metabolome and root exudation dynamics of three phylogenetically distinct plant species. Nat Commun 14(1):1649. https://doi.org/10.1038/s41467-023-37164-x
Mitra D, Guerra Sierra BE, Khoshru B, De Los Santos Villalobos S, Belz C, Chaudhary P, Shahri FN, Djebaili R, Adeyemi NO, El-Ballat EM, El-Esawi MA, Moradi S, Mondal R, Senapati A, Panneerselvam P, Mohapatra PKD (2021) Impacts of arbuscular mycorrhizal fungi on rice growth, development, and stress management with a particular emphasis on strigolactone effects on root development. Commun Soil Sci Plant Anal 52(14):1591–1621. https://doi.org/10.1080/00103624.2021.1892728
Mohapatra R, Naik AKD, Kesh R, Mohanta V, Sahu BB (2024) Exploring plant–microbe interaction in the postgenomic era. In: Microbial diversity in the genomic era. Elsevier, pp 609–624
Mueller CW, Baumert V, Carminati A, Germon A, Holz M, Kögel-Knabner I, Peth S, Schlüter S, Uteau D, Vetterlein D, Teixeira P, Vidal A (2024) From rhizosphere to detritusphere—soil structure formation driven by plant roots and the interactions with soil biota. Soil Biol Biochem 193:109396. https://doi.org/10.1016/j.soilbio.2024.109396

Mulat M, Anand R, Khan F (2020) Defensive role of plant-derived secondary metabolites: indole and its' derivatives. CBIOT 9(2):78–88. https://doi.org/10.2174/2211550109999200728153839

Mutch LA, Young JPW (2004) Diversity and specificity of *Rhizobium leguminosarum* biovar *viciae* on wild and cultivated legumes. Mol Ecol 13(8):2435–2444. https://doi.org/10.1111/j.1365-294X.2004.02259.x

Nazari M (2021) Plant mucilage components and their functions in the rhizosphere. Rhizosphere 18:100344. https://doi.org/10.1016/j.rhisph.2021.100344

Nazari M, Bilyera N, Banfield CC, Mason-Jones K, Zarebanadkouki M, Munene R, Dippold MA (2023) Soil, climate, and variety impact on quantity and quality of maize root mucilage exudation. Plant Soil 482(1–2):25–38. https://doi.org/10.1007/s11104-022-05669-x

Ning X, Lin M, Huang G, Mao J, Gao Z, Wang X (2023) Research progress on iron absorption, transport, and molecular regulation strategy in plants. Front Plant Sci 14:1190768. https://doi.org/10.3389/fpls.2023.1190768

Olanrewaju OS, Babalola OO (2022) The rhizosphere microbial complex in plant health: a review of interaction dynamics. J Integr Agric 21(8):2168–2182. https://doi.org/10.1016/S2095-3119(21)63817-0

Oppenheimer-Shaanan Y, Jakoby G, Starr ML, Karliner R, Eilon G, Itkin M, Malitsky S, Klein T (2022) A dynamic rhizosphere interplay between tree roots and soil bacteria under drought stress. elife 11:e79679. https://doi.org/10.7554/eLife.79679

Orozco-Mosqueda MADC, Fadiji AE, Babalola OO, Glick BR, Santoyo G (2022) Rhizobiome engineering: unveiling complex rhizosphere interactions to enhance plant growth and health. Microbiol Res 263:127137. https://doi.org/10.1016/j.micres.2022.127137

Ourry M, Lebreton L, Chaminade V, Guillerm-Erckelboudt A-Y, Hervé M, Linglin J, Marnet N, Ourry A, Paty C, Poinsot D, Cortesero A-M, Mougel C (2018) Influence of belowground herbivory on the dynamics of root and rhizosphere microbial communities. Front Ecol Evol 6:91. https://doi.org/10.3389/fevo.2018.00091

Paddock K, Robert C, Erb M, Hibbard B (2021) Western corn rootworm, plant and microbe interactions: a review and prospects for new management tools. Insects 12(2):171. https://doi.org/10.3390/insects12020171

Pantigoso HA, Newberger D, Vivanco JM (2022) The rhizosphere microbiome: plant–microbial interactions for resource acquisition. J Appl Microbiol 133(5):2864–2876. https://doi.org/10.1111/jam.15686

Parasar BJ, Sharma I, Agarwala N (2024) Root exudation drives abiotic stress tolerance in plants by recruiting beneficial microbes. Appl Soil Ecol 198:105351. https://doi.org/10.1016/j.apsoil.2024.105351

Park Y-S, Ryu C-M (2021) Understanding plant social networking system: avoiding deleterious microbiota but calling beneficials. IJMS 22(7):3319. https://doi.org/10.3390/ijms22073319

Pascale A, Proietti S, Pantelides IS, Stringlis IA (2020) Modulation of the root microbiome by plant molecules: the basis for targeted disease suppression and plant growth promotion. Front Plant Sci 10:1741. https://doi.org/10.3389/fpls.2019.01741

Pérez-Jaramillo JE, Mendes R, Raaijmakers JM (2016) Impact of plant domestication on rhizosphere microbiome assembly and functions. Plant Mol Biol 90(6):635–644. https://doi.org/10.1007/s11103-015-0337-7

Poudel M, Mendes R, Costa LAS, Bueno CG, Meng Y, Folimonova SY, Garrett KA, Martins SJ (2021) The role of plant-associated bacteria, fungi, and viruses in drought stress mitigation. Front Microbiol 12:743512. https://doi.org/10.3389/fmicb.2021.743512

Preece C, Peñuelas J (2020) A return to the wild: root exudates and food security. Trends Plant Sci 25(1):14–21. https://doi.org/10.1016/j.tplants.2019.09.010

Prescott CE, Grayston SJ, Helmisaari H-S, Kaštovská E, Körner C, Lambers H, Meier IC, Millard P, Ostonen I (2020) Surplus carbon drives allocation and plant–soil interactions. Trends Ecol Evol 35(12):1110–1118. https://doi.org/10.1016/j.tree.2020.08.007

Quiroga G, Erice G, Ding L, Chaumont F, Aroca R, Ruiz-Lozano JM (2019) The arbuscular mycorrhizal symbiosis regulates aquaporins activity and improves root cell water permeability in maize plants subjected to water stress. Plant Cell Environ 42(7):2274–2290. https://doi.org/10.1111/pce.13551

Rafique S (2023) Physiological and biochemical responses in maize under drought stress. In: Wani SH, Dar ZA, Singh GP (eds) Maize improvement. Springer, Cham, pp 117–136

Rasmann S, Köllner TG, Degenhardt J, Hiltpold I, Toepfer S, Kuhlmann U, Gershenzon J, Turlings TCJ (2005) Recruitment of entomopathogenic nematodes by insect-damaged maize roots. Nature 434(7034):732–737. https://doi.org/10.1038/nature03451

Richardson AE, Kawasaki A, Condron LM, Ryan PR, Gupta VVSR (2021) Root microbiome structure and microbial succession in the rhizosphere. In: Gupta VVSR, Sharma AK (eds) Rhizosphere biology: interactions between microbes and plants. Springer Singapore, Singapore, pp 109–128

Rodrigues M, Ganança JFT, Da Silva EM, Dos Santos TMM, Slaski JJ, Zimny J, Pinheiro De Carvalho MÂA (2019) Evidences of organic acids exudation in aluminium stress responses of two Madeiran wheat (*Triticum aestivum* L.) landraces. Genet Resour Crop Evol 66(4):857–869. https://doi.org/10.1007/s10722-019-00754-0

Rolfe SA, Griffiths J, Ton J (2019) Crying out for help with root exudates: adaptive mechanisms by which stressed plants assemble health-promoting soil microbiomes. Curr Opin Microbiol 49:73–82. https://doi.org/10.1016/j.mib.2019.10.003

Roy SJ, Tucker EJ, Tester M (2011) Genetic analysis of abiotic stress tolerance in crops. Curr Opin Plant Biol 14(3):232–239. https://doi.org/10.1016/j.pbi.2011.03.002

Sabbioni G, Forlani G (2022) The emerging role of proline in the establishment and functioning of legume-rhizobium symbiosis. Front Plant Sci 13:888769. https://doi.org/10.3389/fpls.2022.888769

Sardans J, Lambers H, Preece C, Alrefaei AF, Penuelas J (2023) Role of mycorrhizas and root exudates in plant uptake of soil nutrients (calcium, iron, magnesium, and potassium): has the puzzle been completely solved? Plant J 114(6):1227–1242. https://doi.org/10.1111/tpj.16184

Schulz-Bohm K, Gerards S, Hundscheid M, Melenhorst J, De Boer W, Garbeva P (2018) Calling from distance: attraction of soil bacteria by plant root volatiles. ISME J 12(5):1252–1262. https://doi.org/10.1038/s41396-017-0035-3

Shah AA, Gupta A (2023) Secondary metabolite basis of elicitor- and effector-triggered immunity in pathogen elicitation amid infections. In: Singh R, Kumar N (eds) Genetic manipulation of secondary metabolites in medicinal plant. Springer Nature Singapore, Singapore, pp 225–251

Sharma I, Kashyap S, Agarwala N (2023) Biotic stress-induced changes in root exudation confer plant stress tolerance by altering rhizospheric microbial community. Front Plant Sci 14:1132824. https://doi.org/10.3389/fpls.2023.1132824

Sindhu SS, Sehrawat A, Glick BR (2022) The involvement of organic acids in soil fertility, plant health and environment sustainability. Arch Microbiol 204(12):720. https://doi.org/10.1007/s00203-022-03321-x

Solanki MK, Joshi NC, Singh PK, Singh SK, Santoyo G, Basilio De Azevedo LC, Kumar A (2024) From concept to reality: transforming agriculture through innovative rhizosphere engineering for plant health and productivity. Microbiol Res 279:127553. https://doi.org/10.1016/j.micres.2023.127553

Spooren J, Van Bentum S, Thomashow LS, Pieterse CMJ, Weller DM, Berendsen RL (2024) Plant-driven assembly of disease-suppressive soil microbiomes. Annu Rev Phytopathol. https://doi.org/10.1146/annurev-phyto-021622-100127

Stringlis IA, Yu K, Feussner K, De Jonge R, Van Bentum S, Van Verk MC, Berendsen RL, Bakker PAHM, Feussner I, Pieterse CMJ (2018) MYB72-dependent coumarin exudation shapes root microbiome assembly to promote plant health. Proc Natl Acad Sci USA 115(22). https://doi.org/10.1073/pnas.1722335115

Stringlis IA, De Jonge R, Pieterse CMJ (2019) The age of coumarins in plant–microbe interactions. Plant Cell Physiol 60(7):1405–1419. https://doi.org/10.1093/pcp/pcz076

Tak Y, Kumar M (2020) Phenolics: a key defence secondary metabolite to counter biotic stress. In: Lone R, Shuab R, Kamili AN (eds) Plant phenolics in sustainable agriculture. Springer Singapore, Singapore, pp 309–329

Taye ZM, Noble K, Siciliano SD, Helgason BL, Lamb EG (2022) Root growth dynamics, dominant rhizosphere bacteria, and correlation between dominant bacterial genera and root traits through *Brassica napus* development. Plant Soil 473(1–2):441–456. https://doi.org/10.1007/s11104-022-05296-6

Tian T, Reverdy A, She Q, Sun B, Chai Y (2020) The role of rhizodeposits in shaping rhizomicrobiome. Environ Microbiol Rep 12(2):160–172. https://doi.org/10.1111/1758-2229.12816

Tiku AR (2020) Antimicrobial compounds (phytoanticipins and phytoalexins) and their role in plant defense. In: Mérillon J-M, Ramawat KG (eds) Co-evolution of secondary metabolites. Springer, Cham, pp 845–868

Tilgam J, Sreeshma N, Priyadarshini P, Bhavyasree RK, Choudhury S, Bharati A, Ashajyothi M (2022) Rhizosphere engineering for systemic resistance/tolerance to biotic and abiotic stress. In: Singh UB, Rai JP, Sharma AK (eds) Re-visiting the rhizosphere eco-system for agricultural sustainability. Springer Nature Singapore, Singapore, pp 271–300

Tiwari S, Sharma B, Bisht N, Tewari L (2023) Role of beneficial microbial gene pool in mitigating salt/nutrient stress of plants in saline soils through underground phytostimulating signalling molecules. Pedosphere 33(1):153–171. https://doi.org/10.1016/j.pedsph.2022.06.029

Tiziani R, Miras-Moreno B, Malacrinò A, Vescio R, Lucini L, Mimmo T, Cesco S, Sorgonà A (2022) Drought, heat, and their combination impact the root exudation patterns and rhizosphere microbiome in maize roots. Environ Exp Bot 203:105071. https://doi.org/10.1016/j.envexpbot.2022.105071

Trivedi P, Batista BD, Bazany KE, Singh BK (2022) Plant–microbiome interactions under a changing world: responses, consequences and perspectives. New Phytol 234(6):1951–1959. https://doi.org/10.1111/nph.18016

Unnikrishnan BV, Pradeepkumar T, Sidharthan PP, Mohan M (2023) Microbe mediated abiotic stress tolerance in cucurbitaceous vegetables. Biologia 78(10):2863–2873. https://doi.org/10.1007/s11756-023-01456-w

Upadhyay SK, Srivastava AK, Rajput VD, Chauhan PK, Bhojiya AA, Jain D, Chaubey G, Dwivedi P, Sharma B, Minkina T (2022) Root exudates: mechanistic insight of plant growth promoting rhizobacteria for sustainable crop production. Front Microbiol 13:916488. https://doi.org/10.3389/fmicb.2022.916488

Upadhyay R, Saini R, Shukla PK, Tiwari KN (2024) Role of secondary metabolites in plant defense mechanisms: a molecular and biotechnological insights. Phytochem Rev. https://doi.org/10.1007/s11101-024-09976-2

Van Den Berg N, Swart V, Backer R, Fick A, Wienk R, Engelbrecht J, Prabhu SA (2021) Advances in understanding defense mechanisms in *Persea americana* against *Phytophthora cinnamomi*. Front Plant Sci 12:636339. https://doi.org/10.3389/fpls.2021.636339

Vimal SR, Singh JS, Prasad SM (2024) Crop microbiome dynamics in stress management and green agriculture. In: Microbiome drivers of ecosystem function. Elsevier, pp 341–366

Vives-Peris V, De Ollas C, Gómez-Cadenas A, Pérez-Clemente RM (2020) Root exudates: from plant to rhizosphere and beyond. Plant Cell Rep 39(1):3–17. https://doi.org/10.1007/s00299-019-02447-5

Wahab A, Muhammad M, Munir A, Abdi G, Zaman W, Ayaz A, Khizar C, Reddy SPP (2023) Role of arbuscular mycorrhizal fungi in regulating growth, enhancing productivity, and potentially influencing ecosystems under abiotic and biotic stresses. Plants 12(17):3102. https://doi.org/10.3390/plants12173102

Wang J, Hou Q, Li P, Yang L, Sun X, Benedito VA, Wen J, Chen B, Mysore KS, Zhao J (2017) Diverse functions of multidrug and toxin extrusion (MATE) transporters in citric acid efflux and metal homeostasis in *Medicago truncatula*. Plant J 90(1):79–95. https://doi.org/10.1111/tpj.13471

Wang N, Kong C, Wang P, Meiners SJ (2021a) Root exudate signals in plant–plant interactions. Plant Cell Environ 44(4):1044–1058. https://doi.org/10.1111/pce.13892

Wang P, Chai YN, Roston R, Dayan FE, Schachtman DP (2021b) The *Sorghum bicolor* root exudate sorgoleone shapes bacterial communities and delays network formation. mSystems 6(2):e00749-20. https://doi.org/10.1128/mSystems.00749-20

Williams A, Sinanaj B, Hoysted GA (2024) Plant–microbe interactions through a lens: tales from the mycorrhizosphere. Ann Bot 133(3):399–412. https://doi.org/10.1093/aob/mcad191

Xing X, Koch AM, Jones AMP, Ragone D, Murch S, Hart MM (2012) Mutualism breakdown in breadfruit domestication. Proc R Soc B 279(1731):1122–1130. https://doi.org/10.1098/rspb.2011.1550

Xu L, Coleman-Derr D (2019) Causes and consequences of a conserved bacterial root microbiome response to drought stress. Curr Opin Microbiol 49:1–6. https://doi.org/10.1016/j.mib.2019.07.003

Xu X, Fang P, Zhang H, Chi C, Song L, Xia X, Shi K, Zhou Y, Zhou J, Yu J (2019) Strigolactones positively regulate defense against root-knot nematodes in tomato. J Exp Bot 70(4):1325–1337. https://doi.org/10.1093/jxb/ery439

You MP, Lamichhane JR, Aubertot J-N, Barbetti MJ (2020) Understanding why effective fungicides against individual soilborne pathogens are ineffective with soilborne pathogen complexes. Plant Dis 104(3):904–920. https://doi.org/10.1094/PDIS-06-19-1252-RE

Yuan A, Kumar SD, Wang H, Wang S, Impa S, Wang H, Guo J, Wang Y, Yang Q, Liu XJA, Jagadish Sv K, Shao R (2024) Dynamic interplay among soil nutrients, rhizosphere metabolites, and microbes shape drought and heat stress responses in summer maize. Soil Biol Biochem 191:109357. https://doi.org/10.1016/j.soilbio.2024.109357

Zhao G (2023) Plant compensatory growth: its mechanisms and implications to agricultural sustainability under global environmental changes. Int J Sustain Dev World Ecol:1–13. https://doi.org/10.1080/13504509.2023.2289475

Ziegler J, Schmidt S, Strehmel N, Scheel D, Abel S (2017) Arabidopsis transporter ABCG37/PDR9 contributes primarily highly oxygenated coumarins to root exudation. Sci Rep 7(1):3704. https://doi.org/10.1038/s41598-017-03250-6

Zolobowska L, Van Gijsegem F (2006) Induction of lateral root structure formation on petunia roots: a novel effect of GMI1000 *Ralstonia solanacearum* infection impaired in Hrp mutants. MPMI 19(6):597–606. https://doi.org/10.1094/MPMI-19-0597

Secondary Metabolites in Plants: Role in Combating Environmental Stress

9

Varsha K. Singh, Sapana Jha, Palak Rana, Neha Kumari, Garvita Singh, and Rajeshwar P. Sinha

Abstract

Plants are a source of numerous secondary metabolites (SMs) that are required by them to adapt and survive in the extremities. These metabolites being the by-products of primary metabolites help plants to withstand multiple biotic and abiotic stresses like pathogens, predators, drought, salinity, UV radiation, thermal, temperature, etc. Plant growth and productivity are largely affected by the endogenous levels of these metabolites. The type and endogenous levels of these metabolites depends on several factors like species, genotype, physiology, and environmental factors, etc. SMs comprise of three main groups: phenolics, nitrogen-containing compounds, and terpenes. To combat the detrimental effects of abiotic stressors, plants undergo rapid shifts in SMs biosynthesis and accumulation. As a result, SMs have a strong association with plants' ability to survive environmental stressors. Several chemical compounds, including phytohormones have been utilized as exogenous treatments to improve plant tolerance to various stressors. This chapter highlights the SMs mediated defense responses and details the probable mechanism behind the crosstalk between phytohormones/plant growth regulators (PGRs) and various environmental stresses.

Keywords

Abiotic stress · Defense responses · Environmental stressors · PGRs · SMs

V. K. Singh · S. Jha · P. Rana · N. Kumari · R. P. Sinha (✉)
Laboratory of photobiology and Molecular Microbiology, Centre of Advanced Study in Botany, Institute of Science, Banaras Hindu University, Varanasi, India

G. Singh (✉)
Department of Botany, Gargi College, University of Delhi, Delhi, India
e-mail: garvita.singh@gargi.du.ac.in

U. Pankaj et al. (eds.), *Plant-microbiome Interactions for Climate-resilient Agriculture*, https://doi.org/10.1007/978-981-96-3534-4_9

1 Introduction

Secondary metabolites (SMs) are tiny molecules with a variety of biological functions and chemical configurations. They lack high-energy bonds and are not required for core life processes, in contrast to primary metabolites, which are the primary forces behind vital life functions, such as cell development. SMs perform crucial secondary physiological and biochemical roles that aid in plant survival and fitness, especially in relation to how well plants interact with their surroundings and manage biotic and abiotic stress (Twaij and Hassan 2022). Environmental stressors have a substantial impact on worldwide agricultural productivity and nutritional stability. It is estimated that 90% of cropping regions will be affected by factors like as drought, salt, temperature, heavy metals (HMs), and flooding extremities, which can result in average yield losses of up to 70% (Leisner 2020; Joshi et al. 2021; Nazir et al. 2021; Ngugi et al. 2022). Although the SMs produced by plants are produced in trace amounts and do not significantly contribute to the continuation of fundamental plant life activities, they are essential for the adaptation of plants to unfavorable environments (Aguirre-Becerra et al. 2021). Alkaloids, flavonoids, anthocyanins, lignans, quinones, peptides, phenolics, terpenoids, and amines are just a few examples of the many categories of molecules that make up SMs. These compounds are mostly utilized in biopesticides, agrochemicals, food additives, and perfumes (Patil 2020). Moreover, they have a direct or indirect relationship to the flavor, color, and aroma of plants. SMs in plants are often produced by various metabolite pathways and play a direct role in the defense response of plants. The physiology and growth conditions of a plant species determine how much of SMs are produced, and this is mostly because varied environmental growth conditions have different effects on the metabolic pathways linked to their biosynthesis. Numerous studies have demonstrated the wide range of ecological roles that are attributed to plant secondary metabolism, including defense against pathogen, insect, and herbivore attacks as well as resistance against environmental stresses (Pang et al. 2021). Recent developments in SMs' study of plant stress physiology indicate that it may be involved in reducing the harmful effects of a variety of stressors. Apart from their defensive roles, SMs also serve as molecules that shield primary metabolites like proteins and nucleic acids from damage caused by stress (Al-Khayri et al. 2023). Consequently, physiological changes occur in the plant body, such as adjustments to secondary metabolism, ion and water balance, reduction of oxidative damage, etc., which can either directly or indirectly produce the phenotypic response of stress resistance. Depending on how many and what kinds of genes are up- or down-regulated, the stress response may potentially result in growth inhibition or cell death (Salam et al. 2023; Sun and Fernie 2023). But in the majority of instances, secondary metabolism adjustment plays a significant role in defensive mechanisms, either by boosting or lowering their production in the plant body. This chapter covers the role of SMs in the regulation of a variety of biotic and abiotic stressors in plants.

2 SMs: Types and Diversity

SMs are generally produced by all plants for defense, communication, and stress mediation (Isah 2019). For instance, VOCs are defensive chemicals produced by plants that possess antibacterial and insect-repelling properties (Kesselmeier and Staudt 1999). With more than 391,000 plant species known worldwide (Willis 2017), there is still room for new discoveries despite the identification of more than 200,000 PSMs (Neilson et al. 2013). Certain PSMs are unique to closely related plant species, and depending on plant ontogeny and tissue type, their concentrations can differ between populations and individual plants (Acamovic and Brooker 2005). These variations in SMs can be a result of genetic variability, but environmental abiotic factors (growth conditions) such as heat stress, drought, UV radiation, and O_3- which are predicted to become more intense with climate change as well as herbivore and pathogen attacks have an impact on their concentrations (Ryde et al. 2021). PSMs have been broadly classified into three major groups based on a chemical structure and biosynthetic pathway: (i) phenolic compounds (flavonoids, phenolic acids, lignin, lignans, coumarins, stilbenes, and tannins); (ii) terpenoids (plant volatiles, sterols, carotenoids, saponins, and glycosides); and (iii) nitrogen-containing compounds (alkaloids, glucosinolates, and cyanogenic glycosides) (Fig. 9.1) (Jogawat et al. 2021; Yadav et al. 2021).

2.1 Phenolic Compounds

Among the phenolic group compounds, flavonoids are the most abundant ones while stilbenes and lignans are the rarest ones. Flavonoids are a broad class of SMs

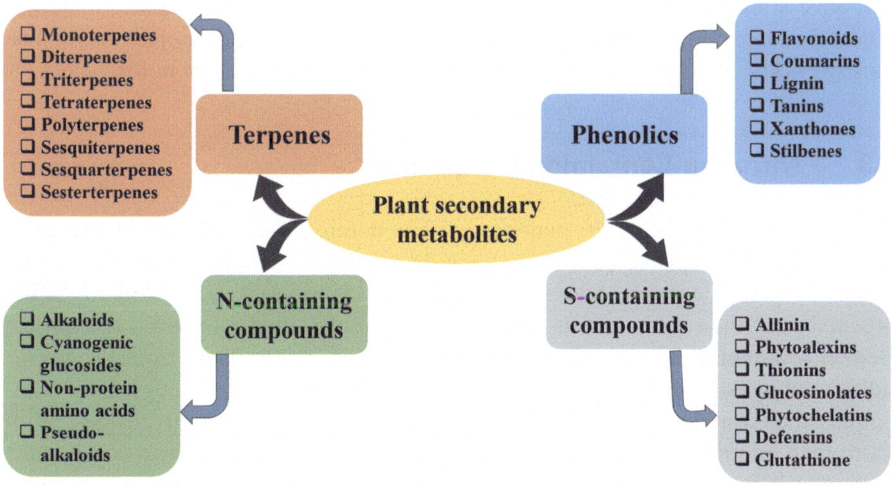

Fig. 9.1 The major types of secondary metabolites produced by plants

that serve a variety of purposes, one of which is stress defense. The seven subgroups of flavonoids are flavones, flavonols, flavanones, isoflavonoids, flavan-3-ols or catechins, and anthocyanins. These subgroups are classified according to the degree of oxidation and unsaturation of their C-ring as well as the carbon C-ring to which their B-ring is linked (Panche et al. 2016). Flavones have a ketone functional group at position 4 of the C-ring and a double bond between positions 2 and 3. In contrast, flavonols occasionally undergo glycosylation and include a hydroxy group at position 3 of the C-ring. Flavonones, in contrast to flavones, have a double bond saturated between positions 2 and 3 of the C-ring. Although the exact role of phenolics in protecting plants is unknown, it is thought that they do so by scavenging free radicals and filtering harmful UV radiation (Agati and Tattini 2010; Takahashi and Ohnishi 2004). The most well-known UV-B attenuators in plants are ferulic acid, caffeic acid, and p-coumaric acid (derivatives of hydroxycinnamic acid) (Harborne and Williams 2000).

2.2 Terpenoids

One of the most structurally varied naturally occurring PSMs is terpenoids, also known as isoprenoids. Their primary structural component is a five-carbon isopentyl unit known as 2-methyl-1, 3-butadiene, or isoprene. While terpenoids have additional functional groups like hydroxyl and heterocyclic rings, terpenes exclusively contain isoprene units. Terpenoids can be divided into two categories based on their structural makeup: cyclic terpenoids (like limonene) and aliphatic terpenoids (like geraniol). In plants, a number of terpenoids have demonstrated protective functions against biotic and abiotic stressors. Dahham et al. (2015) and Porres-Martínez et al. (2016) documented the antioxidant properties of terpenes (monoterpenes 1,8-cineole and α-pinene) and sesquiterpene b-caryophyllene, indicating their role in mitigating oxidative stress caused by external stimuli. By promoting photorespiration, terpenoids are also said to shield plants from oxidative stress and photodamage. The most well-known terpenoids involved in photoprotection are carotenoids.

2.3 Nitrogen-Containing Compounds

Alkaloids, which make up the majority of plant defense chemicals, are known to be present in 20% of plant species and include one or more nitrogen atoms that are obtained by the decarboxylation of amino acids (Song et al. 2014). Based on the precursors of their amino acids, alkaloids can be classified into seven categories: Benzylisoquinoline, indolequinoline, quinolizidine, piperidine, tropane, pyrrolidine, and pyrrolizidine alkaloids (Kurek 2019). Other categories of defensive compounds that contain nitrogen are glucosinolates and cyanogenic glycosides. With more than a hundred chemicals identified from each group, these two compounds share similarities with amino acid precursors but differ greatly in terms of structural diversity (Jamwal et al. 2018).

3 SMs-Mediated Plant Defense Responses

3.1 Abiotic Stress

Several metabolites are produced by plants to overcome the stressful restraints which arise during adverse environmental conditions (Edreva et al. 2008). Various SMs are synthesized from primary molecules and are amassed in cells of plants. In vitro production of these metabolites can be done by providing treatment of abiotic factors and signaling molecules to the cell cultures (Ramakrishna et al. 2011; Rejeb et al. 2014; Caretto et al. 2015; Narayani and Srivastava 2018). The SMs produced by plants are involved in several functions as they protect plants from various environmental stresses and provide defense against insects, pathogens, and herbivores (Hartmann 2004; Freeman and Beattie 2008; Kim et al. 2010). They also play an important role as antioxidant and anti-radical molecules to cope up with oxidative stress induced in the adverse environmental conditions by involving compounds having hydroxyl or thiol group. Various compounds such as ascorbic acid, lipoic acid, flavonoids having o-dihydroxy group like arylamines, aliphatic and unsaturated fatty acids, quercetin, carotenoids (Edreva 2005), SMs play significant role during particular stress conditions in plant cells (Mazid et al. 2011; Koricheva and Barton 2012; Moore et al. 2014; Rejeb et al. 2014; Gobbo-Neto et al. 2017). Sesquiterpenes are linked with defense and are found in members of solanaceae family, glucosinolates-myrosinase are found in brassicaceae family, stilbenes are related with Vitaceae family, isoflavones are produced by Fabaceae family and limonoids are found in members of Rutaceae and Meliaceae (Freeman and Beattie 2008; Goyal et al. 2012).

3.1.1 Drought

Drought is an important abiotic stress that affects growth and development of plant and causes photoinhibition and temperature stresses. Various physiology and biochemistry of plants are altered such as cell growth are arrested, photosynthesis with increased respiration (Mashilo et al. 2017). Due to the introduction of ionic or osmotic stress, plants exposed to drought demonstrated increased synthesis of many kinds of SMs (SMs), such as terpenes, complex phenols, and alkaloids, during both in vitro and in vivo growth (de Matos Nunes et al. 2014; Niinemets 2015; Quan et al. 2016; Afzal et al. 2017; Piasecka et al. 2017). When cultivated under drought stress, the concentration and quantity of phenolic compounds biosynthesized in *Hypericum brasilience* and *Pisum sativum* were much higher than in the control plants (Nogués et al. 1998; de Abreu and Mazzafera 2005). *Salvia officinalis* produce terpenes and loss of more biomass was related with increased number of monoterpenes (Nowak et al. 2010). Production of flavonoids was increased during oxidative stress (Nakabayash et al. 2014) and provides protection to the plants which were exposed to toxic metals like aluminum (Sharma and Dubey 2005; Poschenrieder et al. 2008; González-Santana et al. 2012; Nakabayashi et al. 2014; Al Hassan et al. 2017; Bojórquez-Quintal et al. 2017), and tocopherol, shikonin, and digitoxin were biosynthesized in plant cell cultures which were treated with Fe^{2+}, cadmium, Ca^{2+} and its chloride, and $MnSO_4$ (Cetin et al. 2014; Yazaki 2017). Increased biosynthesis of SMs was seen in leaves of willow plants during drought condition (Larson 1988) and biosynthesis of saponins was decreased in *Chenopodium quinoa* when grown under water deficient condition (Soliz-Guerrero et al. 2002). Because of the increasing stress conditions, the strength of the culture medium nutrients affected the profile of phenolic compounds in *Bellis perennis* callus via inducing the antioxidant system (Cingöz and Karakaş 2016). $AgNO_3$ or $CdCl_2$ stress increased the amount of tropane alkaloids such as scopolamine and hyoscyamine in *Brugmansia candida* treated hairy root cultures (Angelova et al. 2006).

3.1.2 Temperature

Low temperature is one of the detrimental abiotic stimuli that plants in temperate climates must withstand. Species that are acclimated to this situation modify their metabolic pathways to produce higher concentrations of cryo-protectants, which are necessary for tolerance in the fall (Karwasara and Dixit 2013). *Helicteres isora* callus cultures demonstrated a synergistic effect of light and temperature on the formation of SMs (Joshi 2015). It has been demonstrated that when beans and tobacco are exposed to heat shock and water stress, polyamines are produced and then phenylamides are formed. The phenylamides are then thought to have a ROS-scavenging role during the stress (Edreva et al. 2008). The Brassicaceae members' synthesis of carotenoids was somewhat reduced by thermal treatments (Morison and Lawlor 1999), while in *Panax quinquefolius*, high levels encouraged root SM and leaf senescence (Jochum et al. 2007). While trees that adapted to a cold climate produced significant amounts of chlorogenic acid, cold stress encouraged the development of phenolics and their subsequent inclusion in plant cell walls as suberin or lignin (Griffith and Yaish 2004; Pérez-Ilzarbe et al. 1997). *Eleutherococcus senticosus* somatic embryos were exposed to both low and high temperatures, which had an impact on the formation of SMs. The higher temperature caused oxidative stress, which was more pronounced than the lower temperature (Shohael et al. 2006). The generation of ginsenosides in the hairy root culture of Panax ginseng was impacted by temperature and light quality whereas *Melastoma malabathricum* cell cultures cultivated at low temperatures yielded greater levels of anthocyanins and biomass than those grown at higher temperatures (Yu et al. 2005; Chan et al. 2010). Antioxidant enzyme activity revealed that temperature changes during incubation affected the formation of SMs in *Brassica napus* callus cultures via causing oxidative stress (Hura et al. 2015). In the hairy root cultures of *Silybum marianum*, it affected the accumulation of flavolignans; when paired with the treatments, acid pH proved to be the most effective (Rahimi and Hasanloo 2016). An increase in incubation temperature had an impact on the development of hairy root cultures of *Stevia rebaudiana*, but it also partially increased stevioside synthesis (Kumari and Chandra 2016).

3.1.3 Light

UV-A, UV-B, photosynthetic active light, and infrared radiation make up the solar radiation that reaches the earth's surface; however, plants only use a small portion of UV-B, which is the most energetic component of the daytime light spectrum, for growth and development. This is dependent on the exposure wavelength and how the radiation interacts with environmental signals (Caldwell et al. 2003). Depending on the species or genotype, stage of development, kind of light, and duration of exposure, light is also considered one of the limiting elements that affects plant growth and development in both in vivo and in vitro environments. This can have an impact on the synthesis of SMs (Ghosh et al. 2018; Isah and Umar 2018). In *Plantago lanceolata* progeny, phenotypic plasticity related to light and nutritional conditions affected the accumulation of iridoid glycosides and biomass (Miehe-Steier et al. 2015). UV-B light exposure in *Catharanthus roseus* had a major influence on the production of vincristine and vinblastine, which are currently used for anti-lymphoma and anti-leukemia medications (Binder et al. 2009). The biomass and CPT content production of *C. acuminata* seedlings were impacted by the intensity and duration of light exposure, as demonstrated by the expression of genes involved in its biosynthesis (Hu et al. 2016). In *Scutellaria lateriflora* shoot cultures, light conditions had a significant impact on SM; blue light was shown to be the most effective, and a link with plant growth regulators (PGRs) was identified (Kawka et al. 2017). Based on the time of growth of *Artemisia absinthium* cell suspension cultures, light and dark conditions of incubation have shown a significant impact on biomass and SMs production (Ali and Abbasi 2014). In callus cultures of *Zingiber officinale*, it promoted the formation of zingiberol and zingiberene (Anasori and Asghari 2009), while in *C. acuminata* seedlings, it affected CPT biosynthesis (Dai et al. 2004; Hu et al. 2016). Light irradiation altered the biosynthesis of artemisinin in the hairy root cultures of *Artemisia annua*, while white light had an impact on the formation of taxol and baccatin III in the cell cultures of *Taxus cuspidata* (Fett-Neto et al. 1995). When *Eurycoma longifolia* calli were exposed to UV light, compact calli with higher alkaloids biosynthesis levels were produced compared to the control (Parikrama and Esyanti 2014).

3.1.4 Salinity

Drought can also trigger salinity-induced stress that triggers the development of secondary metabolic pathways by causing solutes to accumulate at greater concentrations through osmotic adjustment. This might result in water loss from osmotic stress brought on by salt buildup at greater concentrations in the soil and plants (Xu et al. 2016; Munns 2002; Adak et al. 2019). Physiological alterations that disrupt the functional integrity of membranes, redox equilibrium, and nutritional balance, ultimately impacting primary metabolism, which supplies building blocks for SM pathways, and stomatal function (Chaves et al. 2002; Shabala and Cuin 2007; Gupta and Huang 2014; Xu et al. 2016). Commonly occurring physiological alterations include salt exclusion and sequestration, resistance to accumulated ions and limited K^+ loss, control over water homeostasis and osmotic adjustment, and regulation of growth and enlargement through biochemical expression (Sanchez et al. 2008; Munns and Tester 2008; Hamanishi et al. 2015). For extremophiles that have acclimated to saline growth circumstances, this pre-adaptation involves raising the amounts of SMs biosynthesis triggered by salinity stress; a drop in these levels and salt stress has an adverse effect on physiological functions.

A considerable amount of PSMs, categorized as alkaloids, flavonoids, phenolics, and steroids and have been documented to be generated, engaged in, or activated in defense response function in response to cellular stress, which is regulated by the salty conditions of plant development physiology (Harborne 1999; Bourgaud et al. 2001; Sytar et al. 2018). Salinity stress and free-radical scavengers are thought to mediate the production of aromatic compounds (such as alkaloids, isoprenoids, and phenols) and phenylpropanoids-derived compounds (such as tannins, flavonoids, and hydroxycinnamate esters) at higher levels in SMs-producing plants (Sytar et al. 2018; Selmar 2008). Important physiological changes that affect a plant species' ability to survive and produce metabolites in saline growth conditions in vitro or in vivo include osmotic adjustment, which involves the synthesis and accumulation of soluble sugars (such as glutamate, sorbitol, mannitol, oligosaccharides, fructans, and sucrose) and cellular osmolytes (such as polyols, proline, sugar alcohols, pinitol, glucosinolates, and glycine betaine, among others) and ovaryl osmolytes (Parihar et al. 2015; Omamt et al. 2006; Sujata and Kshitija 2013). The buildup of cellular osmolytes, such as polyamines, proline, soluble sugars, and glycine betaine, is one of the biochemical indicators for salinity stress tolerance. This is partially due to their function in preserving the integrity of the membrane and other cellular structures (Parihar et al. 2015; Martinez et al. 2005). Many plant species manufacture phenolics to protect themselves from biotic or abiotic stressors during growing conditions, and the accumulation of these compounds is correlated with the antioxidant capacity of a variety of plant species (Yi et al. 2010; Fu et al. 2011; Abideen et al. 2015). Stressful salinity environments can activate SM, which protects cells from oxidative damage brought on by ion buildup at the cellular and subcellular levels. This lessens the toxicity of SMs (Hossain et al. 2017). Salinity increased the biosynthesis of proline and solasodine in calli of *Solanum nigrum* for the tolerance in Table 9.1 (Šutković et al. 2011).

3.2 Biotic Stresses

Due to their sessile nature, plants are extremely vulnerable to environmental stressors. In addition to abiotic stressors that have been explained in the previous sections of this chapter, plants are susceptible to various biotic stress as well. The various biotic factors that can have detrimental effect on plant growth includes pathogens microbes (bacteria, fungi, viruses), pests (nematodes, arthropods etc.), parasitic plants, weeds and the attack of various herbivores (Dresselhaus and Huckelhoven 2018; Ashapkin et al. 2020). Plants have evolved highly advanced defensive mechanisms to counteract the detrimental impacts of biotic and environmental stresses (Fig. 9.2). Certain plants, like those found in deserts, have a limited lifespan and are ephemeral; whereas others, are vulnerable and perish as soon as they come into contact with these unfavorable circumstances. Those that exhibit resistance adapt complex, tightly controlled systems to protect themselves from a variety of pressures. These organisms have evolved primary defense mechanisms that involve modifications in photosynthetic rates, stomatal conductance, transpiration, cell wall

Table 9.1 Application of secondary metabolites produced by different plant families on exposure to different stress conditions

Environmental stress condition	Plant species family	Metabolite produced	Class of secondary metabolite	Action of metabolite/ pharmaceutical application	References
Cold stress	*Catharanthus roseus* (Apocynaceae)	Vindoline	Alkaloids	Antidiabetic	Goboza et al. (2019)
Herbivory (*Helicoverpa armigera*)	*Cajanus platycarpus* (Fabaceae)	t-ferulic acid	Flavonoids	Resistance to herbivory	Al-Khayr et al. (2023)
Herbivory	*Chrysanthemum indicum* (Asteraceae)	Pyrethrins	Terpenes	Neurotoxic to parasitic and herbivores	Tiku (2018)
Fungal infection (*Microcyclus ulei*)	*Hevea brasiliensis* (Euphorbiaceae)	Scopoletin	Terpenes	Fungistatic effect	Lemos et al. (2020)
Environmental competition	*Artemesia annua* (Asteraceae)	Artemisinin	Sesquiterpene lactones	Antiparasitic/allelopathic herbicide	Grech-Baran and Pietrosiuk (2012)
Cold stress	*Withania somnifera* (Solanaceae)	Withanolide A, withaferin A	Terpenes	Neuroprotective and anticancerous	Hahm et al. (2011)
Drought	*Amaranthus tricolor* (Amaranthaceae)	Hydroxycinnamic acids	Phenolics	Antisickling activity	Akojie and Fung (1992)
Salinity stress	*Capsicum* sp. Solanaceae	Capsaicin	Capsaicinoids	Analgesic	Othman et al. (2011)
Herbivory (*Plutella xylostella*)	*Brassica oleracea* (Cruciferae)	β-ocimene	Monoterpenes	Attraction of parasitoids	Nouri-Ganbalani et al. (2018)
Heat stress	*Solanum lycopersicon* (Solanaceae)	β-phellandrene	Terpenoids	Antifungal	Zhang et al. (2017)

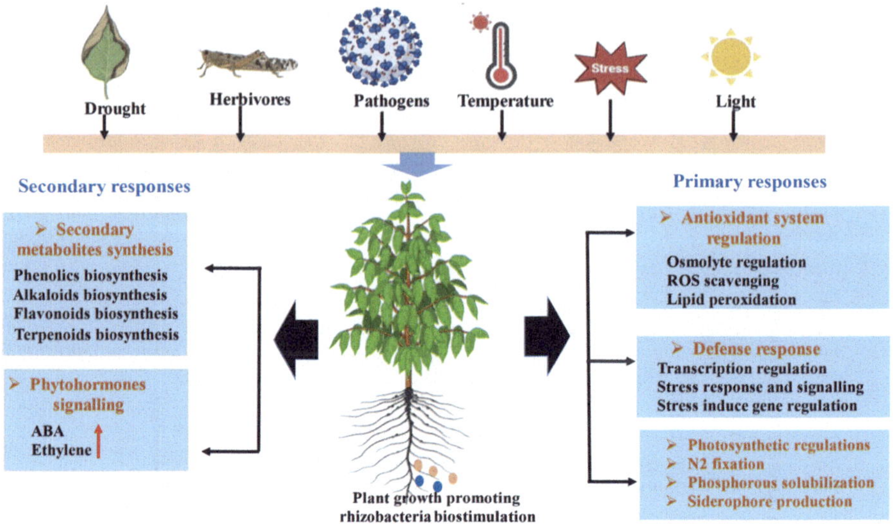

Fig. 9.2 Plant on exposure to various biotic and abiotic stress, such as high temperature, scarcity of water, high intensity of light, and other stressors including microorganisms, herbivores pathogens, etc. produces various kinds of defence responses. This fig depicts different defence responses exhibited by the plant in response to various biotic and abiotic stress

architecture, membrane structure remodeling, cell cycle, and division rates. These adaptations have a cumulative effect on general growth and are essential for fine-tuning the physiology and metabolism of bioactive compounds (Akhi et al. 2021). Plants also adapt many cellular, molecular, and biochemical alterations to withstand changing environmental conditions. These modifications aid plants in warding off pathogens without interfering with their regular physiology. In the entire life cycle, crop plants are vulnerable to attacks from several noxious plant diseases that impede their growth and development. Therefore, plants secrete several SMs, including terpenoids, flavonoids, alkaloids, saponins, steroids, and amino acids, which help to lessen the negative impacts of disease on agricultural plants. Crop plants produce a diverse range of metabolites, most of which have been discovered to be potent promoters of plant defense against biotic and environmental stressors (Kumar et al. 2023). The two main types of metabolites manufactured by the plant body are primary and SMs, among the metabolites produced by plants. Primary metabolites, which include the basic biomolecules such as carbohydrates, lipids, and proteins, have a direct role in the cellular metabolism, growth, and development of plants, while SMs offer distinctive additional advantages to the plant that are not directly related to metabolism but nevertheless aid in the plant's ability to survive. The ability to cope with biotic and abiotic stressors, developing adaptation, acting as pollen attractants and defence against stressors are the beneficial of these metabolites. SMs primarily consist of three types: nitrogen (N)-containing chemicals (glucosinolates, glycosides, and alkaloids), phenolics (flavonoids, lignins, and tannins), and terpenes (isoprenoids and terpenoids). Primary metabolism involves the biomolecules amino acids, chemical compounds, and nucleosides. These molecules follow alternate metabolic pathways diverging from primary metabolism and results in creation of plant SMs (Pott et al. 2019). Plants produce large amount phenolic substances, like flavonoids and allied phenolics compounds, these chemicals change the cellular permeability of plant cells, preventing pathogen growth in the apoplast. These compounds alter the topology of casing proteins, which alters membrane permeability and leads to pH imbalance and ATP generation. Saponin such as α-tomatine in tomato, later bind to plant cell membrane and by inducing the stimulates phosphotyrosine kinase and G-protein pathways, causes ROS burst and Ca^{2+}increase in *Fusarium oxysporum*, leading to the death of the fungi (Ito et al. 2007). Terpenes have multiple purposes in plants. They serve a significant role in stress mitigation, ensuring plant survival, persistence, and competitiveness in facing a variety of stress from both biotic (living organisms) and abiotic (environmental) stressors. They are effective in quenching reactive oxygen species (antioxidant molecules), providing defense against viruses and herbivores, regulators of plant growth, and substances that affect the growth and development of nearby plants, either directly or indirectly (Masyita et al. 2022). Plants use terpenes for a variety of purposes generally. In addition to serving as substances that affect surrounding plants' growth and development (directly or indirectly), they also serve as defense molecules against infections and herbivores. Phytoalexins such as rishitin, gossypol, and capsidiol are major tarpenes that are crucial in agricultural crops of malvaceae (cotton) and solanaceous vegetables. Similarly, the rice plant produces a range of diterpenoid phytoalexins that strengthen its defenses against serious fungal diseases, such as *Magnaporthe oryzae*'s rice blast (Kumar et al. 2023). Various other terpenes including menthol, camphor, pyrethrins, artemisinins, and farnesol are effective in restricting the growth of protozoa, bacteria along with fungi. Capsaicin, a common terpene found in *Capsicum annuum* shows bactericidal properties and is effective for successful growth of the flora. Hemiterpenes also improve photosynthetic efficiency and thermotolerance in certain plant species. It has been reported that when fumigating the leaves of *Quercus ilex* with monoterpenes, the thermo-tolerance of the plant improves many folds (Tiku 2018; Price et al. 1989).

Numerous SMs exhibit antimicrobial activity. Exudates from *Arabidopsis* roots infected with *Pseudomonas syringae* have shown antibacterial action by preventing the growth of the bacteria. Seven of the eight strains are incapable of infecting plants, and when plants were exposed to non-pathogenic strains, they produced more SMs. The non-pathogenic bacteria exhibited a moderate level of antibacterial activity against the non-infecting strains. The infecting strains' antibacterial activity in the exudate from the roots was negligible. An important vegetable crop grown worldwide, *Brassica rapa* can acquire resistance against the infection of *Xanthomonas campestris* pv. *campestris* by exudating some SMs of phenolic chemicals and glucosinolates. The chloroform extracts of *Zingiber officinale* and methanol extracts of *Polyalthia longifolia* and *Terminalia chebula* were more effective in vitro against *Xanthomonas campestris* pv. *campestris*. Another secondary metabolite, caffeine, has been shown effective antibacterial agent against *X. campestris* subsp. as well as various other bacterial strains including *Ralstonia solancearum*,

Clavibacter michiganesis subsp. *sepedonicus*, *Dickeya solani*, *Pectobacterium carotovorum* subsp. *carotovorum*, *P. atrosepticum*. Caffeine affects DNA, RNA, and protein synthesis in addition to lengthening the bacterial generation time of all species studied and causing morphological alterations in cells. *Satureja montana* contains an essential oil that has antibacterial properties against *Clostridium perfringens*, a bacterium known to induce cell lysis and structural damage (Zaynab et al. 2018). Several SMs (alkaloids, phenolics, and flavonoids) also show antiviral properties and were used for curing viral infections in ancient Chinese herbs. *Hosta plantaginea* has the alkaloid 7-deoxytrans-dihydronarciclasine, which has an anti-TMV impact with the lowest IC50 value of 1.80 μM. Additionally, the extract of *Brassica javanica* contains brucine-D, which has an inhibitory effect on PVY, CMV, and TMV. A study has shown that applying a 100 μg/L Bruceine-D embedded extract of *Chenopodium amaranticolor*, or white goose foot, for 15 min prevented over 90% of PVY and CMV infection.

4 Role of PGRs in Secondary Metabolite Production

Plant growth regulators (PGRs) are important regulators of many physiological processes in plants, including the synthesis of SMs. Organic substances known as SMs are vital for defense against pathogens, herbivores, and environmental stressors as well as for interactions with other species. Specifications, age, varieties, growth stage, environmental factors, nutritional status, endogenous hormone balance, and physiological state can all affect how a plant responds to PGRs. Even though PGRs prevent plant growth, they have been shown to have some positive effects on plants, particularly in response to abiotic stressors. Under conditions of environmental stress, PGRs control a variety of physiological responses and growth. Even though a variety of PGR functions under various environmental stress conditions have been thoroughly investigated, their function in mediating the production of SM is poorly recognized. Application of various growth regulators, according to Winkel-Shirley (2001), stimulates the formation of SMs and increases anthocyanin content, which is beneficial in protecting plants from UV radiation damage. The following conditions can be used to induce the production of anthocyanins in plant sources: low sugar and nutrition levels, pathogen invasion, wounds, intense light, and ultraviolet and blue light radiation. To determine the impact of distinct PGRs on the production of SMs, numerous investigations have been carried out and validated in large plants (Naeem et al. 2011). Over their entire lives, plants naturally produce primary metabolites (PMs) and SMs. Environmental stress management requires SMs, even though PMs have an immediate role in the fundamental processes of metabolism (Raza et al. 2022). By triggering different signaling pathways and coordinating adaptive responses, SMs play critical roles in regulating plant defense against environmental stressors. Numerous studies have emphasized the growing significance of classical PGRs as mediators of environmental stress, such as ethylene, gibberellic acid (GA), cytokinins (CK), auxins, and abscisic acid (ABA). These PGRs have drawn a great deal of attention due to their potential roles in regulating SM production to protect plant growth and productivity in the face of numerous stressful conditions, including drought, salinity, heavy metal toxicity, high temperatures, and flooding. Plantlets of *Hypericum sampsonii* and *Hypericum perforatum* exhibit enhanced levels of SMs in response to CK (Liu et al. 2007). According to research by Kalt et al. (2001), utilizing ABA in *Orthosiphon stimaneus* led to oxidative stress while also increasing the production of SMs. Terpenes, phenolics, compounds containing nitrogen, and sulfur are some of the most prominent categories of SMs. Units of 5-C isopentanoid make up terpenes. The main source of phenols is the byproduct of the shikimic acid pathway. Compounds comprising nitrogen and sulfur belong to the third major group and are primarily synthesized from common amino acids (VanEtten et al. 2001). Many insects and mammals use plants as food and most of the various terpene structures produced from plants as SMs are thought to be complex defense mechanisms against toxic compounds and consuming limitations. Both the quantity and the quality of terpenoids are enhanced by PGRs, which also promote terpenoid biosynthesis and growth in several aromatic plants. PGRs application may have an impact on essential oils because of its effects on the enzymatic pathways of terpenoid biosynthesis (Sangwan et al. 2001). After being treated with gibberellic acid, 2, 4-D, and BA (benzyl adenine), *Salvia officinalis* showed a higher number of essential oils and more terpenoids. Reda et al. (2005) noted that the application of GA resulted in a higher yield of essential oils on *Thymus vulgaris* plants. An aromatic ring's hydroxyl functional group and phenol group are components of a wide range of secondary products that are produced by plants. Fruits, vegetables, teas, cocoa, and other plants with specific health benefits produce one of the biggest groups of secondary plant components. They may offer defense against certain diseases and oxidative stress due to their anti-inflammatory, antioxidant, and anti-carcinogenic qualities, among other significant biological attributes (Park et al. 2001). *Orthosiphon stamineus* produced more phenolic compounds as the concentration of ABA increased. To promote flavonoid biosynthesis, GA increases growth and decreases CHS (chalcone synthase) activity in *Ginkgo biloba*. In *Taraxacum officinale*, the application of ethephon contributed to a decrease in total flavonoid concentration, whereas GA3, SA, and KIN considerably enhanced it (Kim et al. 2009). JA and its equivalent methyl jasmonate (MeJA) caused numerous plants, including *Nicotiana tabacum*, *Picea abies*, and *Rubus* sp., to produce SMs like flavonoids (Zayed and Wink 2009). It has been observed that exogenously applied ABA increases the amount of phenolic compounds in *Phragmites communis* and *Vitis rotundifolia* (Ding et al. 2010). The hairy roots of *Salvia miltiorrhiza* were found to be responsive to ABA, GA, and ethylene in terms of inducing the production of phenolic acids and increasing the activities of TAT (tyrosine amino transferase) and PAL (phenylalanine ammonia lyase) (Liang et al. 2013). In *Coleus blumei*, auxin and gibberellin both control the biosynthesis of lignin (Aloni et al. 1990). In nature, flavonoids are polyphenolic compounds that are widely distributed. Numerous flavonoids are found in fruits, vegetables, and drinks such as tea, coffee, and fruit drinks (Masondo et al. 2014). According to the variety of chemical structures and mechanisms of action, naturally occurring sulfur-containing products offer plants a broad spectrum of chemical defenses against a wide range of possible challenges.

5 Biochemical Pathways and Regulation

Many intracellular second messengers have protein kinases as their target enzymes; these enzymes control the phosphorylation of intracellular proteins, which in turn transmits signals further (Goldsmith and Cobb 1994). Protein kinases, which impact various plant processes like photosynthesis, self-incompatibility, cold and light resistance, and cell division are known to exist in over 30 different varieties (Meng and Zhang 2013; Mo et al. 2021). Three primary forms of MAPK, MAPKKs, MAPKKKs, and MAPKs are known to exist among the protein kinases that are Ser/Thr kinases. External signals are amplified and transferred through a sequence of reactions they create. MAPKKKs are found upstream of the chain reaction, and through phosphorylation, they connect outside signals with regulators downstream (Lin et al. 2021). MAPKK is activated by an activated MAPKKK, and MAPKK then phosphorylates the Thr/Tyr phosphorylation sites on the MAPK, causing dual phosphorescence. When MAPK enters the nucleus, it phosphorylates transcription factors to control the expression of specific genes. Alternatively, it remains in the cell to activate a sequence of other protein kinases (Banerjee et al. 2020). In reaction to pathogen infection, MAPK signaling mediates several defense mechanisms and actively engages in PTI and ETI responses (Liu et al. 2018a; Lang and Colcombet 2020).

Two different types of pathogen-derived chemicals start the plant's defensive response at the infection site. Plants have a basic resistance to most infections due to a pathogen-associated molecular pattern (PAMP) that is detected by cell-surface-localized pattern-recognition receptor (PRR)-triggered immunity (PTI). This PAMP triggers the first line of defense in host-induced defense responses. In the pathophysiology of resistant host-pathogenic microbes, resistant hosts directly trigger PTI, such as calcium ion (Ca^{2+}) influx, the formation of reactive oxygen species (ROS), and the activation of mitogen-activated protein kinase (MAPK), after identifying the PAMPs generated by pathogens. Subsequently, the downstream cascade of salicylic acid (SA) or jasmonic acid (JA)/ethylene (ET) signaling is triggered, leading to the manufacture of defense-associated proteins and the initiation of plant disease resistance (Fig. 9.3). Effector proteins, such as AvrPto, AvrPtoB, and AvrPphB, are another type of initiator that are mostly delivered to plant cells by the pathogen type III secretion system (Nicaise and Candresse 2017; Schreiber et al. 2021). The second line of host-induced defense responses can be initiated by effector-triggered immunity (ETI), which is caused by the interaction between pathogen effectors and plant resistance (R) proteins. It triggers a number of defense reactions, including systemic acquired resistance (SAR) and HR/programmed cell death (PCD). Some biotrophic or hemi-biotrophic infections can be strongly resistant to these responses, which can be triggered by SA (Lee et al. 2015). Necrotrophic pathogen resistance is induced by the JA/ET-mediated pathway (Li et al. 2019).

Several signaling channels are used by plants to activate their defense mechanisms against pathogen invasion (Overmyer et al. 2018). Plant MAPKKs group D includes *Arabidopsis* MKK7. Plant defense responses also involve important roles

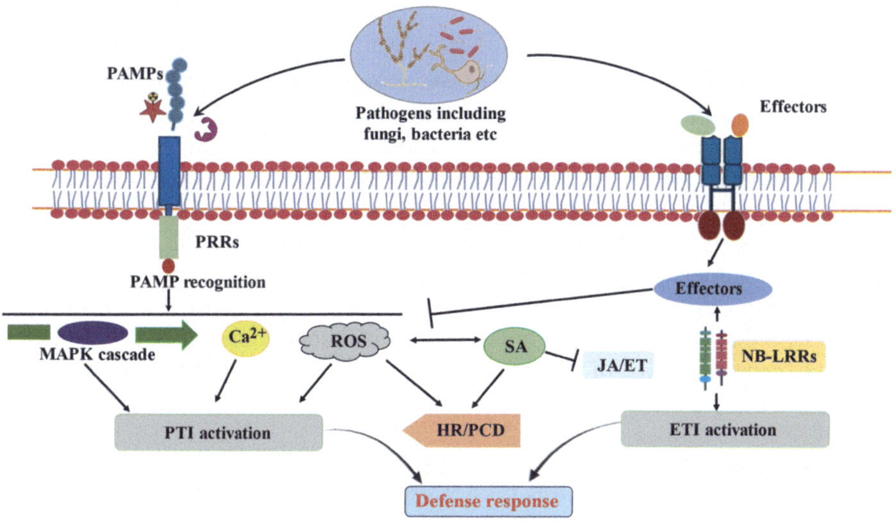

Fig. 9.3 An immunological response model for plant-pathogen interactions. Plants have PTI and ETI as part of their innate immune system. Most infections can be inhibited in growth by PTI that is generated when PRRs recognize PAMPs. Subsequently, signaling pathways downstream are triggered, including ROS signaling, the MAPK cascade, and Ca^{2+} signaling. To hinder PTI, several infections can release pathogenic effectors, which can lead to effector-triggered susceptibility. A pathogen-infection site frequently develops an HR due to the action of disease R proteins with conserved NB-LRR, which can either directly or indirectly recognize certain effectors to induce ETI and therefore restrict the infections growth. Additionally, the SA and JA/ET signaling pathways stimulate downstream transcription factors and start plant defensive responses by activating PTI and ETI and triggering the resistance response to pathogen infections. PRRs pattern recognition receptors, PCD programmed cell death, NB-LRR nucleotide-binding leucine-rich repeat, HR hypersensitive response, R resistance, ROS reactive oxygen species, SA salicylic acid, JA/ET jasmonic acid/ethylene, PAMPs pathogen-associated molecular pattern, MAPK mitogen-activated protein kinase, ETI effector-triggered immunity, PTI PAMP-triggered immunity

for members of Group D from other species. In tomato leaves, for instance, overexpression of MKK4 causes cell death and activates MPK2 and MPK3, two proteins involved in plant defense responses (Ekengren et al. 2003). Using the regulation of SA synthesis, MKK7 overexpression in *Arabidopsis* can positively control SAR and basic resistance. This is distinguished by increased pathogen resistance and constitutive PR gene expression (Zhang et al. 2007). The phosphorylation of *MPKs* is necessary for their transcriptional activation. *Verticillium dahlia* resistance is enhanced in *Arabidopsis* by overexpressing cotton *GhNTF6*, the only MPK family member that needs a phosphorylation reaction during a pathogen attack (Zhou et al. 2022). Many plants have been shown to exhibit resistance to WRKY, a downstream transcription factor in the MAPK pathway. According to an RNA-Seq analysis, rice WRKY67 can cause the transcription of genes linked to defense, and overexpressing the gene in rice can make it more resistant to bacterial blight (Liu et al. 2018b).

6 Conclusion

Although not necessary for the metabolism or growth of plants, SMs are essential for the plants' ability to withstand biotic and abiotic stresses. Transmembrane recognition and the development of SMs are two examples of the stress detection mechanisms under which the plant defense system has evolved. These mechanisms eventually enable plants to tolerate these challenging or stressful environments. In order for plants to interact with their surroundings and adapt to both biotic and abiotic stress situations, their secondary products must have a great diversity and wide range of composition. SMs communicate and crosstalk with PGRs and other organisms to help them adapt in changing environmental conditions. Also, the attributes like high gene plasticity of SMs help the plants to better adapt under different stressors.

Acknowledgments Varsha K. Singh (09/0013(12862)/2021-EMR-1) and Palak Rana (09/0013 (16603)/2023-EMR-I) are thankful to the Council of Scientific & Industrial Research (CSIR), New Delhi, India, for financial assistance in the form of Junior Research Fellowship (JRF). Sapana Jha (No. R/Dev. /Sch./UGC Non-NET Fello./2022-23/52561) is thankful to BHU for providing institutional fellowship. The incentive grant received from IoE (Scheme no. 6031), Banaras Hindu University, Varanasi, India, to Rajeshwar P. Sinha is highly acknowledged.

References

Abideen Z, Qasim M, Rasheed A (2015) Antioxidant activity and poly phenolic content of Phragmites karka under saline conditions. Pak J Bot 47(3):813–818

Acamovic T, Brooker JD (2005) Biochemistry of plant SMs and their effects in animals. Proc Nutr Soc 64:403–412

Adak S, Roy A, Das P, Mukherjee A, Sengupta S, Majumder AL (2019) Soil salinity and mechanical obstruction differentially affects embryonic root architecture in different rice genotypes from West Bengal. Plant Physiol Rep 24:192–209

Afzal SF, Yar AK, Ullah RH (2017) Impact of drought stress on active secondary metabolite production in *Cichorium intybus* roots. J Appl Environ Biol Sci 7(7):39–43

Agati G, Tattini M (2010) Multiple functional roles of flavonoids in photoprotection. New Phytol 186:786–793

Aguirre-Becerra H, Vazquez-Hernandez MC, Saenz de la OD, Alvarado-Mariana A, Guevara-Gonzalez RG, Garcia-Trejo JF, Feregrino-Perez AA (2021) Role of stress and defense in plant SMs production. In: Bioactive natural products for pharmaceutical applications, pp 151–195. https://doi.org/10.1007/978-3-030-54027-2_5

Akhi MZ, Haque MM, Biswas MS (2021) Role of SMs to attenuate stress damages in plants. In: Antioxidants-benefits, sources, mechanisms of action. IntechOpen

Akojie FOB, Fung LM (1992) Antisickling activity of hydroxybenzoic acids in *Cajanus cajan*. Planta Med 58(04):317–320

Al Hassan M, Chaura J, Donat-Torres MP, Boscaiu M, Vicente O (2017) Antioxidant responses under salinity and drought in three closely related wild monocots with different ecological optima. AoB Plants 9(2):plx009

Al Othman ZA, Hadj Ahmed YB, Abdelaty Habila M, Abdel Ghafar A (2011) Determination of capsaicin and dihydrocapsaicin in capsicum fruit samples using high performance liquid chromatography. Molecules 16:8919–8929

Ali M, Abbasi BH (2014) Light-induced fluctuations in the biomass accumulation, SMs production and antioxidant activity in the cell suspension cultures of *Artemisia absinthium* L. J Photochem Photobiol B Biol 140:223–227

Al-Khayri JM, Rashmi R, Toppo V, Chole PB, Banadka A, Sudheer WN, Nagella P, Shehata WF, Al-Mssallem MQ, Alessa FM, Almaghasla MI (2023) Plant SMs: the weapons for biotic stress management. Metabolites 13(6):716

Aloni R, Tollier MT, Monties B (1990) The role of auxin and gibberellin in controlling lignin formation in primary phloem fibers and in xylem of *Coleus blumei* stems. Plant Physiol 94(4):1743–1747

Anasori P, Asghari G (2009) Effects of light and differentiation on gingerol and zingiberene production in the callus culture of *Zingiber officinale* Rosc. Planta Med 75(09):59–63

Angelova Z, Georgiev S, Roos W (2006) Elicitation of plants. Biotechnol Biotechnol Equip 20(2):72–83

Ashapkin VV, Kutueva LI, Aleksandrushkina NI, Vanyushin BF (2020) Epigenetic mechanisms of plant adaptation to biotic and abiotic stresses. Int J Mol Sci 21(20):7457

Banerjee G, Singh D, Sinha AK (2020) Plant cell cycle regulators: mitogen-activated protein kinase, a new regulating switch? Plant Sci 301:110660

Binder BY, Peebles CA, Shanks JV, San KY (2009) The effects of UV-B stress on production of terpenoid indole-alkaloids in *Catharanthus roseus* hairy roots. Biotechnol Prog 25:8615

Bojórquez-Quintal E, Escalante-Magaña C, Echevarría-Machado I, Martínez-Estévez M (2017) Aluminum, a friend or foe of higher plants in acid soils. Front Plant Sci 8:1767

Bourgaud F, Gravot A, Milesi S, Gontier E (2001) Production of plant SMs: a historical perspective. Plant Sci 161(5):839–851

Caldwell MM, Ballaré CL, Bornman JF (2003) Terrestrial ecosystems, increased solar ultraviolet radiation and interactions with other climatic change factors. Photochem Photobiol Sci 2:29–38

Caretto S, Linsalata V, Colella G, Mita G, Lattanzio V (2015) Carbon fluxes between primary metabolism and phenolic pathway in plant tis sues under stress. Int J Mol Sci 16(11):26378–26394

Cetin ES, Babalik Z, Hallac-Turk F, Gokturk-Baydar N (2014) The effects of cadmium chloride on secondary metabolite production in *Vitis vinifera* cv. cell suspension cultures. Biol Res 47(1):47

Chan LK, Koay SS, Boey PL, Bhatt A (2010) Effects of abiotic stress on biomass and anthocyanin production in cell cultures of *Melastoma malabathri* cum. Biol Res 43(1):127–135

Chaves MM, Pereira JS, Maroco J (2002) How plants cope with water stress in the field? Photosynthesis and growth. Ann Bot 89(7):907–916

Cingöz G, Karakaş FP (2016) The effects of nutrient and macronutrient stress on certain secondary metabolite accumulations and redox regulation in callus cultures of *Bellis perennis* L. Turk J Biol 40(6):1328–1335

Dahham SS, Tabana YM, Iqbal MA, Ahamed MBK, Ezzat MO, Majid ASA (2015) The anticancer, antioxidant and antimicrobial properties of the sesquiterpene β-Caryophyllene from the essential oil of *Aquilaria crassna*. Molecules 20:11808–11829

Dai SJ, Wang Y, Yan X, Ma M (2004) Effects of color films on growth and camptothecin content in the leaves of *Camptotheca acuminata* seedlings. Acta Ecol Sin 24(5):869–875

de Abreu IN, Mazzafera P (2005) Effect of water and temperature stress on the content of active constituents of *Hypericum brasiliense* Choisy. Plant Physiol Biochem 43(3):241–248

de Matos Nunes J, Bertodo LOO, Da Rosa LMG, Von Poser GL, Rech SB (2014) Stress induction of valuable SMs in *Hypericum polyanthemum* acclimatized plants. S Afr J Bot 94:182–189

Ding W, Song L, Wang X, Bi Y (2010) Effect of abscisic acid on heat stress tolerance in the calli from two ecotypes of *Phragmites communis*. Biol Plant 54(4):607–613

Dresselhaus T, Hückelhoven R (2018) Biotic and abiotic stress responses in crop plants. Agronomy 8(11):267

Edreva A (2005) Generation and scavenging of reactive oxygen species in chloroplasts: a submolecular approach. Agric Ecosyst Environ 106:119–133

Edreva A, Velikova V, Tsonev T (2008) Stress-protective role of SMs: diversity of functions and mechanisms. Gen Appl Plant Physiol 34(1–2):67–78

Ekengren SK, Liu Y, Schiff M, Dinesh-Kumar SP, Martin GB (2003) Two MAPK cascades, NPR1, and TGA transcription factors play a role in Pto-mediated disease resistance in tomato. Plant J 36(6):905–917

Fett-Neto AG, Pennington JJ, DiCosmo F (1995) Effect of white light on taxol and baccatin III accumulation in cell cultures of *Taxus cuspidata* Sieb & Zucc. J Plant Physiol 146(5–6):584–590

Freeman BC, Beattie GA (2008) An overview of plant defenses against pathogens and herbivores. Plant Health Instruct 94. https://doi.org/10.1094/PHI-I-2008-0226-01

Fu XZ, Ullah Khan E, Hu SS, Fan QJ, Liu JH (2011) Overexpression of the betaine aldehyde dehydrogenase gene from *Atriplex hortensis* enhances salt tolerance in the transgenic trifoliate orange (*Poncirus trifoliata* L. Raf.). Environ Exp Bot 74:106–113

Ghosh S, Watson A, Gonzalez-Navarro OE (2018) Speed breeding in growth chambers and glasshouses for crop breeding and model plant research. Nat Protoc 13(12):2944–2963

Gobbo-Neto L, Bauermeister A, Sakamoto HT (2017) Spatial and temporal variations in SMs content of the Brazilian Arnica leaves (*Lychnophora ericoides* Mart., Asteraceae). J Braz Chem Soc 28(12):2382–2390

Goboza M, Aboua YG, Chegou N, Oguntibeju OO (2019) Vindoline effectively ameliorated diabetes-induced hepatotoxicity by docking oxidative stress, inflammation and hypertriglyceridemia in type 2 diabetes-induced male Wistar rats. Biomed Pharmacother 112:1–11

Goldsmith EJ, Cobb MH (1994) Protein kinases. Curr Opin Struct Biol 4:833–840

González-Santana IH, Márquez-Guzmán J, Cram-Heydrich S, Cruz Ortega R (2012) *Conostegia xalapensis* (Melastomataceae): an aluminum accumulator plant. Physiol Plant 144(2):134–145

Goyal S, Lambert C, Cluzet S, Mérillon JM, Ramawat KG (2012) SMs and plant defense. In: Mérillon JM, Ramawat KG (eds) Plant defence: biological control. Springer, pp 109–138

Grech-Baran M, Pietrosiuk A (2012) *Artemisia* species in vitro cultures for production of biologically active SMs. J Biotechnol Comput Biol Bionanotechnol 93(4):371–380

Griffith M, Yaish MW (2004) Antifreeze proteins in overwintering plants: a tale of two activities. Trends Plant Sci 9(8):399–405

Gupta B, Huang B (2014) Mechanism of salinity tolerance in plants: physiological, biochemical, and molecular characterization. Intl J Genomics 2014:701596

Hahm ER, Moura MB, Kelley EE, Van Houten B, Shiva S, Singh SV (2011) Withaferin A-induced apoptosis in human breast cancer cells is mediated by reactive oxygen species. PLoS One 6:1–12

Hamanishi ET, Barchet GL, Dauwe R, Mansfield SD, Campbell MM (2015) Poplar trees reconfigure the transcriptome and metabolome in response to drought in a genotype and time-of-day-dependent manner. BMC Genomics 16(1):329

Harborne JB (1999) Classes and functions of secondary products from plants. In: Walton NJ, Brown DE (eds) Chemicals from plants, perspectives on secondary plant products. Imperial College Press, London, pp 1–25

Harborne JB, Williams CA (2000) Advances in flavonoid research since 1992. Phytochemistry 55:481–504

Hartmann T (2004) Plant-derived SMs as defensive chemicals in herbivorous insects: a case study in chemical ecology. Planta 219:1–4

Hossain MS, Persicke M, ElSayed AI, Kalinowski J, Dietz KJ (2017) Metabolite profiling at the cellular and subcellular level reveals metabolites associated with salinity tolerance in sugar beet. J Exp Bot 68:5961–5976

Hu Y, Yu W, Song L (2016) Effects of light on the production of camptothecin and expression of key enzyme genes in seedlings of *Camptotheca acuminate* Decne. Acta Physiol Plant 38:65

Hura K, Rapacz M, Hura T, Żur I, Filek M (2015) The effect of cold on the response of *Brassica napus* callus tissue to the secondary metabolites of *Leptosphaeria maculans*. Acta Physiol Plant 37(2):13

Isah T (2019) Stress and defense responses in plant SMs production. Biol Res 52:1–25

Isah T, Umar S (2018) Influencing in vitro clonal propagation of *Chonemorpha fragrans* (moon) Alston by culture media strength, plant growth regulators, carbon source and photoperiodic incubation. J For Res 1:27–43
Ito SI, Ihara T, Tamura H, Tanaka S, Ikeda T, Kajihara H, El-Sayed MA (2007) α-Tomatine, the major saponin in tomato, induces programmed cell death mediated by reactive oxygen species in the fungal pathogen *Fusarium oxysporum*. FEBS Lett 581(17):3217–3222
Jamwal K, Bhattacharya S, Puri S (2018) Plant growth regulator mediated consequences of SMs in medicinal plants. J Appl Res Med Aromat Plant 9:26–38
Jochum GM, Mudge KW, Thomas RB (2007) Elevated temperatures increase leaf senescence and root secondary metabolite concentration in the understory herb *Panax quinquefolius* (Araliaceae). Am J Bot 94:819–826
Jogawat A, Yadav B, Chhaya LN, Singh AK, Narayan OP (2021) Crosstalk between phytohormones and SMs in the drought stress tolerance of crop plants: a review. Physiol Plant 172:1106–1132
Joshi N (2015) Influence of light and temperature on secondary metabolite accumulation in callus cultures of *Helicteres isora* L. IOSR J Environ Sci Toxicol Food Technol 1(1):2319–2399
Joshi S, Thoday-Kennedy E, Daetwyler HD, Hayden M, Spangenberg G, Kant S (2021) High-throughput phenotyping to dissect genotypic differences in safflower for drought tolerance. PLoS One 16(7):e0254908
Kalt W, Ryan DA, Duy JC, Prior RL, Ehlenfeldt MK, Vander Kloet SP (2001) Interspecific variation in anthocyanins, phenolics, and antioxidant capacity among genotypes of high bush and lowbush blueberries (Vaccinium section *Cyanococcus* spp.). J Agric Food Chem 49(10):4761–4767
Karwasara VS, Dixit VK (2013) Culture medium optimization for camptothecin production in cell suspension cultures of *Nothapodytes nimmoniana* (J. Grah.) Mabberley. Plant Biotechnol Rep 7(3):357–369
Kawka B, Kwiecień I, Ekiert H (2017) Influence of culture medium composition and light conditions on the accumulation of bioactive compounds in shoot cultures of *Scutellaria lateriflora* L. (American Skullcap) grown in vitro. Appl Biochem Biotechnol 183:1414–1425
Kesselmeier J, Staudt M (1999) Biogenic volatile organic compounds (VOC): an overview on emission, physiology and ecology. J Atmos Chem 33:23–88
Kim YH, Hamayun M, Khan AL, Na CI, Kang SM, Han HH, Lee I (2009) Exogenous application of plant growth regulators increased the total flavonoid content in *Taraxacum officinale* Wigg. Afr J Biotechnol 8:5727–5732
Kim YS, Choi YE, Sano H (2010) Plant vaccination: stimulation of defense system by caffeine production in planta. Plant Signal Behav 5(5):489–493
Koricheva J, Barton K (2012) Temporal changes in plant secondary metabolite production. In: Iason G, Dicke M, Hartley S (eds) The ecology of plant SMs: from genes to global processes. Cambridge University Press, Cambridge, pp 34–55
Kumar S, Korra T, Thakur R, Arutselvan R, Kashyap AS, Nehela Y, Keswani C (2023) Role of plant SMs in defence and transcriptional regulation in response to biotic stress. Plant Stress 8:100154
Kumari M, Chandra S (2016) Secondary metabolite production in transformed cultures: stevioside glycosides production from *Stevia rebaudiana* hairy root cultures. In: Kumari M, Chandra S (eds) Transgenesis and secondary metabolism: part of the series reference series in phytochemistry. Springer, Berlin, pp 1–19
Kurek J (2019) Alkaloids—their importance in nature and for human life. In: Kurek J (ed) Their importance in nature and for human life. IntechOpen, London, pp 1–7
Lang J, Colcombet J (2020) Sustained incompatibility between MAPK signaling and pathogen effectors. Int J Mol Sci 21:7954
Larson RA (1988) The antioxidants of higher plants. Phytochemistry 27:969–978
Lee HJ, Park YJ, Seo PJ, Kim JH, Sim HJ, Kim SG, Park CM (2015) Systemic immunity requires snrk2.8-mediated nuclear import of NPR1 in *Arabidopsis*. Plant Cell 27:3425–3438

Leisner CP (2020) Climate change impacts on food security-focus on perennial cropping systems and nutritional value. Plant Sci 293:110412

Lemos AS, Florêncio JR, Pinto NC, Campos LM, Silva TP, Grazul RM, Fabri RL (2020) Antifungal activity of the natural coumarin scopoletin against planktonic cells and biofilms from a multidrug-resistant *Candida tropicalis* strain. Front Microbiol 11:1525

Li N, Han X, Feng D, Yuan D, Huang LJ (2019) Signaling crosstalk between salicylic acid and ethylene/jasmonate in plant defense: do we understand what they are whispering? Int J Mol Sci 20:671

Liang Z, Ma Y, Xu T, Cui B, Liu Y, Guo Z, Yang D (2013) Effects of abscisic acid, gibberellin, ethylene and their interactions on production of phenolic acids in *Salvia miltiorrhiza* Bunge hairy roots. PLoS One 8(9):e72806

Lin L, Wu J, Jiang M, Wang Y (2021) Plant mitogen-activated protein kinase cascades in environmental stresses. Int J Mol Sci 22:1543

Liu XN, Zhang XQ, Sun JS (2007) Effects of cytokinins and elicitors on the production of hypericins and hyperforin metabolites in *Hypericum sampsonii* and *Hypericum perforatum*. Plant Growth Regul 53:207–214

Liu X, Li J, Xu L, Wang Q, Lou Y (2018a) Expressing *OsMPK4* Impairs plant growth but enhances the resistance of rice to the striped stem borer *Chilo suppressalis*. Int J Mol Sci 19:1182

Liu Q, Li X, Yan S, Yu T, Yang J, Dong J, Zhang S, Zhao J, Yang T, Mao X (2018b) *OsWRKY67* positively regulates blast and bacteria blight resistance by direct activation of PR genes in rice. BMC Plant Biol 18:257

Martinez JP, Kinet JM, Bajji M, Lutts S (2005) NaCl alleviates polyethylene glycol-induced water stress in the halophyte species *Atriplex halimus* L. J Exp Bot 56:2421–2431

Mashilo J, Odindo AO, Shimelis HA, Musenge P, Tesfay SZ, Magwaza LS (2017) Drought tolerance of selected bottle gourd [*Lagenaria siceraria* (Molina) Standl.] landraces assessed by leaf gas exchange and photosynthetic efficiency. Plant Physiol Biochem 120:75–87

Masondo NA, Aremu AO, Finnie JF, Van Staden J (2014) Plant growth regulator induced phytochemical and antioxidant variations in micropropagated and acclimatized *Eucomis autumnalis* subspecies *autumnalis* (Asparagaceae). Acta Physiol Plant 36:2467–2479

Masyita A, Sari RM, Astuti AD, Yasir B, Rumata NR, Emran TB, Simal-Gandara J (2022) Terpenes and terpenoids as main bioactive compounds of essential oils, their roles in human health and potential application as natural food preservatives. Food Chem 13:100217

Mazid M, Khan TA, Mohammad F (2011) Role of SMs in defense mechanisms of plants. Biol Med 3(2):232–249

Meng X, Zhang S (2013) MAPK cascades in plant disease resistance signaling. Annu Rev Phytopathol 51:245–266

Miehe-Steier A, Roscher C, Reichelt M, Gershenzon J, Unsicker SB (2015) Light and nutrient-dependent responses in SMs of *Plantago lanceolata* offspring are due to phenotypic plasticity in experimental grasslands. PLoS One 10(9):e0136073

Mo S, Qian Y, Zhang W, Qian L, Wang Y, Cailin G, Ding H (2021) Mitogen-activated protein kinase action in plant response to high-temperature stress: a mini review. Protoplasma 258:477–482

Moore BD, Andrew RL, Külheim C, Foley WJ (2014) Explaining intraspecific diversity in plant SMs in an ecological context. New Phytol 201(3):733–750

Morison JIL, Lawlor DW (1999) Interactions between increasing CO_2 concentration and temperature on plant growth. Plant Cell Environ 22:659–682

Munns R (2002) Comparative physiology of salt and water stress. Plant Cell Environ 25(2):239–250

Munns R, Tester M (2008) Mechanisms of salinity tolerance. Annu Rev Plant Biol 59:651–681

Naeem MMMA, Khan MMA, Moinuddin IM, Aftab T (2011) Triacontanol-mediated regulation of growth and other physiological attributes, active constituents and yield of *Mentha arvensis* L. Plant Growth Regul 65:195–206

Nakabayashi R, Mori T, Saito K (2014) Alternation of flavonoid accumulation under drought stress in *Arabidopsis thaliana*. Plant Signal Behav 9(8):e29518

Narayani M, Srivastava S (2018) Elicitation: a stimulation of stress in in vitro plant cell/tissue cultures for enhancement of secondary metabolite production. Phytochem Rev 16(6):1227–1252

Nazir F, Fariduddin Q, Hussain A, Khan TA (2021) Brassinosteroid and hydrogen peroxide improve photosynthetic machinery, stomatal movement, root morphology and cell viability and reduce Cu-triggered oxidative burst in tomato. Ecotoxicol Environ Saf 207:111081

Neilson EH, Goodger JQD, Woodrow IE, Møller BL (2013) Plant chemical defense: at what cost? Trends Plant Sci 18:250–258

Ngugi MM, Gitari HI, Muui CW, Gweyi-Onyango JP (2022) Growth tolerance, concentration, and uptake of heavy metals as ameliorated by silicon application in vegetables. Int J Phytoremediation:1–14. https://doi.org/10.1080/15226514.2022.2045251

Nicaise V, Candresse T (2017) Plum pox virus capsid protein suppresses plant pathogen-associated molecular pattern (PAMP)-triggered immunity. Mol Plant Pathol 18:878–886

Niinemets Ü (2015) Uncovering the hidden facets of drought stress: SMs make the difference. Tree Physiol 36(2):129–132

Nogués S, Allen DJ, Morison JIL, Baker NR (1998) Ultraviolet-B radiation effects on water relations, leaf development and photosynthesis in droughted pea plants. Plant Physiol 117:173–181

Nouri-Ganbalani G, Borzoui E, Shahnavazi M, Nouri A (2018) Induction of resistance against *Plutella xylostella* (L.) (Lep.: Plutellidae) by jasmonic acid and mealy cabbage aphid feeding in *Brassica napus* L. Front Physiol 9:859

Nowak M, Manderscheid R, Weigel H-J, Kleinwächter M, Selmar D (2010) Drought stress increases the accumulation of monoterpenes in sage (*Salvia officinalis*), an effect that is compensated by elevated carbon dioxide concentration. J Appl Bot Food Qual 83:133–136

Omamt EN, Hammes PS, Robbertse PJ (2006) Differences in salinity tolerance for growth and water-use efficiency in some amaranth (*Amaranthus* spp.) genotypes. N Z J Crop Hortic Sci 34:11–22

Overmyer K, Vuorinen K, Brosché M (2018) Interaction points in plant stress signaling pathways. Physiol Plant 162:191–204

Panche AN, Diwan AD, Chandra SR (2016) Flavonoids: an overview. J Nutr Sci 5:1–5

Pang Z, Chen J, Wang T, Gao C, Li Z, Guo L, Xu J, Cheng Y (2021) Linking plant SMs and plant microbiomes: a review. Front Plant Sci 12:621276

Parihar P, Singh S, Singh R, Singh VP, Prasad SM (2015) Effect of salinity stress on plants and its tolerance strategies: a review. Environ Sci Pollut Res 22(6):4056–4075

Parikrama R, Esyanti RR (2014) Effect of UV elicitation on callus growth, alkaloid and terpenoid contents in *Eurycoma longifolia* Jack. Int J Adv Chem Eng Biol Sci 1(1):12–15

Park ES, Moon WS, Song MJ, Kim MN, Chung KH, Yoon JS (2001) Antimicrobial activity of phenol and benzoic acid derivatives. Int Biodeterior Biodegradation 47:209–214

Patil AS (2020) Plant SMs: isolation, characterization & biological properties. Studera Press

Pérez-Ilzarbe J, Hernández T, Estrella I, Vendrell M (1997) Cold storage of apples (cv. Granny Smith) and changes in phenolic compounds. Z Lebensm Unters Forsch A 204(1):52–55

Piasecka A, Sawikowska A, Kuczyńska A (2017) Drought-related SMs of barley (*Hordeum vulgare* L.) leaves and their metabolomic quantitative trait loci. Plant J 89(5):898–913

Porres-Martinez M, Gonzalez-Burgos E, Carretero ME, Gomez-Serranillos MP (2016) In vitro neuroprotective potential of the monoterpenes alpha-pinene and 1,8-cineole against H_2O_2-induced oxidative stress in PC12 cells. Z Nat C J Biosci 71:191–199

Poschenrieder C, Gunsé B, Corrales I, Barceló J (2008) A glance into aluminum toxicity and resistance in plants. Sci Total Environ 400(1–3):356–368

Pott DM, Osorio S, Vallarino JG (2019) From central to specialized metabolism: an overview of some secondary compounds derived from the primary metabolism for their role in conferring nutritional and organoleptic characteristics to fruit. Front Plant Sci 10:454686

Price AH, Atherton NM, Hendry GA (1989) Plants under drought-stress generate activated oxygen. Free Radic Res Commun 8(1):61–66

Quan NT, Anh LH, Khang DT (2016) Involvement of SMs in response to drought stress of rice (*Oryza sativa* L.). Agriculture 6(2):23

Rahimi S, Hasanloo T (2016) The effect of temperature and pH on biomass and bioactive compounds production in *Silybum marianum* hairy root cultures. Res J Pharmacogn 3(2):53–59

Ramakrishna A, Giridhar P, Ravishankar GA (2011) Phytoserotonin: a review. Plant Signal Behav 6:800–809

Raza A, Su W, Jia Z, Cheng Y, Zou X (2022) Mechanistic insights into trehalose-mediated cold stress tolerance in rapeseed (*Brassica napus* L.) seedlings. Front Plant Sci 13:857980

Reda F, Abdel-Rahim EA, El-Baroty GSA, Ayad HS (2005) Response of essential oils, phenolic components and polyphenol oxidase activity of thyme (*Thymus vulgaris* L.) to some bioregulators and vitamins. Int J Agric Biol 7:735–739

Rejeb IB, Pastor V, Mauch-Mani B (2014) Plant responses to simultaneous biotic and abiotic stress: molecular mechanisms. Plan Theory 3(4):458–475

Ryde I, Li T, Rieksta J, Dos Santos BM, Neilson EHJ, Gericke O, Jepsen JU, Bork LRH, Holm HS, Rinnan R (2021) Seasonal and elevational variability in the induction of specialized compounds from mountain birch (*Betula pubescens* var. pumila) by winter moth larvae (*Operophtera brumata*). Tree Physiol 41:1019–1033

Salam U, Ullah S, Tang ZH, Elateeq AA, Khan Y, Khan J, Khan A, Ali S (2023) Plant metabolomics: an overview of the role of primary and SMs against different environmental stress factors. Life 13(3):706

Sanchez DH, Lippold F, Redestig H, Hannah M, Erban A, Krämer U, Kopka J, Udvardi MK (2008) Integrative functional genomics of salt acclimatization in the model legume *Lotus japonicus*. Plant J 53(6):973–987

Sangwan NS, Farooqi AHA, Shabih F, Sangwan RS (2001) Regulation of essential oil production in plants. Plant Growth Regul 34:3–21

Schreiber KJ, Chau-Ly IJ, Lewis JD (2021) What the wild things do: mechanisms of plant host manipulation by bacterial type III-secreted effector proteins. Microorganisms 9:1029

Selmar D (2008) Potential of salt and drought stress to increase pharmaceutical significant secondary compounds in plants. Landbauforschung Volkenrode Agric For Res 58(1/2):139–144

Shabala S, Cuin TA (2007) Potassium transport and plant salt tolerance. Physiol Plant 133(4):651–669

Sharma P, Dubey RS (2005) Modulation of nitrate reductase activity in rice seedlings under aluminium toxicity and water stress: role of osmolytes as enzyme protectant. J Plant Physiol 162(8):854–864

Shohael AM, Ali MB, Yu KW, Hahn EJ, Paek KY (2006) Effect of temperature on SMs production and antioxidant enzyme activities in *Eleutherococcus senticosus* somatic embryos. Plant Cell Tissue Organ Cult 85(2):219–228

Soliz-Guerrero JB, de Rodriguez DJ, Rodriguez-Garcia R, Angulo Sanchez JL, Mendez-Padilla G (2002) Quinoa saponins: concentration and composition analysis. In: Janick J, Whipkey A (eds) Trends in new crops and new uses. ASHS Press, Alexandria, p 110

Song MC, Kim EJ, Kim E, Rathwell K, Nam SJ, Yoon YJ (2014) Microbial biosynthesis of medicinally important plant SMs. Nat Prod Rep 31:1497–1509

Sujata B, Kshitija S (2013) Drought stress adaptation: metabolic adjustment and regulation of gene expression. Plant Breed 132(1):21–32

Sun Y, Fernie AR (2023) Plant secondary metabolism in a fluctuating world: climate change perspectives. Trends Plant Sci 29(5):560–571

Šutković J, Lerl D, Ragab MGA (2011) In vitro production of solasodine alkaloid in *Solanum nigrum* under salinity stress. J Phytology 3(1):43–49

Sytar O, Barki S, Zivcak M, Brestic M (2018) The involvement of different SMs in salinity tolerance of crops. In: Kumat V (ed) Salinity responses and tolerance in plants, vol 2. Springer, Berlin, pp 21–48

Takahashi A, Ohnishi T (2004) The significance of the study about the biological effects of solar ultraviolet radiation using the exposed facility on the international space station. Biol Sci Space 18:255–260

Tiku AR (2018) Antimicrobial compounds and their role in plant defense. In: Singh A, Singh IK (eds) Molecular aspects of plant-pathogen interaction. Springer, Singapore, pp 283–307

Twaij BM, Hasan MN (2022) Bioactive SMs from plant sources: types, synthesis, and their therapeutic uses. Int J Plant Biol 13(1):4–14

VanEtten H, Temporini E, Wasmann C (2001) Phytoalexin (phytoanticipin) tolerance as a virulence trait: why is it not required by all pathogens? Physiol Mol Plant Pathol 59(2):83–93

Willis KJ (2017) State of the world's plants. Available online https://www.kew.org/about-us/press-media/state-of-the-worlds-plants-2017. Accessed on 9 Nov 2021

Winkel-Shirley B (2001) Flavonoid biosynthesis: a colourful model for genetics biochemistry, cell biology, and biotechnology. Plant Physiol 126:485–493

Xu C, Tang X, Shao H, Wang H (2016) Salinity tolerance mechanism of economic halophytes from physiological to molecular hierarchy for improving food quality. Curr Genomics 17(3):207–214

Yadav B, Jogawat A, Rahman MS, Narayan OP (2021) SMs in the drought stress tolerance of crop plants: a review. Gene Rep 23:1–14

Yazaki K (2017) *Lithospermum erythrorhizon* cell cultures: present and future aspects. Plant Biotechnol 34(3):131–142

Yi G, Lei Z, Zhong-Ji S (2010) Stomatal clustering, a new marker for environmental perception and adaptation in terrestrial plants. Bot Stud 51:325–336

Yu KW, Murthy HN, Hahn EJ, Paek KY (2005) Ginsenoside production by hairy root cultures of Panax ginseng: influence of temperature and light quality. Biochem Eng J 23(1):536

Zayed R, Wink M (2009) Induction of pyridine alkaloid formation in transformed root cultures of *Nicotiana tabacum*. Z fur Naturforsch 64(11–12):869–874

Zaynab M, Fatima M, Abbas S, Sharif Y, Umair M, Zafar MH, Bahadar K (2018) Role of SMs in plant defense against pathogens. Microb Pathog 124:198–202

Zhang X, Dai Y, Xiong Y, DeFraia C, Li J, Dong X, Mou Z (2007) Overexpression of *Arabidopsis* MAP kinase 7 leads to activation of plant basal and systemic acquired resistance. Plant J 52:1066–1079

Zhang JH, Sun HL, Chen SY, Zeng L, Wang TT (2017) Anti-fungal activity, mechanism studies on alpha-phellandrene and nonanal against *Penicillium cyclopium*. Bot Stud 58:1–9

Zhou J, Wu Y, Zhang X, Zhao L, Feng Z, Wei F, Zhang Y, Feng H, Zhou Y, Zhu H (2022) MPK homolog *GhNTF6* was involved in cotton against *Verticillium* wilt by interacted with *VdEPG1*. Int J Biol Macromol 195:456–465

Omics Techniques to Decipher Plant Root-Microbe Interactions

10

Alok Ranjan, Shruti Kaushik, and Jyotsana Tilgam

Abstract

The next generation technologies have enabled in-depth investigation of plant root-microbe interactions through various omics approaches, such as genomics, metagenomics, transcriptomics, translatomics, proteomics, and metabolomics. These omics technologies decipher gene functions and its structures, reveal insights into the microbial diversity, transcript differences and its differential expressions, provide ideas for protein-protein interaction networks, and several models for metabolic processes and network. Collectively, these techniques enhance our understanding of the role of omics in plant root-microbe interactions. This chapter provides a comprehensive overview of different omics approach in elucidating beneficial role of plant-microbe interactions, which helps to the development of sustainable agricultural practices.

Keywords

Metabolomics · Omics · Plant root-microbes interactions · Rhizosphere · Root

A. Ranjan (✉)
Department of Biotechnology, Patna Women's College, Patna, Bihar, India

S. Kaushik
ICAR-Central Soil Salinity Research Institute, Karnal, Haryana, India

J. Tilgam
ICAR-National Bureau of Agriculturally Important Micro-organisms, Maunath Bhanjan, Uttar Pradesh, India

U. Pankaj et al. (eds.), *Plant-microbiome Interactions for Climate-resilient Agriculture*, https://doi.org/10.1007/978-981-96-3534-4_10

1 Introduction

Recent advancements in different omics techniques provide a promising opportunity to overcome crop productivity challenges by elucidating the benefits of plant-microbe interactions (Pang et al. 2022; Korra et al. 2023). The advent of next-generation omics technologies has revolutionized and provided robust high-throughput data of genomics, transcriptomics proteomes, and metabolomes (Meng et al. 2016). These omics techniques have proven invaluable in elucidating the physiological, molecular, cellular, and biochemical aspects of plant-microbe interactions under various biotic and abiotic stresses (Shafi et al. 2021; Doni et al. 2022). Using multi-omics data from different sources of databases together is essential for effective utilization and application of omics technologies. The multi-omics data integration helps in visualization, interpretation, and data analysis which facilitates a deeper understanding of biological processes and plant-microbes interaction (Mcgowan et al. 2020; Bodein et al. 2022). Plant rhizosphere are colonized by various microbes, with mutualistic benefits that assist each other in obtaining nutritional resources from soil (Orozco-Mosqueda et al. 2018; Ranjan et al. 2022). The complex mutualistic or symbiotic interaction between plants and microbes is crucial for ecosystem dynamics and sustainable agriculture (Rane et al. 2022). Plant-microbe interaction is affected by various factors including host-related elements, soil properties, soil pH, moisture content, biotic and abiotic stresses (Fadiji et al. 2023). Elucidating these factors by using multi-omics studies could deepen our understanding of plant-microbe interaction. Multi-omics approaches reveal the molecular complexity of plant microbe interactions. Genomics analysis reveal key and novel genes (Arigoni et al. 1998), transcriptomics approach identifies expression pattern and dynamics of thousands of gene (Schenk et al. 2012), proteomics analysis identifies functions of essential proteins (Jayaraman et al. 2012), similarly metabolomics analysis provides deep understanding of the functions of metabolites and small molecules (Barkal et al. 2016). Integrated omics techniques facilitated the elucidation of various interactions between plant and microbes such as symbiotic, non-symbiotic, and pathogenic relationships (Kimotho and Maina 2024). These insights highlight the transformative potential of omics technologies in advancing sustainable agriculture by providing a deeper understanding of beneficial plant-microbe relationships. In this context, we evaluate our current understanding of recent advancements in omics approaches by citing relevant examples. We explore how these developments, when integrated with modern omics techniques and data analysis are enhancing our understanding to plant microbe's interaction and crop improvement landscape. The multi-omics data analyzed by using bioinformatics tools has demonstrated that plant-microbe interactions in polluted rhizospheres trigger the release of several antioxidants and phytohormones, consequently it activates plant defense mechanisms (Kimotho and Maina 2024). Phytohormones act as key messengers' molecules and regulate plant growth and induce the synthesis of secondary metabolites. Omics techniques including metatranscriptomics and metabolomics used in the detection and analysis of different phytohormones which deepen our understanding of their functional roles in plant microbes' interactions (Eichmann et al. 2021). Recent advancements in analytical technologies have facilitated the analysis to trace chemical substances, which helps in the study of allelochemical biosynthesis and plant responses, thereby increasing our understanding of rhizosphere chemistry (Hickman et al. 2021). Each omics technique has its certain limitations that may impact on the data analysis and results. However, integrating different omics approaches can help overcome some of these limitations. The integral multi-omics data analysis is inherently difficult because transcription, translation, and metabolism: All processes occur on different timescales (Kimotho and Maina 2024). In plant sciences research, the trend towards ever-larger datasets with multi-omics approaches will leverage machine learning and artificial intelligence to classify and interpret key relationships across various variables, including plant host genotype, microbiome composition and function, environment, time, etc. (Kimotho and Maina 2024). In this concise chapter, we have attempted to present the important recent multi-omics application in the field of plant-microbe interaction at the interface to better understand the mechanism, which is indispensable for researchers working in plant-microbe interactions.

2 Multi-Omics Strategy for Plant-Microbes Interactions: An Overview

The suffix "omics" denotes a suite of technologies designed to analyze extensive and complete data sets of specific classes of biomolecules in defined conditions within cells, tissues, organs, or entire organisms (Beck et al. 2022). In recent years, plant molecular biology focuses on exploring cellular, molecular, and biological processes, their genetic foundations, and their interactions with environmental changes. This multifaceted and thorough investigation requires large-scale experiments that target entire genetic, structural, or functional components. These extensive studies are referred to as "omics" (Beck et al. 2022). The leading sub-branches of omics in fundamental systems biology studies and modern plant-microbe interaction interventions include genomics, epigenomics, transcriptomics, metatranscriptomics, proteomics, metaproteomics, metabolomics, and systems biology (Kimotho and Maina 2024). These fields primarily involve comprehensive investigations of the genome, transcriptome, proteome, and metabolome (Table 10.1). Each of these omics branches is closely connected to systems biology and bioinformatics (Rodriguez et al. 2019). The omics sub-branches constitute different layers of the omics cascade, each of which defines a biosystem or an organism at different biomolecular levels. Genomics has significantly advanced our understanding of various aspects of plant pathosystems (Arigoni et al. 1998). The genomics approach study of plant-microbe interactions has provided information of entire genomes sequence and understanding of genes and their annotations (Levy et al. 2018). The genetic variation between species on a whole-genome scale. Genomics allow for the comparison and correlation of candidate genes with specific phenotypes, like beneficial and harmful traits (Arigoni et al. 1998). Metagenomics systematically provide genetic sequence information from the entire microbial community in

Table 10.1 Different omics strategy for studying plant-microbe interactions

Strategy	Target molecule	Description	Tools and technologies used	References
Genomics	DNA	Provide insight of entire set of genetic information contained within an organism, analyze genome sequences, uncover DNA protein interactions	*Technologies:* Next generation sequencing like Illumina, Ion Torrent, and pacific Bioscience, Third generation sequencing Oxford nanopore and PacBio Sequel. *Software tools:* Alignment and assembly tolls (Bowtie, BWA, SAMtools, Variant calling software) (GATK (Genome analysis Toolkit) and VarScan Annotation tools, (ANNOVAR and Ensembl), Pathways analysis tools (Ingenuity Pathway Analysis and Reactome)	Hardwick et al. (2017) and Jiao and Schneeberger (2017)
Metagenomics	DNA	Outperform traditional culture-based methods, identify novel genes and function, offer insight into both individual units and ecosystem functions		Quince et al. (2017) and Hassa et al. (2018)
Epigenomics	DNA	Explore heritable alteration in gene expression independent of DNA sequence changes, understand how modification enable environmental adaptation	*Technologies:* Bisulfite Sequencing, Whole Genome Bisulfite Sequencing (WGBS), Reduced Representation Bisulfite Sequencing (RRBS), Chromatin Immunoprecipitation (ChIP), ATAC-Seq (Assay for transposase-Accessible Chromatin using Sequencing). *Software tools:* DNA methylation analysis tools	Füllgrabe et al. (2023)
Transcriptomics	RNA	Comprises the study of the entire RNA molecules, including mRNA, rRNA, tRNA, snRNAs, and other ncRNA of cells. Involves gene expression profiling and analysis, targeted sequencing, and single-cell analysis	*Technologies:* RNA-Seq technology enable the high throughput sequencing, Single cell RNA sequencing. *Software tools:* MegaHIT, MetaVelvet-SL, MetaPhlAn, KEGG, DESeq2, edgeR, limma, Microarray, qPCR	Satam et al. (2023)
Metatranscriptomics	RNA	Illuminate specific transcribed genes enable functional analysis of plant associated microbial communities, helps understand microbial physiology		Hassa et al. (2018)

Proteomics	Protein	Characterization of the entire compliment of proteins an organism possess at any given time. Encompasses analysis of protein abundance, diversity, functions, compartmentalization, and interactions with other proteins, as well as post translational modifications	*Technologies:* SDS-PAGE, Liquid Chromatography- tandem mass Spectrometry (LC-MS/MS), Matrix Assisted Laser Desorption/Ionization (MALDI-MSI), Protein Microarrays. *Software Tools:* Mascot, SEQUEST, MaxQuant, SILAC (Stable Isotopes Labelling by Amino acod in Cell Culture) and iTRAQ (Isobaric Tags for relative and Absolute Quantification) STRING and DAVID	Nobori et al. (2020)
Metaproteomics	Protein	The large-scale characterization of the entire protein complements of environmental microbiota. Link specific proteins to microbial utilization of particular carbon substrates, especially in the phyllosphere, identify proteins involved in defence mechanism		Starr et al. (2018)
Metabolomics	Metabolites	Involves the identification, quantitation, and characterization of a wide array of low molecular weight metabolites in a given cell, tissue, organ, or whole organism at a specific developmental stage at a given time	*Technologies:* Gas Chromatography-Mass Spectrometry (GC-MS), Liquid Chromatography-Mass Spectrometry (LC-MS), Direct Infusion Mass spectroscopy (DIMS), Nuclear Magnetic Resonance (NMR), Spectroscopy, High Performance Liquid Chromatography (HPLC), Imaging Mass Spectroscopy (IMS). *Software Tools:* Mzmine, XCMS, METLIN, MetaboLight	Castro-moretti et al. (2020)
Systems biology	Combinational	It is premised on integration of approaches to establish a meaningful relationship between the genotype and phenotype. Encompasses component integration and analysis of dynamic regulatory networks, in order to develop a meaningful interpretation of complex physiological and cellular processes	Microarray data base, protein interaction data base, Pathways data base	Rodriguez et al. (2019)

rhizosphere. These microbes form intricate communities, known as the rhizobiome, which are vital to the host plant health, growth, plant yield, and survival (Olanrewaju and Babalola 2022). Transcriptome profiling has enabled accurate and absolute quantifications of transcript expression and the sequence of all transcripts in a given sample. The metatranscriptomics strategy enables the assessment of gene expression and better understanding of the specific functions of genes or characterizes the microbiome members that play a crucial role in plant-microbe interactions (Guo et al. 2021). Unlike genomics and transcriptomics, proteomics and metaproteomics have enabled to detect and quantify expressed protein and thus provide deep understanding of the biological mechanisms that regulate cellular functional and determine composition of microbial communities in plant-microbes interactions (Starr et al. 2018). The significance of specific metabolites in plant-microbe interactions has been recognized for quite some time. By use of metabolomics methods, a set of metabolites identify and quantify accurately together with understanding the mechanistic roles of metabolites which play important role in plant-microbe interactions (Gupta et al. 2022). However, the complexity of plant microbes' interactions means that dynamic edaphic factors and spatio-temporal molecular interactions do not actually follow simple biological and physiological mechanism therefore cannot be studied by a single high throughput technique. Hence, a system's biology approach offers a comprehensive method for dissecting the genetic, molecular, and biochemical mechanisms that govern plant-microbes interactions.

2.1 Genomics

Recent introduction of first to third-generation sequencers, high throughput amplicon sequencing approaches have been widely used to reveal the bacterial diversity, archaeal, fungal, and micro eukaryotic communities on plants and helping to understand plant microbe's interaction (Levy et al. 2018). Recently molecular characterization and analysis of microbes has enhanced our understanding of molecular plant-microbe interactions. The first and important step in the genomic approach to study plant–microbe interactions is the sequencing of the genome of microbes which associated with plant or rhizosphere (Priya et al. 2021). The whole genome of microbe gives information of the complete set of genes present in the microbe and other associated biochemical pathways, thus enhancing our understanding of the biogeochemical cycle (Fuentes-Pardo and Ruzzante 2017). Biochemical pathways are usually interconnected and regulated by feedforward and feedback mechanisms at several levels. Deeper understanding of these biological mechanism in plant-microbe interactions will further help in agricultural implications, including application of beneficial microbial-derived products to enhance crop yields for sustainable agriculture. With advancements in genome-wide association studies (GWAS), deciphering complex interactions between plants and microbes to achieve desirable traits through genetic manipulation (Bartoli and Roux 2017). Genome-wide association studies have driven efforts to understand the molecular factors that influence both beneficial and harmful aspects of plant-microbe interactions, such as signaling pathways, immune responses, and genetic changes caused by heavy metal toxicity and agrochemicals (Beilsmith et al. 2019). Comparative analysis of plants and microbes, several researchers have identified numerous effector genes and proteins, observed pathogen-induced pathogenesis, and noted significant variations in plants' defense mechanisms (Yang et al. 2017). This method harnesses a comprehensive recombination repository to correlate traits within a genetically diverse population. Therefore, GWAS holds promise as a tool to understand plant-microbe interactions and identify novel genes and proteins for sustainable agricultural practices. In summary, genomics has been pivotal in transforming our understanding of plant-microbe interactions, while also challenging traditional biological assumptions.

2.2 Epigenomics

Epigenomics, used to study plant-microbe interactions, focuses on examining alterations in epigenetic modifications and genetic material across both plants and microbes (Ramos-Cruz et al. 2021). This approach investigates inheritable changes in gene expression that are independent of changes in the DNA sequence such as, DNA methylation patterns, histone protein modifications such as acetylation and methylation, and the roles of non-coding RNAs such as microRNAs. Recently it is shown that DNA demethylation positively affects plant resistance to pathogens such as *Pseudomonas syringae* (Bonnet et al. 2023). The PAV (proline-alanine-valine) gene is linked to distinct epigenomic characteristics in fungi, providing valuable predictive insights into how fungal plant pathogens adapt to hosts and assisting in the formulation of improved disease prevention strategies (Dubey and Jeon 2017). Epigenomic analysis showed that hypermethylation of the xyloglucan endotransglycosylase protein could be influenced by *Neobacillus*, a prevalent and abundant species uniquely found in Naxos tubers (Boutsika et al. 2023). This combined approach identifies key microbial species, their pathways, and epigenetic markers that influence host-microbe interactions, providing new insights into agriculture.

2.3 Metagenomics

Metagenomics systematically gathers genetic sequence information from entire microbial communities, offering more comprehensive insights than traditional culture-based methods (Xu et al. 2021). A microbial genome consists of a unique set of genes that can be identified through metagenomics techniques. Metagenomic studies have the potential to uncover fundamental aspects of plant-microbe interactions by identifying and categorizing microbial genes that unexpectedly influence their plant hosts. The novel nitrogen-fixing genes in Rhizobium (Regalado et al. 2020) and mycotoxin biosynthetic genes in *Alternaria alternate* and its host response in apples (Bhargava et al. 2019) have been identified by using metagenomics approach. Metagenomics has demonstrated a remarkable ability to discover and analyze unknown microbial proteins and related sequences, surpassing traditional culture-based and single-cell genome sequencing techniques (Jansson and Hofmockel 2018). Furthermore, metagenomics provides a chance to study microbes within their natural ecosystems, enabling detailed analyses of microbial community numbers, diversity, and composition. This technique focuses on three main objectives: (i) correlating phylogeny with function, (ii) identifying genes that encode novel enzymes, and (iii) bulk sequencing of environmental samples. Metagenomics methods allow us to study plant-microbe interactions, revealing that certain microbes can slow plant growth, while others enhance plant immunity by inhibiting pathogen growth or blocking harmful invaders (Bhargava et al. 2019). Metagenomic analysis can be conducted through amplicon targeted sequencing or shotgun sequencing to provide valuable insights into plant-associated microbial communities and were implemented in two ways: (i) 16 S rRNA and (ii) Random shotgun sequencing. These two methods allowed novel gene identification, their role in interactions with host plant, and also provide their taxonomic origins. Although, the complexity and strain diversity of plant-associated communities pose a significant challenge, often resulting in lower coverage and poorer assembly of specific genomes. Therefore, it is essential to create a clear framework, guided by specific objectives, to generate a metagenomic dataset and determine the necessary sequence depth based on the type of analysis to be performed. As a result, there is need to advance the field of metagenomics to effectively decipher and interpret the vast existing and emerging multi-omics plant-microbiome data.

2.4 Transcriptomics and Metatranscriptomics

Transcriptomics is used to detect and quantify RNA molecules in biological samples, helping to link gene functions to specific conditions. The next generation sequencing based transcriptome profiling enabled absolute quantifications of transcript expression and the nucleotide sequence of all transcripts in a given sample. Transcriptomics technique has ability to analyze potentially important genes, signaling pathways and network which regulate biological functions (Johnson et al. 2019). Metatranscriptomics strategies enable gene expression assessment in entire community. This approach is used to better understand specific functions and identify the genes or microbiome members that play crucial roles in plant-microbe interactions (Shakya et al. 2019). Metatranscriptomics has been principally used to discover novel microbial interactions and characterize active microbial communities and their function in the context of gene expression and regulation and associated pathways. There are several available bioinformatics tools for metatranscriptomics analysis such as COMAN (Terrón-Camero et al. 2022), FMAP (Kim et al. 2016), HUMAnN2 (Franzosa et al. 2018), and MetaTrans (Martinez et al. 2016). These bioinformatics tools are used based on the types of need for metatranscriptomics data set analysis. Comparative metatranscriptomics analyses of the microbial expression levels enable to decipher mechanism of phytoremediation strategies.

2.5 Proteomics and Metaproteomics

Proteomic studies have characterization of the set of proteins, revealing the role of proteins in microbial metabolic processes (Jayaraman et al. 2012). Proteomics approach provides a comprehensive insight into the complexity of plant-microbe interactions and their association (Jayaraman et al. 2012). In plant-microbe interactions, proteins play a key role in cellular homeostasis and signaling networks. Proteomic analysis provides practical insights into these interactions. Several methods and bioinformatics tools have been developed to functional analysis of proteins, such as yeast two-hybrid system (Y2H) to investigate protein-protein interactions (PPIs) by employing "Prey" and "Bait" proteins (Altmann et al. 2018). Mass spectrometric data analysis tools, such as MASCOT (Brosch et al. 2009), PeptIdent (Wilkins and Williams 1997), and SEQUEST (Eng et al. 2008) facilitate proteomic analysis. In addition, programs such as ProFound, (Zhang and Chait 2000) and MS-Tag offer specialized analyses of MALDI-TOF and MS/MS spectra, enhancing the depth of proteomic investigation. Electrophoresis gel-based analysis, such as two-dimensional gel electrophoresis (2DE) and differential gel electrophoresis (DIGE) coupled with mass spectrometry (MS), enable the identification and quantification of proteins. Proteomic analysis is instrumental in unravelling the pathogenicity, stress-related responses, and antioxidant mechanisms involved in plant-microbe interactions, providing valuable insights into physiological and cellular processes. Nodules related proteins have identified in *M. truncatula* by using 2D-LC/MS/MS (Larrainzar et al. 2007). Similarly, comparative proteomic analyses have elucidated the protein secretion patterns of the symbiont *S. meliloti* and phytopathogen *P. syringae* in response to root exudates from different host plants (Cheng et al. 2010). These findings highlight the significance of proteomic approaches in unraveling the complex interactions between plants and microbes. Metaproteomic studies are crucial for examining soil fertility, plant-microbe interactions, nutrient cycling, and bioremediation (Wang et al. 2011; Khatabi et al. 2019). A metaproteomics study of the phyllosphere across various plants revealed consistent dominant bacterial taxa and identified proteins common to different species (Wang et al. 2011). Metaproteomics can link specific proteins to microbial utilization of particular carbon substrates, especially in the phyllosphere. However, the wide range of protein abundances caused by posttranslational modifications in both plant tissues and microbial communities, along with the challenge of distinguishing between plant and microbial proteins, has been a major obstacle for proteomics-based approaches, especially when species-specific databases are unavailable.

2.6 Metabolomics

Metabolomics is the detailed exploration of an organism's metabolites. This method involves identifying and measuring all the metabolites within a cell or whole microbial community (Gupta et al. 2022). In the past decade, metabolomics has become a trusted method for studying plant-microbe interactions. It offers both qualitative and quantitative insights into how microbes interact with plants (Gupta et al. 2022). In plant-microbe interaction, metabolites play a crucial role as the immediate effector molecules of recognition and communication between host plant and microbes. Results were obtained from genomics, transcriptomics, and proteomics analysis. We know that microbes associated with plants can produce thousands of molecules that interact with and influence ecosystems, especially in the rhizosphere. Nowadays, mass spectrometry-based metabolomics is an important method for analyzing and understanding these diverse and complex chemical molecules (Saito and Matsuda 2010). In many crops by using technologies such as: Gas Chromatography-Mass Spectrometry (GC-MS), Liquid Chromatography-Mass Spectrometry (LC-MS), Direct Infusion Mass spectroscopy (DIMS), Nuclear Magnetic Resonance (NMR), Spectroscopy, High Performance Liquid Chromatography (HPLC), Imaging Mass Spectroscopy (IMS) and bioinformatics tools such as; Mzmine, XCMS, METLIN, and MetaboLight facilitate metabolites profiling. Easier access to metabolomics and the knowledge it provides has been essential in understanding how plant metabolites shape their microbial communities and how metabolites molecules produced by microbes influence plant-microbe interactions. Understanding the metabolic connections in plant-microbe interactions is very important. Metabolomics provides deeper insights into how metabolites produced by microbes influence plant traits. This can occur through specific interactions between individual plants and microbes or through broader changes driven by the plant's microbiome (Aharoni et al. 2023). Currently, there is still little to no effort to distinguish the origins of metabolites analyzed in plant-microbe co-cultures. This makes it difficult to connect metabolome data to specific plant traits. Differentiating between metabolites will enhance our understanding of plant-microbe and microbe-microbe metabolic interactions. Using stable isotope labeling is seen as a promising technical advancement in this area (Chokkathukalam et al. 2014).

2.7 Systems Biology

Systems biology is a comprehensive approach to study of genes, proteins, metabolites, and their interactions within a cell, tissue, or whole organism by integrating multi-omics datasets to establish qualitative and quantitative associations among different biological components (Rodriguez et al. 2019). By using system biology approach, it enables us to understand a complex biological system and modelling it with the help of computational techniques (Fig. 10.1). The significant progress and understanding of the complexity in plant-microbe interactions has been achieved. The use of multi-omics tools combined with bioinformatics software for data analysis and visualization has enabled us to investigate plant microbe interactions in detail (Fondi and Liò 2015). Using multi-omics technologies can be challenging because of the huge amount of data, the lack of complementation between different data types, and the varying scales at which each omics technology measures plant-microbe interactions. Advancements in computational tools and deep-learning applications, which can handle the increasing and varied data types, are making it

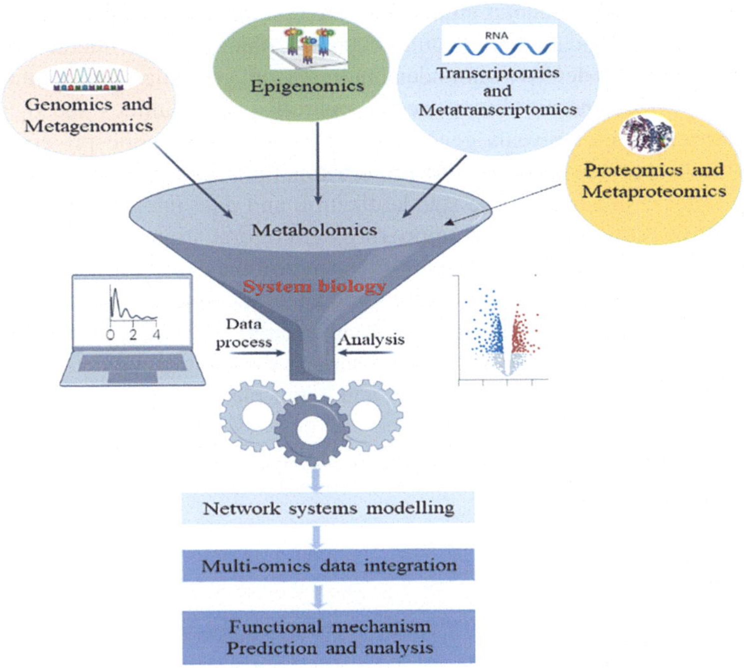

Fig. 10.1 Overview of omics-based approaches to understand plant-microbe interactions

easier to interpret this information. Meanwhile, network analyses remain a valuable method for integrating multiple data sets (Tolani et al. 2021). Systems biology requires a "top-down" approach, integrating various omics techniques with systems analyses. This means measuring and establishing relationships between multiple components, such as genes, RNA, proteins, and metabolites, using both qualitative and quantitative data. Additionally, practical mathematical and computational models are developed to provide formal descriptions of the components. These models help illustrate and calculate the relationships between the system's layers, describing both the local and global properties of the system. These models can then be experimentally tested to create new hypotheses, which help refine and improve the models. Based on the data and these refined models, computational algorithms should be efficient enough to quickly analyze the behavior of complex systems (Ali and Mohammed 2023). This process ultimately deepens our understanding of biological systems and aids in discovering and reconstructing new biological traits.

3 Conclusions

Integrating multi-omics technologies—like genomics, transcriptomics, proteomics, and metabolomics—marks a ground-breaking advance in understanding the complexities of plant-microbe interactions. By exploring the intricate factors that shape these associations, this chapter emphasizes the crucial role of omics approaches in understanding biological and molecular details of plant microbe interactions. Each omics approach adds depth to our understanding, from genomic insights into genetic variations to metagenomic discoveries of microbial communities and their functions. Transcriptomics reveals gene expression dynamics, while proteomics and metabolomics provide insights into protein functions and metabolic landscapes. Despite ongoing challenges like standardization and data integration, omics technologies offer great potential for improving agricultural strategies and strengthening resilient plant-microbe partnerships. Furthermore, exploring how plants and microbes interact in diverse environments and understanding epigenomics' role will be crucial. Overall, ongoing advances in omics technologies offer promising opportunities to enhance our understanding and harness these interactions for sustainable agriculture, environmental conservation, and better plant health.

Acknowledgments AR express their gratitude to all laboratory colleagues and research staff members for their assistance.

References

Aharoni A, Goodacre R, Fernie AR (2023) Plant and microbial sciences as key drivers in the development of metabolomics research. Proc Natl Acad Sci USA. https://doi.org/10.1073/pnas.2217383120

Ali AM, Mohammed MA (2023) A comprehensive review of artificial intelligence approaches in omics data processing: evaluating progress and challenges. Int J Math Stat Comput Sci. https://doi.org/10.59543/ijmscs.v2i.8703

Altmann M, Altmann S, Falter C, Falter-Braun P (2018) High-quality yeast-2-hybrid interaction network mapping. Curr Protoc plant Biol. https://doi.org/10.1002/cppb.20067

Arigoni F, Talabot F, Peitsch M et al (1998) A genome-based approach for the identification of essential bacterial genes. Nat Biotechnol. https://doi.org/10.1038/nbt0998-851

Barkal LJ, Theberge AB, Guo CJ et al (2016) Microbial metabolomics in open microscale platforms. Nat Commun. https://doi.org/10.1038/ncomms10610

Bartoli C, Roux F (2017) Genome-wide association studies in plant pathosystems: toward an ecological genomics approach. Front Plant Sci 8:763

Beck AE, Kleiner M, Garrell AK (2022) Elucidating plant-microbe-environment interactions through omics-enabled metabolic modelling using synthetic communities. Front Plant Sci. https://doi.org/10.3389/fpls.2022.910377

Beilsmith K, Thoen MPM, Brachi B et al (2019) Genome-wide association studies on the phyllosphere microbiome: embracing complexity in host–microbe interactions. Plant J. https://doi.org/10.1111/tpj.14170

Bhargava P, Verma A, Singh A et al (2019) Metagenomics as a tool to explore new insights from plant-microbe interface. In: Plant microbe interface. Springer. https://doi.org/10.1007/978-3-030-19831-2

Bodein A, Scott-Boyer MP, Perin O et al (2022) Interpretation of network-based integration from multi-omics longitudinal data. Nucleic Acids Res. https://doi.org/10.1093/nar/gkab1200

Bonnet DMV, Tirot L, Grob S, Jullien PE (2023) Methylome response to proteasome inhibition by pseudomonas syringae virulence factor syringolin A. Mol Plant-Microbe Interact. https://doi.org/10.1094/MPMI-06-23-0080-R

Boutsika A, Michailidis M, Ganopoulou M et al (2023) A wide foodomics approach coupled with metagenomics elucidates the environmental signature of potatoes. iScience. https://doi.org/10.1016/j.isci.2022.105917

Brosch M, Yu L, Hubbard T, Choudhary J (2009) Accurate and sensitive peptide identification with mascot percolator. J Proteome Res. https://doi.org/10.1021/pr800982s

Castro-moretti FR, Gentzel IN, Mackey D, Alonso AP (2020) Metabolomics as an emerging tool for the study of plant–pathogen interactions. Metabolites. https://doi.org/10.3390/metabo10020052

Cheng Z, McConkey BJ, Glick BR (2010) Proteomic studies of plant-bacterial interactions. Soil Biol Biochem. https://doi.org/10.1016/j.soilbio.2010.05.033

Chokkathukalam A, Kim DH, Barrett MP et al (2014) Stable isotope-labeling studies in metabolomics: new insights into structure and dynamics of metabolic networks. Bioanalysis. https://doi.org/10.4155/bio.13.348

Doni F, Suhaimi NSM, Mispan MS et al (2022) Microbial contributions for rice production: from conventional crop management to the use of 'omics' technologies. Int J Mol Sci. https://doi.org/10.3390/ijms23020737

Dubey A, Jeon J (2017) Epigenetic regulation of development and pathogenesis in fungal plant pathogens. Mol Plant Pathol. https://doi.org/10.1111/mpp.12499

Eichmann R, Richards L, Schäfer P (2021) Hormones as go-betweens in plant microbiome assembly. Plant J. https://doi.org/10.1111/tpj.15135

Eng JK, Fischer B, Grossmann J, MacCoss MJ (2008) A fast SEQUEST cross correlation algorithm. J Proteome Res. https://doi.org/10.1021/pr800420s

Fadiji AE, Yadav AN, Santoyo G, Babalola OO (2023) Understanding the plant-microbe interactions in environments exposed to abiotic stresses: an overview. Microbiol Res. https://doi.org/10.1016/j.micres.2023.127368

Fondi M, Liò P (2015) Multi -omics and metabolic modelling pipelines: challenges and tools for systems microbiology. Microbiol Res. https://doi.org/10.1016/j.micres.2015.01.003

Franzosa EA, McIver LJ, Rahnavard G et al (2018) Species-level functional profiling of metagenomes and metatranscriptomes. Nat Methods. https://doi.org/10.1038/s41592-018-0176-y

Fuentes-Pardo AP, Ruzzante DE (2017) Whole-genome sequencing approaches for conservation biology: advantages, limitations and practical recommendations. Mol Ecol. https://doi.org/10.1111/mec.14264

Füllgrabe J, Gosal WS, Creed P et al (2023) Simultaneous sequencing of genetic and epigenetic bases in DNA. Nat Biotechnol. https://doi.org/10.1038/s41587-022-01652-0

Guo L, Yu H, Wang B et al (2021) Metatranscriptomic comparison of endophytic and pathogenic fusarium-arabidopsis interactions reveals plant transcriptional plasticity. Mol Plant-Microbe Interact. https://doi.org/10.1094/MPMI-03-21-0063-R

Gupta S, Schillaci M, Roessner U (2022) Metabolomics as an emerging tool to study plant-microbe interactions. Emerg Top Life Sci. https://doi.org/10.1042/ETLS20210262

Hardwick SA, Deveson IW, Mercer TR (2017) Reference standards for next-generation sequencing. Nat Rev Genet. https://doi.org/10.1038/nrg.2017.44

Hassa J, Maus I, Off S et al (2018) Metagenome, metatranscriptome, and metaproteome approaches unraveled compositions and functional relationships of microbial communities residing in biogas plants. Appl Microbiol Biotechnol. https://doi.org/10.1007/s00253-018-8976-7

Hickman DT, Rasmussen A, Ritz K et al (2021) Review: allelochemicals as multi-kingdom plant defence compounds: towards an integrated approach. Pest Manag Sci. https://doi.org/10.1002/ps.6076

Jansson JK, Hofmockel KS (2018) The soil microbiome—from metagenomics to metaphenomics. Curr Opin Microbiol. https://doi.org/10.1016/j.mib.2018.01.013

Jayaraman D, Forshey KL, Grimsrud PA, Ané JM (2012) Leveraging proteomics to understand plant-microbe interactions. Front Plant Sci. https://doi.org/10.3389/fpls.2012.00044

Jiao WB, Schneeberger K (2017) The impact of third generation genomic technologies on plant genome assembly. Curr Opin Plant Biol. https://doi.org/10.1016/j.pbi.2017.02.002

Johnson JS, Spakowicz DJ, Hong BY et al (2019) Evaluation of 16S rRNA gene sequencing for species and strain-level microbiome analysis. Nat Commun. https://doi.org/10.1038/s41467-019-13036-1

Khatabi B, Gharechahi J, Ghaffari MR et al (2019) Plant–microbe symbiosis: what has proteomics taught us? Proteomics. https://doi.org/10.1002/pmic.201800105

Kim J, Kim MS, Koh AY et al (2016) FMAP: functional mapping and analysis pipeline for metagenomics and metatranscriptomics studies. BMC Bioinf. https://doi.org/10.1186/s12859-016-1278-0

Kimotho RN, Maina S (2024) Unraveling plant–microbe interactions: can integrated omics approaches offer concrete answers? J Exp Bot. https://doi.org/10.1093/jxb/erad448

Korra T, Perumal TN, Thera UK (2023) Multiomics analysis of rhizosphere and plant health. In: Rhizobiome: ecology, management and application. https://doi.org/10.1016/B978-0-443-16030-1.00022-5

Larrainzar E, Wienkoop S, Weckwerth W et al (2007) *Medicago truncatula* root nodule proteome analysis reveals differential plant and bacteroid responses to drought stress. Plant Physiol. https://doi.org/10.1104/pp.107.101618

Levy A, Salas Gonzalez I, Mittelviefhaus M et al (2018) Genomic features of bacterial adaptation to plants. Nat Genet. https://doi.org/10.1038/s41588-017-0012-9

Martinez X, Pozuelo M, Pascal V et al (2016) MetaTrans: an open-source pipeline for metatranscriptomics. Sci Rep. https://doi.org/10.1038/srep26447

Mcgowan T, Johnson JE, Kumar P et al (2020) Multi-omics visualization platform: an extensible galaxy plug-in for multi-omics data visualization and exploration. Gigascience. https://doi.org/10.1093/gigascience/giaa025

Meng C, Zeleznik OA, Thallinger GG et al (2016) Dimension reduction techniques for the integrative analysis of multi-omics data. Brief Bioinform. https://doi.org/10.1093/bib/bbv108

Nobori T, Wang Y, Wu J et al (2020) Multidimensional gene regulatory landscape of a bacterial pathogen in plants. Nat Plants. https://doi.org/10.1038/s41477-020-0690-7

Olanrewaju OS, Babalola OO (2022) The rhizosphere microbial complex in plant health: a review of interaction dynamics. J Integr Agric. https://doi.org/10.1016/S2095-3119(21)63817-0

Orozco-Mosqueda M del C, Rocha-Granados M del C, Glick BR, Santoyo G (2018) Microbiome engineering to improve biocontrol and plant growth-promoting mechanisms. Microbiol Res. https://doi.org/10.1016/j.micres.2018.01.005

Pang Z, Mao X, Xia Y et al (2022) Multiomics reveals the effect of root rot on polygonati rhizome and identifies pathogens and biocontrol strain. Microbiol Spectr. https://doi.org/10.1128/spectrum.02385-21

Priya P, Aneesh B, Harikrishnan K (2021) Genomics as a potential tool to unravel the rhizosphere microbiome interactions on plant health. J Microbiol Methods. https://doi.org/10.1016/j.mimet.2021.106215

Quince C, Walker AW, Simpson JT et al (2017) Shotgun metagenomics, from sampling to analysis. Nat Biotechnol. https://doi.org/10.1038/nbt.3935

Ramos-Cruz D, Troyee AN, Becker C (2021) Epigenetics in plant organismic interactions. Curr Opin Plant Biol. https://doi.org/10.1016/j.pbi.2021.102060

Rane NR, Tapase S, Kanojia A et al (2022) Molecular insights into plant–microbe interactions for sustainable remediation of contaminated environment. Bioresour Technol. https://doi.org/10.1016/j.biortech.2021.126246

Ranjan A, Sinha R, Singla-Pareek SL et al (2022) Shaping the root system architecture in plants for adaptation to drought stress. Physiol Plant. https://doi.org/10.1111/ppl.13651

Regalado J, Lundberg DS, Deusch O et al (2020) Combining whole-genome shotgun sequencing and rRNA gene amplicon analyses to improve detection of microbe–microbe interaction networks in plant leaves. ISME J. https://doi.org/10.1038/s41396-020-0665-8

Rodriguez PA, Rothballer M, Chowdhury SP et al (2019) Systems biology of plant-microbiome interactions. Mol Plant. https://doi.org/10.1016/j.molp.2019.05.006

Saito K, Matsuda F (2010) Metabolomics for functional genomics, systems biology, and biotechnology. Annu Rev Plant Biol. https://doi.org/10.1146/annurev.arplant.043008.092035

Satam H, Joshi K, Mangrolia U et al (2023) Next-generation sequencing technology: current trends and advancements. Biology (Basel). https://doi.org/10.3390/biology12070997

Schenk PM, Carvalhais LC, Kazan K (2012) Unraveling plant-microbe interactions: can multi-species transcriptomics help? Trends Biotechnol. https://doi.org/10.1016/j.tibtech.2011.11.002

Shafi A, Zahoor I, Habib H (2021) Omics technologies to unravel plant-microbe interactions. In: Plant-microbe dynamics: recent advances for sustainable agriculture. https://doi.org/10.1201/9781003106784

Shakya M, Lo CC, Chain PSG (2019) Advances and challenges in metatranscriptomic analysis. Front Genet. https://doi.org/10.3389/fgene.2019.00904

Starr AE, Deeke SA, Li L et al (2018) Proteomic and metaproteomic approaches to understand host-microbe interactions. Anal Chem. https://doi.org/10.1021/acs.analchem.7b04340

Terrón-Camero LC, Gordillo-González F, Salas-Espejo E, Andrés-León E (2022) Comparison of metagenomics and metatranscriptomics tools: a guide to making the right choice. Genes (Basel). https://doi.org/10.3390/genes13122280

Tolani P, Gupta S, Yadav K et al (2021) Big data, integrative omics and network biology. Adv Protein Chem Struct Biol. https://doi.org/10.1016/bs.apcsb.2021.03.006

Wang HB, Zhang ZX, Li H et al (2011) Characterization of metaproteomics in crop rhizospheric soil. J Proteome Res. https://doi.org/10.1021/pr100981r

Wilkins MR, Williams KL (1997) Cross-species protein identification using amino acid composition, peptide mass fingerprinting, isoelectric point and molecular mass: a theoretical evaluation. J Theor Biol. https://doi.org/10.1006/jtbi.1996.0346

Xu L, Dong Z, Chiniquy D et al (2021) Genome-resolved metagenomics reveals role of iron metabolism in drought-induced rhizosphere microbiome dynamics. Nat Commun. https://doi.org/10.1038/s41467-021-23553-7

Yang Y, Wang N, Guo X et al (2017) Comparative analysis of bacterial community structure in the rhizosphere of maize by highthroughput pyrosequencing. PLoS One. https://doi.org/10.1371/journal.pone.0178425

Zhang W, Chait BT (2000) ProFound: an expert system for protein identification using mass spectrometric peptide mapping information. Anal Chem. https://doi.org/10.1021/ac991363o

Metabolomics: Towards Understanding of Plant-Microbe Interaction for Improved Productivity

11

Neha Saini, Sonam Kumari, Hemlata Kumari, Antresh Kumar, and Neelam S. Sangwan

Abstract

Plant metabolomics is a rapidly developing field in the area of biological sciences that has the potential to understand various aspects of plant-microbe interaction. Metabolomics involves extensive investigations of metabolites occurring in the tissues and cells of plants. In natural environments, the plants closely interact with microbes through plant's roots, shoots, stems, and leaves. The plant-microbe interaction may be advantageous with respect nutrition, development, and stress tolerance of the plant. Mycorrhizal fungi play a vital role in sustaining plant growth and development by supplying nutrients, shielding plants from pathogens, thereby enhancing crop production. Metabolic profiling of both plants and their associated microbes helps researchers to understand expression of their complex biochemical pathways in relation to symbiosis, disease resistance, and stimulating growth. This chapter provides insight into the molecular mechanism linked with this interaction as well as the crucial metabolites affecting microbial colonization, plant immune responses, and the exchange of nutrients specially under mycorrhizal plant-interaction. This interaction helps achieve sustainable agriculture practices, improving plant productivity, soil fertility, and better disease resistance. This chapter discusses the aspects of modulation of metabolomics under abiotic stress, the beneficial microbes, and plant resistance to abiotic stress. This chapter also includes the scopes of metabolomics integration in crop improvement strategies to enhance beneficial plant-microbe interaction and hence crop productivity.

N. Saini · S. Kumari · H. Kumari · A. Kumar · N. S. Sangwan (✉)
Department of Biochemistry, School of Interdisciplinary and Applied Sciences, Central University of Haryana, Mahendergarh, Haryana, India
e-mail: nsangwan@cuh.ac.in

U. Pankaj et al. (eds.), *Plant-microbiome Interactions for Climate-resilient Agriculture*, https://doi.org/10.1007/978-981-96-3534-4_11

Keywords

Metabolomics · Primary and secondary metabolites · Abiotic stress · Plant-microbe interaction · Mycorrhiza

1 Introduction

1.1 Metabolomics and Its Importance in Plant Research

Metabolomics is a complete study of small molecules or metabolites within biological systems (Fiehn, 2002). For the purpose of understanding the function of metabolites, key processes of metabolomics, such as identification, quantification, and further analyses, are necessary (Klassen et al. 2017). Metabolomics is established as an omics technology that used to study complex interactions between genes, environment, and metabolism in plants (Hall et al. 2002). Metabolomics has essentially help to comprehending plants because it detects alterations in metabolic levels that reveal how plants respond to stress and other environmental changes. Plants diseases can now be identified and diagnosed early due to metabolomics, which finds specific metabolites and biomarkers associated with a variety of disorders (Terentey and Dolzhenko 2023). The identification of polymorphic genetic markers linked to advantageous traits like stress tolerance could facilitate effective marker-assisted breeding programs improving crop breeding (Ferrnie and Schauer 2009). Thus, metabolomics enhances our understanding of plant interactions with their environment (Allwood et al. 2021).

1.2 Plant-Microbe Interaction

The interaction between microbes and plants is a complex and dynamic phenomenon that has existed for as long as plant colonization. Plants are often invaded by both pathogenic and beneficial microorganisms, especially fungi and bacteria, in both natural and agricultural environments (Cheng et al. 2019). Plant-microbe interactions are crucial for disease resistance in crops. Microorganisms aid in stress tolerance, growth, and nutrient uptake, benefiting both plants and microbes in a symbiotic relationship (Satheesh et al. 2024). Bacteria-bacteria interactions within the plant microbiota play an important function in plant health and productivity, offering the potential for sustainable agricultural strategies without relying on agrochemicals (Barone et al. 2024). Plants and microbes interaction occurs in a variety of ways at different levels (epiphytic as well as endophytic) (Fig. 11.1) (Schirawski and Perlin 2018). The beneficial interactions can be characterized by a number of indirect and direct process, including the transfer of nutrients by mycorrhizal fungi and rhizobia, which bind to roots and supply plants with fixed nitrogen and mineral nutrients, respectively; direct growth stimulation through phytohormones; antagonistic relationships with pathogenic microorganisms; and

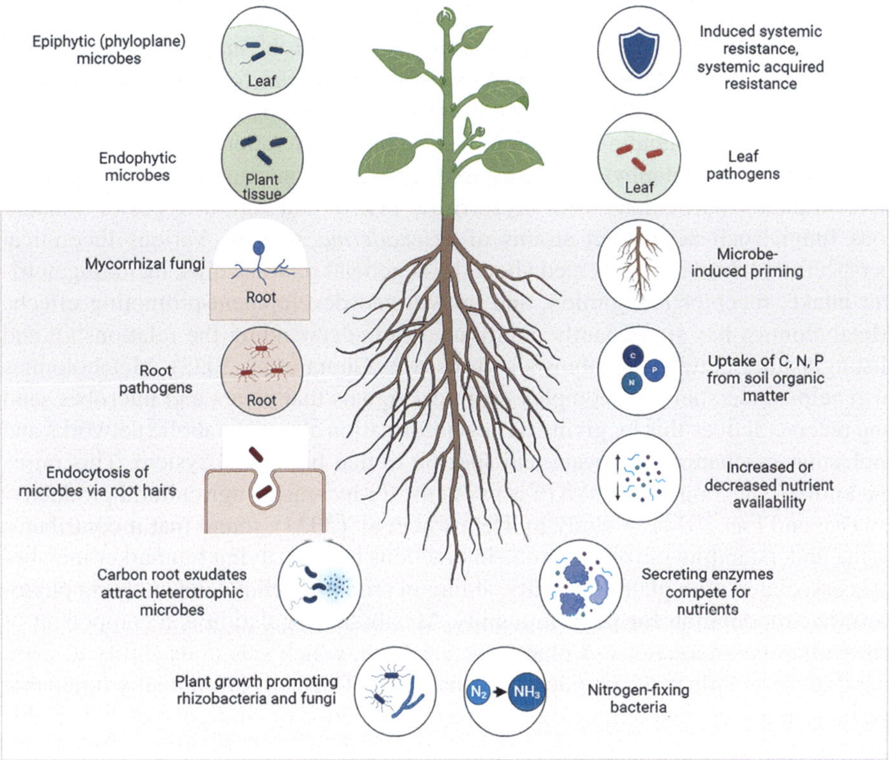

Fig. 11.1 A systematic representation of plant - microbe interaction (BioRender.com)

stress mitigation. The invasive bacteria may be saprophytic, which causes the colonizing plants to necrotrophy, which makes these detrimental interactions hazardous for the plants. Therefore, understanding plant-microbe interaction is essential for recognizing both the positive and negative effect of microbes on plants. Proteomics techniques, next-generation sequencing (NGS) technologies, metagenomics, metabolomics, genomics data analysis, and transcriptomics have all shown to be useful tools and methods for studying plant-pathogen interactions and their interactions (Imam et al. 2016).

1.2.1 Role of Metabolome in Plant-Microbe Interactions

Abiotic factors like drought and salt stress are not the only cause of crop losses in the world; plant diseases also contribute significantly to crop losses. Pathogen outbreaks that cause severe losses can have a profound effect on the agrarian benefits. The fungal pathogen *Phytophthora infestans* caused the Irish potato famine in the 1840s, while the bacterial pathogen *Pantoea stewartii subsp. stewartii* caused the widespread loss of sweet corn in the USA during the early 1900s. Such losses served as a catalyst for study into the mechanisms by which microbial infections cause diseases and the mitigation strategies for future outbreaks (Castro-Moretti et al. 2020). One well-known method for studying plant-microbe interactions is metabolomics. As Fan (2024) suggests, metabolomics is crucial for understanding how plants and microbes interact because it finds metabolic changes that happen when pathogens attack, signaling metabolites in symbiotic relationships, and bioactive compounds from endophytes. There have been studies on significant biotic interactions in plants, particularly with mycorrhiza, PGPB, and various types of filamentous fungi, such as distinct strains of *Trichoderma species*. Various theoretical perspectives have been presented about the beneficial relationships, including nutrient intake, receptor recognition, and growth and development-promoting effects. Metabolomics has significantly contributed to understanding the relationship and distinguishing between symbiosis and disease (Gupta et al. 2022). Metabolomics also helps understand the complex chemical signals that plants and microbes send and receive. It does this by giving us new information about metabolic networks and molecular mechanisms relevant and specific to that biological system. This raises the knowledge about the health of ecosystems for increasing agricultural productivity (Xu and Fan 2024). A study by Demiwal et al. (2024), found that it contributes to the understanding of root-microbe interactions by identifying biomarker metabolites associated with plant immunity, aiding in crop protection, and studying phytohormone modulation for plant immunity. Metabolite regulation is a component of plant-microbe interactions. A plant's metabolome, which aids in its ability to withstand stress, is influenced by microbes (Jha 2024). Thus, it is prominently found that the plant-microbe interaction effect on the accumulation of different metabolites differentially (Table 11.1). In tobacco plant, the interaction of *Bacillus subtilis* with plant enhances the amount of flavonoids (Nazari et al. 2017). In a similar study of *Chrysochus auratus and Asclepias species,* interaction enhance the cardenolides secondary metabolites, which enhance the plant tolerance ability.

2 Primary Metabolites

Plants produce essential substances, called primary metabolites, as part of their normal metabolic processes. They are essential to the growth and development of plants and perform a number of biological processes like synthesis of proteins, cell walls, and energy production (Zaynab et al. 2019) (Fig. 11.2). Different types of primary metabolites present in plants such as sugars, for example, fructose, sucrose, and glucose serve as both the energy sources and the building block of cell walls, amino acids like glutamate, alanine, and asparate etc are the building block of protein synthesis and the precursor of some secondary metabolites (Forde and Lea 2007). Nucleotides are also the primary metabolite, which are the building blocks of DNA and RNA synthesis and the energy currency (Mofatt and Ashihara 2002).

Table 11.1 The role of plant-microbe interaction on the metabolites

Plants	Microbes	Metabolites effect	References
Rice (*Oryza sativa*)	*Fusarium fujikuroi*	Increase in sakuranetin (flavanone)	Siciliano et al. (2015)
Tobacco (*Nicotiana*)	*Bacillus subtilis*	Accumulation of flavonoids	Nazari et al. (2017)
Millet (*Sorghum bicolor*)	*Colletotrichum sublineola*	Increase in 3-deoxyanthocyanidins and luteolin (a flavone)	Du et al. (2010)
Thale cress (*Arabidopsis thaliana*)	*Streptomyces*	Increase in pteridic acid	Yang et al. (2023)
	Pieris rapae Spodoptera	Increase in amino acids	Luo et al. (2023)
Maize (Zea mays)	*Root microbiome*	Decrease of benzoxazinoids	Panz et al. (2021)
Sorghum (*Sorghum bicolor*)	*C. sublineola*	Production of flavonols and anthocyanidins	Wang et al. (2020)
Milkweed (*Asclepias*)	*Chrysochus auratus*	Cardenolides help in tolerance by target site insensitivity	Luo et al. (2023)
Ragwort (*Jacobaea vulgaris*)	*Tyria jacobaeae*	Pyrrolizidine alkaloids help in detoxification by flavin dependent monooxygenase	Luo et al. (2023)
Rhamphospermum nigrum (*Brassica nigra*)	*Brevicoryne brassicae lipaphis erysimi*	Glucosinolates	Luo et al. (2023)

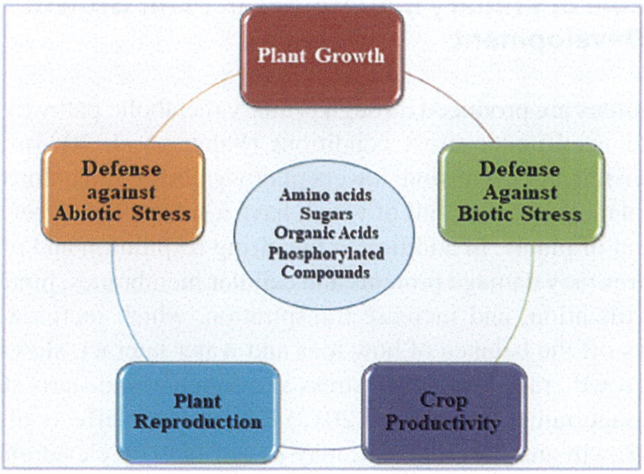

Fig. 11.2 Plant metabolites: Function of different primary plant metabolites in plants

2.1 The Role of Primary Metabolism Under Stress

Plant stress factors are external factors that adversely affect a plant's growth, development, or productivity. Plants are impacted by stress in a variety of ways, including changes in growth rates, yields from crops, which in turn are regulated by cellular metabolism governed by respective gene expression. Both biotic and abiotic environmental stresss have an impact on plants. Biotic stress includes pests and parasites, whereas, abiotic stress caused by factors such as temperature, water deficit or flooding etc, which have negative effect on living organisms (Saini and Sangwan 2024). One of the main elements affecting agricultural productivity and yield is abiotic stress. Plants can be subjected to a variety of abiotic stress, like heavy metals, drought, excessive salt, high and low temperature, and nutritional deficiencies etc (Saini et al. 2024). All of these stresses affect a plant's morphology and physiology, including growth and metabolic functions. These also inhibit important macromolecules, such as enzymes, as well as the transport systems that transport nutrients and large ions. In order to defend against abiotic or biotic stress, plants use a variety of defence mechanisms, either separately or in combination, which can be explored by approaches including transcriptomics, genomics, metabolomics, and proteomics etc (Sangwan et al. 2014, 2015; Salam et al. 2023). Primary metabolites are involved in several aspects of the stress response, such as oxidative stress reduction, osmoprotection and osmotic adjustment, signaling, and stress response regulation. Some primary metabolites that work as osmoprotectants to keep cellular turgor pressure and protect against water stress are proline, glycine betaine, and sugars (Singh et al. 2015). Furthermore, the primary metabolites that have antioxidant activities like ascorbate, phenolic, and glutathione are implicated in the scavenging of reactive oxygen species (ROS) and alleviation of oxidative damage (Khan et al. 2020).

2.2 The Role of Primary Metabolites in Plant Growth and Development

Primary metabolites are produced through primary metabolic pathways of the plants are highly influenced under stress conditions (Salam et al. 2023). Abiotic stress inhibits cell division and expansion, lowers photosynthesis and chlorophyll content, and decreases plant production, all of which have a substantial effect on the growth and development of plants. In addition to impairing respiration and photosynthesis, high temperatures may damage proteins and cellular membranes, hinder pollen generation and fertilization, and increase transpiration, which increase water loss in plants. It throws off the balance of how ions and water interact, slows down photosynthesis and growth, raises oxidative stress and damage, and starts stress signaling pathways (Djanaguiraman and Prasad 2012). Some of the effects of the stress are slowed down growth and development, may be attribute by elevation of stress hormones like ethylene and ABA, stress signaling pathways being activated, antioxidants mechanism being more active, and changes in metabolism and gene expression. Plants also respond to the stress by producing expressing morphological features such as extensive roots system and small leaves to adapt to limited water availability in addition to other morphological and physiological alterations. Molecular expressions are noticeable in terms of modulation of genes involved in several pathways regulated under stress conditions (Dos Santos et al. 2022).

3 Metabolomics Tools and Techniques

3.1 Sample Preparation for Metabolomics Studies

The increasing use of metabolomics allows us to better understand the physiological responses, stress tolerance, and production of secondary metabolites in plants. Accurate metabolomics data require well-prepared plant samples as improper handling might introduce bias, degrade metabolites, and yield inconsistent results (Sangwan et al. 2015; Patel et al. 2021). Compounds with highly complex spectrum having diverse stability, sizes, solubility, amount, and polarity is produced by the structural variety of plant metabolites. Sample preparation is important in plant metabolomics since it may significantly alters the metabolites amounts and hence interpretation. Using this consideration, it is important to collect and store plant samples as quickly as possible to prevent changes to the metabolic processes that occur in the plant cells. A number of critical steps are involved in sample preparation, such as harvesting, quenching, extraction, as well as derivatization. Research on plant metabolomics is most likely to be biased due to improper handling during sample collection. Metabolomics Standards Initiative (MSI) guidelines for sample collection, harvesting, quenching, homogenization, and extract preparation should be followed to support plant metabolomics research (Fig. 11.3) (Sangwan et al. 2015; Patel et al. 2021; Salem et al. 2020).

3.1.1 Sample Collection

In metabolomics research, a large variety of biological samples have been thoroughly analyzed and published. This includes cell cultures (plant, human, algae, animal etc), tissues (derived from animals and plants), and fluids (like saliva, whole blood, urine, amniotic fluid, serum, cerebrospinal fluid, sweat, breast milk, etc.). Different sample preparation techniques are necessary for complex samples since these metabolites are diverse in nature and likely to have distinct matrices (Salem et al. 2020).

3.1.2 Sample Harvesting

In a metabolomics experiment, the precise timing and method of sampling, or harvesting, is critical since it establishes the "metabolic snapshot" of the organism to be studied, which has a direct impact on how the results are biologically interpreted. To prevent diurnal variation and the metabolites loss with high rates of turnover, the harvest process should be performed as soon as the tissues are obtained (Kim and Verpoorte 2010). To prevent biomolecule loss or degradation, biological samples must ideally be frozen in liquid nitrogen (Rodrigues et al. 2019).

3.1.3 Sample Quenching

An important step in metabolomics workflow is quenching, which immediately stops the metabolism and prevents the sample from further changing. This includes metabolite breakdown and changes in concentration, composition, or physical

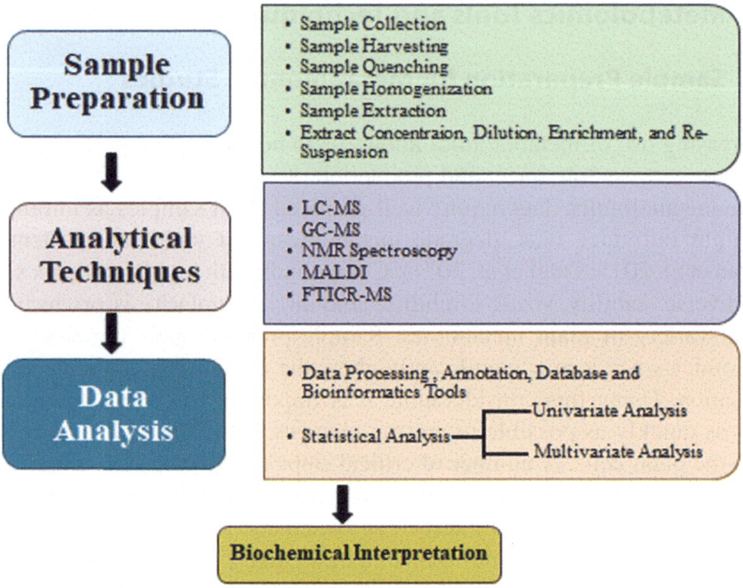

Fig. 11.3 A flow chart of metabolomics studies

characteristics. Freeze-drying and using ice-cold methanol are two more common techniques. Following harvest, freeze-drying is typically the first stage in the processing of the sample and helps retain the biochemical composition of the tissues (Rodrigues et al. 2019).

3.1.4 Sample Homogenization

Homogenization of the sample is accomplished by breaking down the plant cell wall either by homogenizer or grinding by under liquid nitrogen in a chilled mortar and pestle (Rodrigues et al. 2019). Liquid nitrogen is always used during these procedures to keep the tissue safe from thawing; *bead beater*: is especially helpful for hard tissues like seeds and involves mechanically disrupting them with beads, it also offers homogeneity that is quick and consistent; *mortar and pestle*: this conventional approach is popular for lower sample sizes despite being labor-intensive since it ensures from little sample loss and recovers most of it (Kim and Verpoorte 2010) (Table 11.2).

3.1.5 Sample Extraction

The most effective technique for extracting metabolites varies on a number of aspects, such as the physicochemical characteristics of the target metabolites, the system under study's biochemical composition, and the characteristics of the solvent to be utilized. Sonication, solvent extraction, supercritical fluid extraction, and

Table 11.2 Techniques used for metabolite extraction in different plants

Sr. no.	Plant	Tissue	Extraction method	Solvent used	Metabolite extraction	References
1	Tomato (*Solanum lycopersicum*)	Fruit	Solvent extraction, QuEChERS	Acetonitrile, NaCl, $MgSO_4$	Secondary metabolites, phyto-estrogens	Cunha and Fernandes (2011)
2	Potato (*Solanum tuberosum*)	Tubers	Mortar and pestle, solvent extraction	Methanol	Polar metabolites	Kim and Verpoorte (2010)
3	*Arabidopsis thaliana*	Whole plant	Cryogenic grinding, solvent extraction	Methanol: Chloroform	Polar and non-polar metabolites	Kim and Verpoorte (2010)
4	Maize (*Zea mays*)	Seeds	Mortar and pestle, solvent extraction	Methanol (70:30)	Primary and secondary metabolites	Fiehn (2008)
5	Peppermint (*Mentha piperita*)	Leaves	Homogenization with liquid nitrogen, solvent extraction	Methanol	Essential oils, terpenoids	Dettmer et al. (2007)
6	Soybean (*Glycine max*)	Roots	Bead beating, solvent extraction	Methanol, followed by ultrafiltration	Polar and non-polar metabolites	Rochfort (2005)
7	Grapevine (*Vitis vinifera*)	Leaves	Cryogenic grinding, solvent extraction	Methanol: Chloroform	Phenolics and flavonoids	Lisec et al. (2006)

solid-phase extraction are examples of common extraction methods (Tugizimana et al. 2013). Hydrophilic solvents, including hydro-alcoholic solutions, can be used to extract polar and semi-polar metabolites, whereas more hydrophobic solvents can be used to extract lipids (Table 11.2). Liquid-liquid fractionation provides a significant number of process simplifications as compared to single extraction approaches. By removing impurities like more polar metabolites in hydrophilic solvents and hydrophobic compounds concentrated in lipophilic solvents, each fraction is partially purified (Sangwan et al. 2014, 2015; Salem et al. 2020).

3.1.6 Concentration, Dilution, and Enrichment

Other techniques have been employed in addition to liquid-liquid extraction to partially purify samples prior to analytical evaluation. After extracting the sample, solvents are evaporated to concentrate the metabolites. Using a nitrogen flow evaporator is suggested when metabolites, like some lipids, are susceptible to oxidative changes (Salem et al. 2020). Dried samples may go through additional steps before being subjected to GC/MS analysis. Common derivatization reagents are N, O-bis(trimethylsilyl)trifluoroacetamide (BSTFA) and trimethylsilyldiazomethane (TMS) (Wang et al. 2023). Even at extremely low temperatures (like −20 °C), it is not advised to preserve the extracted liquid samples. However, samples can be temporarily stored, if needed, in a dry form before analysis if needed (Salem et al. 2020).

3.2 Analytical Techniques for Plant Metabolomics

Plant metabolomics studies help to understand physiological, developmental, and stress responses (Manickam et al. 2023). Various techniques such as LC-MS, GC-MS, NMR, MALDI, and FTIRC-MS etc., are used to analyze primary and secondary metabolites (Sangwan et al. 2014, 2015; Waris et al. 2022), as shown in (Table 11.3). Plant metabolomics techniques include mass spectrometry, magnetic resonance imaging, and analytical platforms for identifying and quantifying primary and secondary metabolites, aiding in visualizing metabolite localization in plant tissues (Sangwan et al. 2014, 2015; Kusano and Fukushima 2022). In this chapter, we describe recent analytical tools and techniques for studying plant metabolome. We also discuss the significance of sample preparation using targeted and non-targeted methods.

3.2.1 Liquid Chromatography-Tandem Mass Spectrometry (LC-MS)

LC-MS, is one of the most comprehensive analytical techniques, to measure a wide variety of complex compounds. The LC-MS method can be applied for the analysis of high biological mass of plant metabolites, functional groups that are heat-labile, chemical reactive functional groups, and low boiling point compounds (Patel et al. 2021). The crude extracts from simple extraction could be added into the LC-MS, unlike GC-MS, there is no need any sample treatment for their conversion into volatile compounds. The process of obtaining metabolite separation involves the use of columns with specific chemical characteristics, like reversed phase, hydrophobic interaction, and ion exchange. Usually, The development of ultra-performance LC enhanced the method's throughput, sensitivity, and resolution (Patel et al. 2021; Salem et al. 2020). The detection for the phytohormones, folic acid, flavonoids, and other secondary metabolites analysis is accomplished by using the Liquid chromatography-tandem mass spectrometry (LC-MS/MS) (Yi et al. 2022).

Two simple as well as effective methods of treating high-performance liquid chromatography (HPLC) are the gradient elution, where the water-solvent ratio varies for the whole processing and the other being the isocratic elution method where the water-solvent ratio remains constant during the whole processing. Some of the organic solvents that are most commonly used in liquid chromatography include methanol, aceto nitrile, formic acid, and water. Simple samples are best suited under isocratic elution. On the other hand, in changing the gradient elution to the earlier elution technique, it has been shown that there is improved analysis speed, narrower peak, and almost equivalent resolution (Munjal et al. 2022). The

Table 11.3 Analytical techniques used in plant metabolomics

Sr no.	Analytical techniques	Principle	Used in plant metabolomics	Advantages	Limitations	References
1	Liquid Chromatography-Mass Spectrometry (LC-MS)	Separation based on polarity, followed by mass detection	Broad coverage of polar, semi-polar, and non-polar metabolites, such as alkaloids, flavonoids, and phenolic	Wide metabolite coverage, no need for derivatization	Matrix effects, lower reproducibility compared to GC-MS	Dettmer et al. (2007) and Kim and Verpoorte (2010)
2	Gas Chromatography-Mass Spectrometry (GC-MS)	Separation based on volatility and polarity, coupled with mass-based detection	Analysis of volatile and non-volatile metabolites after derivatization (e.g., sugars, amino acids, lipids)	High resolution, sensitivity, and reproducibility	Requires derivatization for non-volatile metabolites	Lisec et al. (2006) and Fiehn (2008)
3	Nuclear Magnetic Resonance (NMR) Spectroscopy	Detection of metabolites based on nuclear spin states in a magnetic field	Identification and quantification of primary metabolites, lipids, and organic acids	Non-destructive, highly reproducible, absolute quantification	Lower sensitivity compared to MS-based techniques	Kim and Verpoorte (2010) and Dunn and Ellis (2005)
4	Matrix-Assisted Laser Desorption/Ionization (MALDI)	Ionizes molecules in a sample using a laser	Identification of plant metabolites like lipids, proteins, and DNA sequences	High resolution, high sensitivity, high analysis scan speed	Matrix absorption	Susniak et al. (2020)

(continued)

Table 11.3 (continued)

Sr no.	Analytical techniques	Principle	Used in plant metabolomics	Advantages	Limitations	References
5	Fourier Transform ion Cyclotron Resonance-Mass Spectrometry (FTICR-MS)	Determines the mass of ions in a magnetic field by examining their cyclotron frequency	Identification of primary and secondary metabolites like amino acids, flavonoids, organic acids	High-resolution fragmented ion detection	High cost, complicated ion-ion interactions, and high magnetic field-related instrumentation	Vinay et al. (2021) and Patel et al. (2021)
6	High-Performance Thin-Layer Chromatography (HPTLC)	Separation based on molecular affinity to the stationary phase	Analysis of pigments, phenolic, alkaloids, and lipids	Simple, cost-effective, can handle multiple samples simultaneously	Limited quantification accuracy, lower resolution	Cunha and Fernandes (2011)
7	Ultra-High-Performance Liquid Chromatography (UHPLC)	Advanced form of LC offering higher pressure and resolution	Improved separation of complex plant metabolite mixtures	High sensitivity and resolution, shorter run times	Requires careful optimization of columns and solvents	Dunn and Ellis (2005)
8	Inductively Coupled Plasma Mass Spectrometry (ICP-MS)	Ionization of samples using plasma, followed by mass detection	Detection of trace metals and metal-containing metabolites in plants	Highly sensitive, accurate quantification	Expensive instrumentation, complex sample preparation	Fiehn (2008)
9	Ion Chromatography (IC)	Separation of ions and charged molecules based on ionic interactions	Analysis of inorganic ions, organic acids, and small polar metabolites	Excellent for ionic metabolites, high sensitivity	Limited to charged metabolites, matrix interference	Dunn and Ellis (2005)

combination of ultra-performance liquid chromatography (UPLC) with high resolution mass analysis like time-of-fight (TOF) mass spectrometry, orbitrap-based mass spectrometer and Fourier transform mass spectrometry has improved the machine performance, which has caused the number of metabolites to rise significantly that can be measured and annotated in recent years (Sangwan et al. 2014, 2015; Salem et al. 2020).

LC-MS is an efficient technique to identify compounds such as alleochemicals and specialized metabolites, under the stress condition (Hazrati et al. 2022). The application of LC-HRMS/MS diagnostic fragmentation filtering technique has shown the ergot alkaloids and indole diterpenoids were found in *Claviceps purpurea* thereby establishing this method as effective for the study of plant metabolome (Uhlig et al. 2021).

3.2.2 Gas Chromatography-Mass Spectrometry (GC-MS)

GC-MS is important technique to plant metabolomics due to its high resolution, sensitivity, repeatability and cost-effectiveness (Singh et al, 2024). It enables the study of genotype, chemotype, genetic mechanisms, stress responses, gene function, and even biotechnological applications. With the help of metabolomics and GC-MS techniques, it has been examined how plant metabolomics can influence these stress responses, and stress resistant crop profiles have been produced, consequently improving their resilience, quality, and productivity (Singh et al 2024; 2021). Moreover, GC-MS allows the profiling of several compounds, including fatty acids, carbohydrates, amino acids which can give more information about plant metabolism and plant stress responses (Shulaev et al. 2008). GC-MS is an important tool for the purpose of identifying and quantifying metabolites as well in plant metabolome study (Ashraf et al. 2022). Furthermore, some of the derivatization methods that can be employed include alkylation, trimethylsilylation, silyation, acylation, and methoxylation. There are two derivatization steps that are required in the extraction as well as the identification of metabolites. Methoxyamine hydrochloride is used in the initial step to convert all carbonyl groups into the appropriate oximes. In the second phase transformation occurs, wherein derivatizing reagents like N-Methyl-N-(trimethylsilyl) trifluoroacetamide (MSTFA) and N, O-bis-(trimethylsilyl)-trifluoroacetamide (BSTFA) are used in order to increase the volatility of the resulting derivative metabolites. TMS-amine, TMS-ester, trimethylsilyl (TMS)-ether, or TMS-sulphide groups are formed as a result of this process; these groups are volatile, less polar, and thermally stable. The hydrogen in some metabolites' -OH, -NH, -COOH, and -SH is substituted with [-Si $(CH_3)_3$] (Patel et al. 2021).

Untargeted GC-MS profiling utilizing full scan collection of information might lead to inaccurate metabolite identifications and require intricate raw data processing. Increased sensitivity, improved specificity, and simpler molecule identification and data processing are all possible outcomes of utilizing GC-MS/MS for targeted analysis. A significant obstacle to the wider application of this technique in metabolomics is the absence of extensive databases containing MRM transitions for endogenous metabolites that are derivatized and those that are not (Choudhury et al. 2022). The potential of *Azadirachta indica* in plant metabolome investigations to identify medicinal secondary metabolites is demonstrated by the discovery of 62 bioactive chemicals by GCMS analysis of leaf extracts (Loganathan et al. 2021).

3.2.3 Nuclear Magnetic Resonance (NMR) Spectroscopy

NMR spectroscopy is very important for plant metabolomics study because it gives us detailed chemical information and helps with metabolite profiling, molecular identification, and biomarker development. As demonstrated in case ofthe *Symphytum anatolicum* study (Kilinc et al. 2023), NMR provides direct quantitative data without chromatographic separation, aiding in depth phytochemical research. Because the technique facilitate the non-destructive analysis of different chemical constituents found in plant sources which contribute toward drug development, metabolomics, and food safety assessments, NMR methods are indispensable tools in plant metabolomics (Chaurasiya et al, 2012; Anand et al. 2022). Nuclear Magnetic Resonance (NMR) based metabolite identification detects samples containing atomic numbers, particularly hydrogen (^{1}H), or carbon (13C) that possess an odd mass. NMR metabolomics tool is more reliable as sample is analyzed in non-destruction mode along with dexterity in the processes of preparation of samples (Patel et al. 2021). There are also no restraints which further supports that NMR spectroscopy has several advantages, such as quick experiment duration, varieties of studies, and excellent repeatability in plant metabolomics (Emwas et al. 2022). Beside mass spectrometry, NMR helps in understanding of metabolic pathways and identification of new compounds, and biosynthetic pathway (Chaurasiya et al, 2012; Nagana Gowda and Raftery 2017). Kumar (2016) finds that in metabolomics profiling, NMR spectroscopy is crucial to the study of plant metabolomics because it has ability to identify the quantitative and qualitative content of various plant sources and products containing endogenous and exogenous biologically active substances of medicinal groups. One study showed that 2D NMR technique like 1H-1H TOCSY or 1H-13C heteronuclear experimental methods are essential for conducting the metabolomics studies and identification of compounds (Mahrous and Farag 2015). *Citrus sinensis* leaf metabolic indicators facilitate the early diagnosis of disease (Pontes et al. 2016). Additionally, researchers examine the primary and secondary metabolites of aromatic Mediterranean plants, providing insights into their molecular composition and seasonal fluctuations (Scognamiglio et al. 2015). Saeidnia and Gohari (2012) found that plant metabolomics techniques using ^{1}H, ^{19}F, ^{31}P, and ^{13}C isotopes to profile various substances like amino acids, lipids, and organic acids, aids in the effective differentiation of metabolites. Although ^{1}H NMR is the most widely used approach for metabolite profiling, it is limited by signal overlapping in the complex combination of plant extracts used in metabolite profiling. ^{1}H NMR is nevertheless rapid and simple to utilize. On the other hand, heteronuclear multiple quantum coherence (HMBC), total correlation spectroscopy (TOCSY), nuclear overhauser effect spectroscopy (NOESY), and two-dimensional (2D) ^{1}H J-resolved NMR are some other advanced techniques. Since chemical extraction is a necessity for both high-resolution magic angle spinning (HRMAS)-NMR, MS and liquid state NMR methods are particularly helpful for solid lyophilized tissue. To identify signals in crowded spectrum regions, J-resolved spectroscopy can be used. Using both J-resolved STOCSY and INADEQUATE C NMR techniques together can speed up large-scale metabolomics research and provide more structural information in lesser time (Uchimiya et al. 2023).

3.2.4 Matrix-Assisted Laser Desorption/Ionization (MALDI)

Metabolome studies in plants employ MALDI, a specialized analytical tool, to generate accurate metabolic quantification and extend our understanding of plant biology by analyzing metabolites. It has been identified that mass spectrometry, including MALDI, can show the identity of plant metabolites under stress (Singh et al. 2021). MALDI MSI is a significant technique of delivering the spatial distribution data without any preinformation about the analyte and applicable in imaging plant metabolites including lipids, proteins, and DNA sequences in tissues (Dong et al, 2020; Susniak et al, 2020). In addition, MALDI-MSI has allowed the identification of the distribution of different proteins and peptides at the same time from a part of plant tissue. The process involved in the MALDI-MSI analysis include sample preparation, acquisition, compound assignment, and image recording (Patel et al. 2021; Alolga et al. 2024) stated that MALDI is the most popular ionization technique used in protein/metabolite imaging when combined with different types of mass analyzers like ToF-ToF, ToF, and QqToF. Some other techniques are in recent use in metabolite identification such as secondary ion mass spectrometry (SIMS), desorption electrospray ionization (DESI), and laser ablation electrospray ionization (LASEI) (Patel et al. 2021).

3.2.5 Fourier Coefficient Cyclotron Resonance-Mass Spectrometry (FTICR-MS)

FTICR-MS, which can detect fragmented ions and provide chemical structural information with high resolution, without the need of chromatographic separation, could give fast metabolite identification in plant metabolomics (Vinay et al. 2021). FTICR-MS is very helpful to look deeply into the metabolites and metabolism of plants (Maia et al. 2023). However, its high cost, intricate ion-ion interactions, and high magnetic field-related apparatus severely hampered its widespread use in plant metabolomics research (Patel et al. 2021). Shahbazy et al. (2020) reported that the time-course metabolic profiling of *Thymus vulgaris* plants under drought stress, could be explored as distinctive metabolites and metabolic pathways. Some other techniques are also used in the detection of plants metabolites such as UHPLC, HPTLC, ICP-MS, IC etc. as shown in Table 11.3.

3.3 Data Processing, Annotation, Database, and Bioinformatics Tools for Plants Metabolome Analysis

With a broad spectrum of analytical tools, metabolomics produces enormous amounts of metabolite data. Different software tools (web-based programs) have been established over the last decades for metabolomics, including mathematical modeling of metabolomics networks and data mining, the processing of raw data, assessment, and interpretation (Patel et al. 2021).

3.3.1 Data Processing and Annotation

There are several applications available for in-silico analysis of spectrum of data of metabolites produced by different analytical devices in large amounts. Processing raw data, mining, and integrating metabolites were all done with web-based tools. Typically, processed data includes baseline shift correction, peak alignment, background noise reduction, and identification, and mass spectrum deconvolution. Many bioinformatics tools are made for pre-processing data; these include MeltDB, MetaboAnalys, MetAlign, AnalyzerPro, MZmine 2, and AMDIS (Automated Mass Spectral Deconvolution and Identification System), for various techniques. One of the most popular programs for processing data from liquid chromatography-mass spectrometry (LC-MS) is called XCMS. It can be found as a cloud-based platform called XCMS, online or as a R package (Domingo-Almenara and Siuzdak 2020). For metabolomics, XCMS is the maximum used peak choosing and deconvolution software (Alboniga et al. 2020). To resource inside the identity of metabolites, the METLIN tandem mass spectrometry (MS/MS) database was developed in 2003. At the time, there was no such database for metabolite identification. The METLIN database, machine learning algorithms, and specific fragmentation information for stable isotopes served as the base for those records (Chen et al. 2022). This database is capable of discovering the metabolites (Hegman 2010). Various techniques exist, including centWave XCMS, which is growth of the centWave set of rules designed for the MZmine2 ADAP wavelet (Chen et al. 2022). MetPA, or Metabolomics Pathway Analysis, is clean-to-use net application that analyzes and visualizes metabolomic records with regards to metabolic pathways. To discover the most applicable metabolic pathways in a given metabolomic study, MetPA combines some of superior pathway enrichment analysis approaches with the evaluation of pathway topological aspects (Xia and Wishart 2010; Chen et al. 2022). The Global Natural Product Social Molecular Networking online mass spectrometry (MS/MS) platform facilitates the processing and annotation of metabolites. GNPS facilitates the identification and discovery of metabolites throughout the data, beginning with data collection and processing, and continuing through post-publication (Patel et al. 2021).

3.3.2 Networks Analysis

The significance of the metabolic pathways can be shown by network analysis, showing connection between these metabolites (Patel et al. 2021). The MetaCyc, small molecular pathway database (SMPDB), AraCyc, and KEGG database are the examples of large-scale metabolic pathway databases. MetaboAnalyst, Paintomics, MPEA, Metscape2, MetaMapRR, ProMeTra, IMPaLA, MBRole, VANTED, and Metabo set enrichment analysis (MSEA) are some of the software that are used for statistical and other metabolite enrichment analyses. Single-sample profiling (SSP), Quantitative enrichment (QEA), and overrepresentation (ORA) are the three types of MSEA analysis. Metscape2, an add-on for the well-known Cytoscape software, allows users to view gene, metabolite, and pathway data in the context of metabolic networks (Patel et al. 2021).

3.4 Statistical Analysis in Metabolomics

3.4.1 Univariate Analysis

Univariate statistical approaches are commonly used to decrease a large number of recorded signals and identify the metabolites that have accumulated differently (Chen et al. 2022). A "univariate analysis" is carried out when just one variable is examined at a time, typically one out of a group of several variables that are evaluated in omics studies. Tests to compare various sample sets, such as the t test or ANOVA, are examples of univariate procedures. When univariate approaches are used in an omics environment, it is typically essential to perform significance testing for tenths to hundreds of metabolites. As a result, multiple test correction is required to protect against an increasing number of false positives (Saccenti et al. 2014). In order to preserve those potentially "information-rich" mzRT traits, univariate techniques are occasionally employed in combination with multivariate models as a filter. After that, only those mzRT features that have shown statistical significance in earlier univariate tests (e.g., p-value <0.05) may be taken into account for the multivariate model (Vinaixa et al. 2012; Chen et al. 2022).

3.4.2 Multivariate Analysis

Multi-omics data usually have specific characteristics that vary depending on phenotypic or experimental conditions; therefore, it is recommended to use multivariate methods, which allow the simultaneous observation and analysis of more than two statistical variables. ANOVA, multiple regression, partial least squares discriminant analysis (PLS), principal component analysis (PCA), cluster analysis, factor analysis, and machine learning (e.g., random forest and SVM) are examples of multivariate analysis. Multivariate analyses are helpful for exploratory data analysis because they employ weighted averages to condense the original variables into fewer variables (Chen et al. 2022). A common unsupervised multivariate linear model in metabolomics studies is principal component analysis (PCA). PCA continues to be the mainstay and industry standard model for handling high-dimensional and complicated data sets in all unsupervised chemo metric techniques (Tugizimana et al. 2013). Partial Least Squares Discriminant Analysis (PLS-DA) is utilized in metabolomics for variable selection in non-targeted metabolomics data, aiding in disease biomarker discovery by classifying relevant variables in complex biological samples (Bujak et al. 2016). PLS-DA can effectively analyze metabolomics data, especially for compositional data in metabolome analysis, aiding in identifying significant metabolites for disease diagnosis (Kalivodova et al. 2015). The use of Random Forests as a classification model in metabolomics data analysis is rare, in contrast to PLS-DA. Although available through freeware tools such as MetaboAnalyst and the Random Forest module on the R platform, its application in metabolomics analysis still requires additional study (Ismail et al. 2017; Lind et al. 2016).

4 Metabolome Profiling in Plant-Microbe Interaction

Metabolomics profiling presents insights into the complex metabolic exchanges that take place at some stage in pathogenic attack, defence responses, mutualism, symbiosis, in plant-microbe interactions. Understanding the complex networks that regulate plant growth, development, and immunity requires an understanding of these interactions at the biochemical and molecular level (Draper et al. 2011). MS and NMR are two examples of analytical techniques that have advanced and made it possible to detect and quantify masses of metabolites (Munjal et al. 2022). Analyzing the chemical change of defensive chemicals and signaling molecules between plants and microbes is a requirement for insights of metabolomics. Researchers may additionally recognize how plants respond to toxic pathogens like bacteria, fungus, and viruses in addition to useful microbes like rhizobia and mycorrhizal fungi by using and analyzing the metabolites of plants and microbes (Mhlongo et al. 2018). This profiling aids in improving our information of how defence mechanisms are triggered by plants and how tolerance of plant is enhanced by beneficial microorganism.

5 Plant-Mycorrhizal Interaction

5.1 Mycorrhizal Fungi: Nutrient Exchange, Protection from Pathogens

Mycorrhiza is a mutualistic symbiotic association between fungi and plants, where fungi obtain fixed carbon sources in exchange for nitrogen and phosphorus sources from plants (Pena and Tibbett 2024). Approximately 50,000 fungal species belong to Basidiomycota, Ascomycota, and Mucoromycota, as well as over 350,000 vascular or non-vascular plants develop mycorrhiza formation for nutrient exchange (Genre et al. 2020). The plant-fungi interactions are categorized as endomycorrhizae, ectomycorrhizae, and ectoendomycorrhizae (Ganugi et al. 2019). Ectomycorrhizal fungi grow on the root surface without penetration of host cells. These are commonly found in cold regions plants (pines and firs), and deciduous forests including poplar and oaks associated with Ascomycetes and Basidiomycetes fungi. However, endomycorrhizal fungi penetrate the cell wall for its growth. Endomycorrhiza establishes a symbiotic relationship with high economic

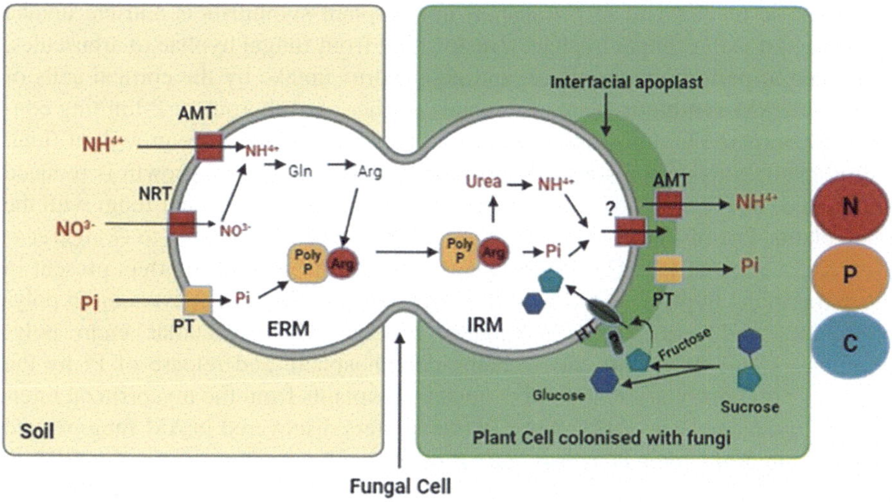

Fig. 11.4 Plants-fungi interactions required for nutrients exchange, growth, and protection

importance. Fungal hyphae enter the plant root system by the epidermis and penetrate the cortical parenchyma of the plant cell. The growth of fungal hyphae from outside the plant roots to inside the soil helps in nutrient uptake, plant growth, stress tolerance, resulting to enhance plant productivity and protection from pathogens (van der Heijden et al. 2015). In fungal-plant symbiosis, plant species invest carbon and gain phosphorous and nitrogen from fungi. (Fig. 11.4; Walder et al. 2012). Out of all the mycorrhizal fungi, ectomycorrhizae and arbuscular mycorrhizae (AM) show high economic and ecological importance in nutrient exchange, stress tolerance, and pathogen protection (Kumar et al. 2017).

5.1.1 Nutrient Exchange Between Plants and Fungi

Nutrient absorption in plants is achieved by plant root hairs and or through the mycorrhizal association (Kumar et al. 2017). The mycorrhizal pathway is a bidirectional that is regulated by both partners via a symbiotic relationship. In the plant-mycorrhizal association, plants ordinarily offer carbon in the form of carbohydrates to the symbiotic fungi in return for nutrients like nitrogen phosphorous, sulfur, potassium, etc. (Xie et al. 2022). Symbiosis association between arbuscular mycorrhizae and plants is the most common mutualism, where arbuscular fungi invade the plant root system and start forming arbuscules in the cortex. The fungal nutrients enter in the plant cell through the peri-arbuscular membrane. The increased nutrients supply to the plants, helps to initiate the carbohydrate supply from plants. Higher the carbon content provided by plants, higher the mineral nutrient offered by the fungi (Wang et al. 2017). Phosphorus (P) and nitrogen (N) are vital macronutrients required for plant growth and development. The available sources of these macronutrients include orthophosphate ($H_2PO_4^-$ and HPO_4^{2-}), nitrite, nitrate, and ammonium ions, etc.

The basic mechanism of P transport in AM-plant symbiosis is nutrient uptake from the soil using fungal hyphae, translocation from fungal hyphae to arbuscules, unloading at periarbuscular space, and phosphorus uptake by the cortical cells of host plant. AM-symbiotic plants can uptake higher phosphorous in P-limiting conditions, compared to non-symbiont plants. Plants also regulate the growth of fungi with P availability in the soil, if P availability is higher, fungal growth is reduced (Johri et al. 2015). P uptake varies with the colonization of different fungi with the same plants. For instance, *G. intraradices* supply more P, compared to *G. aggregatum* (Kiers et al. 2011). The Pi (inorganic orthophosphate) transporters present in the extraradical hyphae of AM help in Pi uptake from the soil, conversion in polyphosphate, and sequestered in tubular vesicles where vacuolar endo polyphosphatase cut down into short chains polyphosphate and release of Pi by the action of polyphosphatase. This Pi is uptaken by plants from the mycorrhizal interface (Nguyen and Saito 2021). Four Pi transporters discovered in AM fungi named as *GvPT* in *Rhizophagus clarus*, *GiPT* in *Rhizophagus irregularis, GigmPT* in *Gigaspora margarita,* and *GmosPT* in *Funneliformis mosseae. GvPT* and *GiPT* are found on extraradical mycelium help in P uptake from the soil, *GmosPT* is found on both intra and extra-radical mycelium, *GigmPT* helps in P acquisition from periarbuscular space (Rui et al. 2022). The phosphate transporters present on periarbuscular membrane are phosphate/H^+ cotransporters which generate a proton gradient for P transport from periarbuscular space to plant cytoplasm. HA1 a H+/ATPase present in *M. truncatula* is essential for P transport during AM symbiosis (Krajinski et al. 2014). Regulation of P homeostasis is a crucial factor in AM symbiosis regulation and establishment.

The mechanism of N transport in AM symbiosis is not completely explored. Ammonium is uptake by mycorrhizae, assimilation in the form of glutamine then in arginine, and transported with phosphate granules, the negative charge on these granules helps in arginine transport from extraradical to intraradical hyphae. The mechanism after intraradical hyphae is still unknown (Wang et al. 2017). *GintAMT1, GintAMT2,* and *GintAMT3* are the AMT (Ammonium transporters) present in *Rhizophagus irregularis,* induced during low nitrogen conditions and help in ammonium uptake from the soil (Wang et al. 2017). The ammonium transporters named *ATM2;3* in Medicago and *GmAMT4.1* in soybean are present on peri-arbuscular membranes suggesting the transport of ammonium in arbuscules (Breuillin-Sessoms et al. 2015; Kobae et al. 2010). Another transporter *LjAMT2;2* (an ammonium transporter) present in *Lotus japonicus* highly upregulated in symbiosis with *G. margarita* (Guether et al. 2009).

5.1.2 Role of Mycorrhizal Fungi in Plant Pathogen Protection

Approximately 45% of crop yield gets lost per year due to different plant diseases which leads to food shortages. However, the pesticides used for the treatment of these pathogens also cause serious health issues (Smms et al. 2021). Mycorrhizal fungi are the suitable alternative and eco-friendly substitute of harmful pesticides used for crop protection from various pathogens. Mycorrhizal fungi symbiotically grow in plant roots and reduce the chances of pathogen attack, this may be the result of physiological and biological changes, host immune system activation, morphology of plant root, nematode parasitism, etc. (Larbi-Koranteng et al. 2022). Mycorrhizal fungi protect the plants from metal toxicity, they can grow in heavy metal-contaminated areas and provide plant resistance to metal toxicity (Hildebrandt et al. 2007). VAM (Vesicular arbuscular mycorrhizae fungi) is used to control the sesame plant root rot disease and wilting caused by *F. oxysporum* (Ziedan et al. 2011). In another study, mycorrhizal fungi help in the strawberry plant to induce resistance against *P. fragariae* by altering the root architecture (Norman et al. 1996).

Mycorrhizal fungi help in the deposition of lignin on plant roots which makes them thicker as compared to non-mycorrhizal plants, thickening of roots hinders the entry of plant pathogens, reduces bacterial blight disease caused by *P. syringae* in mulberry plants by inoculating it with mycorrhizal fungi (*G. mosseae*) (Imad Khrieba 2019). Mycorrhizal fungi play a significant role in maintaining plant growth and development by providing nutrients, protection from pathogens, stress tolerance (cold, drought, salinity, etc.), resistance, crop yield enhancement, soil quality improvement, etc. It's a good alternative to harmful pesticides.

6 Microbial Metabolites and Their Effect on Plants

Plant-microbe interaction is greatly encouraged via microbial metabolites, which might be produced by means of numerous types of soil bacteria. These bioactive materials have a major impact on plant resistance to disease, strain tolerance, and growth. Secondary metabolites, which might be produced by using micro-organism, fungus, and other microorganisms and might be characteristic as growth regulators, signaling molecules, or bio-protectants, are a major component in the symbiotic interaction among vegetation and microbes (Basit et al. 2021). Various microbes produced a number of metabolites which are used against plant disease (Jayaprakashvel and Mathivanan 2011). The growth and reproduction of microorganisms are directly influenced by primary metabolites. These include substances produced through the metabolic processes of microorganisms, such as ethanol, lactic acid, and organic acids. For instance, *Lactobacillus* species produce lactic acid during fermentation, which has significant application in the food and pharmaceutical industries due to its acidic qualities and ability to preserve food. Likewise, *Saccharomyces cerevisiae* and other yeasts create ethanol, which is essential for biofuel production and brewing (Demain 2014).

Although secondary metabolites are not directly involved in growth, they are crucial for an organism's survival in stressful environments. These consist of substances with antibacterial, antifungal, or other pharmacological activities, pigments, alkaloids, and enzymes. Probably the most well-known microbial secondary metabolite, penicillin was the first extensively used antibiotic and revolutionized modern medicine (Demain 2014; Lakshmanaperumalsamy 2009).

Microbial secondary metabolites frequently have unique chemical entities that can be developed into environmentally acceptable, bio rational alternatives to traditional agrochemicals (Saxena 2014). The biosynthesis of secondary metabolites in

Table 11.4 Role of different microbial metabolites on plants

Sr no.	Metabolites/enzyme	Role in Plant interaction	Impact on plants	References
1	Abascisic Acid (ABA)	Stress hormone regulation, ripening signaling	Increase stress tolerance, accelerates fruit ripening, and improves drought resilience	Dresler et al. (2020)
2	1-Aminocyclopropane-1-carboxylate (ACC) deaminase	Ethylene modulation	Reduces ethylene levels in plants under stress, enhancing root growth and stress tolerance	Mir and Karabulut (2025)
3	Exopolysaccharides (EPS)	Biofilm formation, stress mitigation	Protects plants from desiccation and salt stress, enhances root colonization	Bhagat et al. (2021)
4	Flavonoids	Antioxidant activity, stress response signaling	Enhances defense against pathogens and UV stress, regulates auxin transport	Ferreyra et al. (2021)
5	Gibberellins	Hormone-like growth regulation	Promotes stem elongation and flowering, delays senescence	Keswani et al. (2022)
6	Indole-3-acetic acid (IAA)	Phytohormone production, stimulates plant growth	Promotes root development, enhances nutrient uptake, and increases biomass	Timmusk et al. (2023)
7	Phenylpropanoid amides	Phytotoxicity and allelopathy	Inhibits growth of competing plants, improves plant competitiveness	Bertrand et al. (2021)
8	Polyamines	Stress tolerance	Increases abiotic stress tolerance (salinity, drought) and enhances plant development in stressful environments	Sabkia et al. (2021)
9	Siderophores	Iron chelation, nutrient acquisition	Improves plant growth under iron-limiting conditions, enhances stress tolerance	Chandwani et al. (2022)

microbes occurs by different metabolic pathway such as polyketide synthase pathway, shikimate pathway, beta-lactam pathway, and the carbohydrate pathway (Ullah et al. 2020). These microbial metabolites play various roles in plants interaction also effect plant as shown in Table 11.4.

7 Role of Metabolomics for Crop Improvement

Metabolomics is essential to crop development and improvement as it provides key metabolic features which regulate important pathways (Razzaq et al. 2022). For example, metabolomics investigation of wheat types resistant to drought showed higher amounts of osmoprotectants such as proline and glycine betaine, which help protect plants from water, stress (Wahid et al. 2007). Metabolism, identifing critical regulatory points and comprehends metabolic pathways, demonstrates plant response to environmental changes (Shulaev et al. 2008). For instance, analyzing the metabolomes of *Arabidopsis sps* that were exposed to excessive temperatures revealed the role of elevated sugar and tolerance to heat stress (Kaplan et al. 2004). Studies of plant under salt stress, showed adjustments in lipid metabolism that led to an increase in the synthesis of metabolites that protect against stress, such inference help understand mechanism of stress management by plants Metabolism offers insights into plant metabolisc pathways and modulation to environmental responses thus enhancing crop quality and yield, therefore facilitating precision agriculture (Razzaq et al. 2022). Reports by Lecourieux et al. (2020), for example, determined that metabolomics study revealed in flavonoid metabolism in response to temperature alterations, making these helped focus on viticulture strategies feasible.

8 Challenges and Future Direction

A variety of microbial species, such as bacteria, fungi, and archaea, interact with in different ways. Different types of microbes may have different effects on plant metabolism, this microbial diversity adds another level of complexity. A significant challenge is determining how individual microbial taxa contribute to plant metabolomics and comprehending the collective influence of microbial communities. Another challenge is that those metabolites which are present in low concentration are difficult to detect. Similarly, differentiating between highly similar structural metabolites and isomers is highly tricky and difficult. In future, further improvement in analytical techniques, such as high resolution MS, etc. will improve the metabolites detection and quantification, sensitivity and accuracy. A more dynamic overview of plant-microbe interactions will be possible with the development of real-time, non-invasive metabolomics tools that will allow for continuous monitoring of metabolite variations throughout time. Beneficial interactions between plants and microbes, such as nutrient uptake, stress resistance, or disease resistance, may be improved by introducing by microbial consortia or genetically engineering plants with particular metabolic trait features. In order to track and improve these improved features for relationships, metabolicomics will be essential tool.

9 Conclusion

The communication between plants and their microbiota primarily occurs through metabolites. These metabolites are detected with the help of several metabolomics techniques such as NMR, LC-MS, GC-MS, MALDI, FTICR-MS, IMS etc. In order to improve plant-microbe interactions and gain indepth knowledge of the interactions and influences between plants and their associated microbiomes, metabolomics is essential tool. In evaluating the current state of the field, it appears that there is an opportune alignment of research questions with methodological tools because many significant processes in plant-microbe metabolic interactions are underpinned by scantly characterized metabolic exchanges, and newly emerging metabolomic approaches provide a high-throughput platform to identify the relevant molecules. However, studying plant-microbe interactions through metabolomics entails inherent technical challenges, but new approaches like exometabolomics and stable isotope probing seem well-suited to deliver mechanistic insights.

References

Albóniga OE, González O, Alonso RM, Xu Y, Goodacre R (2020) Optimization of XCMS parameters for LC–MS metabolomics: an assessment of automated versus manual tuning and its effect on the final results. Metabolomics 16:1–12

Allwood JW, Williams A, Uthe H, van Dam NM, Mur LA, Grant MR, Pétriacq P (2021) Unravelling plant responses to stress—the importance of targeted and untargeted metabolomics. Metabolites 11(8):558

Alolga RN, Wang SL, Qi LW, Zang H, Huang FQ (2024) MALDI mass spectrometry imaging in targeted drug discovery and development: the pros, the cons, and prospects in global omics techniques. TrAC Trends Anal Chem 178:117860

Anand A, Sharma A, Kaur Saini H, Sharma S, Sharma R, Thakur C et al (2022) Profiling of plant derived natural constituents by using magnetic resonance techniques. Concept Magn Reson A 2022(1):5705637

Ashraf MF, Hou D, Hussain Q, Imran M, Pei J, Ali M et al (2022) Entailing the next-generation sequencing and metabolome for sustainable agriculture by improving plant tolerance. Int J Mol Sci 23(2):651

Barone GD, Zhou Y, Wang H, Xu S, Ma Z, Cernava T, Chen Y (2024) Implications of bacteria–bacteria interactions within the plant microbiota for plant health and productivity. J Zhejiang Univ Sci B 25:1–16

Basit A, Shah ST, Ullah I, Ullah I, Mohamed HI (2021) Microbial bioactive compounds produced by endophytes (bacteria and fungi) and their uses in plant health. In: Plant growth-promoting microbes for sustainable biotic and abiotic stress management. Springer, pp 285–318

Bertrand C, Gonzalez-Coloma A, Prigent-Combaret C (2021) Plant metabolomics to the benefit of crop protection and growth stimulation. In: Advances in botanical research, vol 98. Academic, pp 107–132

Bhagat N, Raghav M, Dubey S, Bedi N (2021) Bacterial exopolysaccharides: insight into their role in plant abiotic stress tolerance. J Microbiol Biotechnol 31(8):1045

Breuillin-Sessoms F, Floss DS, Gomez SK, Pumplin N, Ding Y, Levesque-Tremblay V, Noar RD, Daniels DA, Bravo A, Eaglesham JB (2015) Suppression of arbuscule degeneration in *Medicago truncatula* phosphate transporter4 mutants is dependent on the ammonium transporter 2 family protein AMT2; 3. Plant Cell 27(4):1352–1366

Bujak R, Daghir-Wojtkowiak E, Kaliszan R, Markuszewski MJ (2016) PLS-based and regularization-based methods for the selection of relevant variables in non-targeted metabolomics data. Front Mol Biosci 3:35

Castro-Moretti FR, Gentzel IN, Mackey D, Alonso AP (2020) Metabolomics as an emerging tool for the study of plant–pathogen interactions. Metabolites 10(2):52

Chandwani S, Chavan SM, Paul D, Amaresan N (2022) Bacterial inoculations mitigate different forms of iron (Fe) stress and enhance nutrient uptake in rice seedlings (*Oryza sativa* L.). Environ Technol Innov 26:102326

Chaurasiya, ND, Sangwan NS, Sabir, F, Misra, LN, Sangwan R S (2012) Withanolide biosynthesis recruits both mevalonate and DOXP pathways of isoprenogenesis in Ashwagandha Withania somnifera L. (Dunal). Plant Cell Reports 31: 1889-1897. https://doi.org/10.1007/s00299-012-1302-4
Chen Y, Li EM, Xu LY (2022) Guide to metabolomics analysis: a bioinformatics workflow. Metabolites 12(4):357
Cheng YT, Zhang L, He SY (2019) Plant-microbe interactions facing environmental challenge. Cell Host Microbe 26(2):183–192
Choudhury FK, Pandey P, Meitei R, Cardona D, Gujar AC, Shulaev V (2022) GC-MS/MS profiling of plant metabolites. Methods Mol Biol 2396:101–115
Cunha SC, Fernandes JO (2011) Quantification of phytoestrogens in foods using QuEChERS and gas chromatography–mass spectrometry. Food Chem 126(3):791–798
Demain AL (2014) Importance of microbial natural products and the need to revitalize their discovery. J Ind Microbiol Biotechnol 41(2):185–201
Demiwal P, Tayade S, Yadav SR, Sircar D (2024) A metabolomics perspective on root-derived plant immunity and phytohormone interaction. Physiol Plant 176(1):e14150
Dettmer K, Aronov PA, Hammock BD (2007) Mass spectrometry-based metabolomics. Mass Spectrom Rev 26(1):51–78
Djanaguiraman M, Prasad PV (2012) Effects of salinity on ion transport, water relations and oxidative damage. In: Ecophysiology and responses of plants under salt stress. Springer New York, New York, pp 89–114
Domingo-Almenara X, Siuzdak G (2020) Metabolomics data processing using XCMS. In: Computational methods and data analysis for metabolomics. Humana New York, pp 11–24
Dong Y, Sonawane P, Cohen H, Polturak G, Feldberg L, Avivi SH et al (2020) High mass resolution, spatial metabolite mapping enhances the current plant gene and pathway discovery toolbox. New Phytol 228(6):1986–2002
Dos Santos TB, Ribas AF, de Souza SGH, Budzinski IGF, Domingues DS (2022) Physiological responses to drought, salinity, and heat stress in plants: a review. Stresses 2(1):113–135
Draper J, Rasmussen S, Zubair H (2011) Metabolite analysis and metabolomics in the study of biotrophic interactions between plants and microbes. In: Annual plant reviews volume 43: biology of plant metabolomics, vol 43. Wiley, pp 25–59
Dresler S, Strzemski M, Kováčik J, Sawicki J, Staniak M, Wójciak M, Sowa I, Hawrylak-Nowak B (2020) Tolerance of facultative metallophyte carlina acaulis to cadmium relies on chelating and antioxidative metabolites. Int J Mol Sci 21(8):1–2. https://doi.org/10.3390/ijms21082828
Du Y, Chu H, Wang M, Chu IK, Lo C (2010) Identification of flavone phytoalexins and a pathogen-inducible flavone synthase II gene (SbFNSII) in sorghum. J Exp Bot 61(4):983–994
Dunn WB, Ellis DI (2005) Metabolomics: current analytical platforms and methodologies. Trends Anal Chem 24(4):285–294
Emwas AH, Szczepski K, Poulson BG, McKay R, Tenori L, Saccenti E et al (2022) Nuclear magnetic resonance in metabolomics. In: Metabolomics perspectives. Academic, pp 149–218
Fan K (2024) Unraveling plant-microbe dynamics with metabolomics: pathogens, symbionts, and endophytes. ChemRxiv
Fernie AR, Schauer N (2009) Metabolomics-assisted breeding: a viable option for crop improvement? Trends Genet 25(1):39–48
Ferreyra MLF, Serra P, Casati P (2021) Recent advances on the roles of flavonoids as plant protective molecules after UV and high light exposure. Physiol Plant 173(3):736–749
Fiehn O (2002) Metabolomics—the link between genotypes and phenotypes. Plant Mol Biol 48(1–2):155–171
Fiehn O (2008) Extending the breadth of metabolite profiling by gas chromatography coupled to mass spectrometry. Trends Anal Chem 27(3):261–269
Forde BG, Lea PJ (2007) Glutamate in plants: metabolism, regulation, and signalling. J Exp Bot 58(9):2339–2358

Ganugi P, Masoni A, Pietramellara G, Benedettelli S (2019) A review of studies from the last twenty years on plant–arbuscular mycorrhizal fungi associations and their uses for wheat crops. Agronomy 9(12):840. https://doi.org/10.3390/agronomy9120840

Genre A, Lanfranco L, Perotto S, Bonfante P (2020) Unique and common traits in mycorrhizal symbioses. Nat Rev Microbiol 18(11):649–660. https://doi.org/10.1038/s41579-020-0402-3

Guether M, Neuhäuser B, Balestrini R, Dynowski M, Ludewig U, Bonfante P (2009) A mycorrhizal-specific ammonium transporter from *Lotus japonicus* acquires nitrogen released by arbuscular mycorrhizal fungi. Plant Physiol 150(1):73–83

Gupta S, Schillaci M, Roessner U (2022) Metabolomics as an emerging tool to study plant–microbe interactions. Emerg Top Life Sci 6(2):175–183

Hall R, Beale M, Fiehn O, Hardy N, Sumner L, Bino R (2002) Plant metabolomics: the missing link in functional genomics strategies. Plant Cell 14(7):1437–1440

Hazrati H, Kudsk P, Ding L, Uthe H, Fomsgaard IS (2022) Integrated LC–MS and GC–MS-based metabolomics reveal the effects of plant competition on the rye metabolome. J Agric Food Chem 70(9):3056–3066

Hegeman AD (2010) Plant metabolomics—meeting the analytical challenges of comprehensive metabolite analysis. Brief Funct Genomics 9(2):139–148

Hildebrandt U, Regvar M, Bothe H (2007) Arbuscular mycorrhiza and heavy metal tolerance. Phytochemistry 68(1):139–146

Imad Khrieba M (2019) Mycorrhizae's role in plant nutrition and protection from pathogens. Curr Investig Agric Curr Res 8(1). https://doi.org/10.32474/ciacr.2019.08.000277

Imam J, Singh PK, Shukla P (2016) Plant microbe interactions in post genomic era: perspectives and applications. Front Microbiol 7:1488

Ismail SN, Maulidiani M, Akhtar MT, Abas F, Ismail IS, Khatib A et al (2017) Discriminative analysis of different grades of gaharu (*Aquilaria malaccensis* Lamk.) via 1H-NMR-based metabolomics using PLS-DA and random forests classification models. Molecules 22(10):1612

Jayaprakashvel M, Mathivanan N (2011) Management of plant diseases by microbial metabolites. In: Bacteria in agrobiology: plant nutrient management. Springer Berlin, Heidelberg, pp 237–265

Jha Y (2024) Role of microbes in altering plant metabolome under stressful environments. In: Fungal secondary metabolites. Elsevier, pp 427–437

Johri AK, Oelmüller R, Dua M, Yadav V, Kumar M, Tuteja N, Varma A, Bonfante P, Persson BL, Stroud RM (2015) Fungal association and utilization of phosphate by plants: success, limitations, and future prospects. Front Microbiol 6(OCT):1–13. https://doi.org/10.3389/fmicb.2015.00984

Kalivodová A, Hron K, Filzmoser P, Najdekr L, Janečková H, Adam T (2015) PLS-DA for compositional data with application to metabolomics. J Chemom 29(1):21–28

Kaplan F, Kopka J, Haskell DW, Zhao W, Schiller KC, Gatzke N et al (2004) Exploring the temperature-stress metabolome of Arabidopsis. Plant Physiol 136(4):4159–4168

Keswani C, Singh SP, García-Estrada C, Mezaache-Aichour S, Glare TR, Borriss R et al (2022) Biosynthesis and beneficial effects of microbial gibberellins on crops for sustainable agriculture. J Appl Microbiol 132(3):1597–1615

Khan N, Ali S, Zandi P, Mehmood A, Ullah S, Ikram M et al (2020) Role of sugars, amino acids and organic acids in improving plant abiotic stress tolerance. Pak J Bot 52(2):355–363

Kiers ET, Duhamel M, Beesetty Y, Mensah JA, Franken O, Verbruggen E, Fellbaum CR, Kowalchuk GA, Hart MM, Bago A, Palmer TM, West SA, Vandenkoornhuyse P, Jansa J, Bücking H (2011) Reciprocal rewards stabilize cooperation in the mycorrhizal symbiosis. Science 333(6044):880–882. https://doi.org/10.1126/science.1208473

Kılınc H, D'Urso G, Paolillo A, Alankus O, Piacente S, Masullo M (2023) LC-MS and NMR based plant metabolomics: a comprehensive phytochemical investigation of *Symphytum anatolicum*. Metabolites 13(10):1051

Kim HK, Verpoorte R (2010) Sample preparation for plant metabolomics. Phytochem Anal 21(1):4–13

Klassen A, Faccio AT, Canuto GAB, da Cruz PLR, Ribeiro HC, Tavares MFM, Sussulini A (2017) Metabolomics: definitions and significance in systems biology. In: Metabolomics: from fundamentals to clinical applications. Springer, pp 3–17

Kobae Y, Tamura Y, Takai S, Banba M, Hata S (2010) Localized expression of arbuscular mycorrhiza-inducible ammonium transporters in soybean. Plant Cell Physiol 51(9):1411–1415

Krajinski F, Courty P-E, Sieh D, Franken P, Zhang H, Bucher M, Gerlach N, Kryvoruchko I, Zoeller D, Udvardi M (2014) The H+-ATPase HA1 of *Medicago truncatula* is essential for phosphate transport and plant growth during arbuscular mycorrhizal symbiosis. Plant Cell 26(4):1808–1817

Kumar D (2016) Nuclear magnetic resonance (NMR) spectroscopy for metabolic profiling of medicinal plants and their products. Crit Rev Anal Chem 46(5):400–412

Kumar N, Jaitak V (2024) Recent advancement in NMR based plant metabolomics: techniques, tools, and analytical approaches. Crit Rev Anal Chem 11:1–25

Kumar V, Kumar M, Sharma S, Prasad R (2017) Probiotics in agroecosystem. In: Probiotics in agroecosystem. Springer, pp 1–537. https://doi.org/10.1007/978-981-10-4059-7

Kusano M, Fukushima A (2022) Plant metabolomics: the great potential of plant metabolomics in big data biology. In: Plant omics: advances in big data biology. CABI, pp 50–66

Larbi-Koranteng S, Kankam F, Adomako J, Abdulai M (2022) Role of mycorrhizae in crop protection. In: Arbuscular mycorrhizal fungi in agriculture-new insights. IntechOpen

Lecourieux D, Kappel C, Claverol S, Pieri P, Feil R, Lunn JE et al (2020) Proteomic and metabolomic profiling underlines the stage-and time-dependent effects of high temperature on grape berry metabolism. J Integr Plant Biol 62(8):1132–1158

Lind MV, Savolainen OI, Ross AB (2016) The use of mass spectrometry for analysing metabolite biomarkers in epidemiology: methodological and statistical considerations for application to large numbers of biological samples. Eur J Epidemiol 31:717–733

Lisec J, Schauer N, Kopka J, Willmitzer L, Fernie AR (2006) Gas chromatography mass spectrometry-based metabolite profiling in plants. Nat Protoc 1(1):387–396

Loganathan T, Barathinivas A, Soorya C, Balamurugan S, Nagajothi TG, Jayakumararaj R (2021) GCMS profile of bioactive secondary metabolites with therapeutic potential in the ethanolic leaf extracts of *Azadirachta indica*: a sacred traditional medicinal plant of India. J Drug Deliv Ther 11(4-S):119–126

Luo M, Li B, Jander G, Zhou S (2023) Non-volatile metabolites mediate plant interactions with insect herbivores. Plant J 114(5):1164–1177

Mahrous EA, Farag MA (2015) Two dimensional NMR spectroscopic approaches for exploring plant metabolome: a review. J Adv Res 6(1):3–15

Maia M, Figueiredo A, Cordeiro C, Sousa Silva M (2023) FT-ICR-MS-based metabolomics: a deep dive into plant metabolism. Mass Spectrom Rev 42(5):1535–1556

Manickam S, Rajagopalan VR, Kambale R, Rajasekaran R, Kanagarajan S, Muthurajan R (2023) Plant metabolomics: current initiatives and future prospects. Curr Issues Mol Biol 45(11):8894–8906

Mhlongo MI, Piater LA, Madala NE, Labuschagne N, Dubery IA (2018) The chemistry of plant–microbe interactions in the rhizosphere and the potential for metabolomics to reveal signaling related to defense priming and induced systemic resistance. Front Plant Sci 9:112

Mir MY, Karabulut F (2025) Promoting plant growth by ACC deaminase enzymes producing bacterial endophytes under biotic stress conditions. In: Microbial biostimulants. Apple Academic Press, pp 261–276

Moffatt BA, Ashihara H (2002) Purine and pyrimidine nucleotide synthesis and metabolism. Arabidopsis Book/Am Soc Plant Biol 1:e0018

Munjal Y, Tonk RK, Sharma R (2022) Analytical techniques used in metabolomics: a review. Syst Rev Pharm 13(5):515–521

Nagana Gowda GA, Raftery D (2017) Recent advances in NMR-based metabolomics. Anal Chem 89(1):490–510

Nazari F, Safaie N, Soltani BM, Shams-Bakhsh M, Sharifi M (2017) Bacillus subtilis affects miRNAs and flavanoids production in Agrobacterium-Tobacco interaction. Plant Physiol Biochem 118:98–106

Nguyen CT, Saito K (2021) Role of cell wall polyphosphates in phosphorus transfer at the arbuscular interface in mycorrhizas. Front Plant Sci 12:725939

Norman JR, Atkinson D, Hooker JE (1996) Arbuscular mycorrhizal fungal-induced alteration to root architecture in strawberry and induced resistance to the root pathogen *Phytophthora fragariae*. Plant Soil 185:191–198

Patel MK, Pandey S, Kumar M, Haque MI, Pal S, Yadav NS (2021) Plants metabolome study: emerging tools and techniques. Plan Theory 10(11):2409

Pena R, Tibbett M (2024) Mycorrhizal symbiosis and the nitrogen nutrition of forest trees. Appl Microbiol Biotechnol 108(1). https://doi.org/10.1007/s00253-024-13298-w

Pontes JGM, Ohashi WY, Brasil AJ, Filgueiras PR, Espíndola APD, Silva JS et al (2016) Metabolomics by NMR spectroscopy in plant disease diagnostic: huanglongbing as a case study. ChemistrySelect 1(6):1176–1178

Razzaq A, Wishart DS, Wani SH, Hameed MK, Mubin M, Saleem F (2022) Advances in metabolomics-driven diagnostic breeding and crop improvement. Metabolites 12(6):511

Rochfort S (2005) Metabolomics reviewed: a new "omics" platform technology for systems biology and implications for natural products research. J Nat Prod 68(12):1813–1820

Rodrigues AM, Ribeiro-Barros AI, António C (2019) Experimental design and sample preparation in forest tree metabolomics. Metabolites 9(12):285

Rui W, Mao Z, Li Z (2022) The roles of phosphorus and nitrogen nutrient transporters in the arbuscular mycorrhizal symbiosis. Int J Mol Sci 23(19). https://doi.org/10.3390/ijms231911027

Sabkia MH, Ongb PY, Ibrahimc N, Leea CT, Klemešd JJ, Lie C, Gaoe Y (2021) A review on abiotic stress tolerance and plant growth metabolite framework by plant growth-promoting bacteria for sustainable agriculture. Chem Eng Trans. https://doi.org/10.3303/CET2183062

Saccenti E, Hoefsloot HC, Smilde AK, Westerhuis JA, Hendriks MM (2014) Reflections on univariate and multivariate analysis of metabolomics data. Metabolomics 10:361–374

Saeidnia S, Gohari AR (2012) Application of spectroscopy in herbal metabolomics. DARU J Pharm Sci 20:1–2

Saini N, Sangwan NS (2024) Interplay between nitric oxide and hydrogen sulfide during plant response to abiotic stress. In: H2S in plants. Academic, pp 277–295

Saini N, Modolo LV, Deswal R, Sehrawat A, Sangwan NS (2024) Expanding roles of cross-talk between hydrogen sulfide and nitric oxide under abiotic stress in plants. Plant Physiol Biochem 214:108852

Salam U, Ullah S, Tang ZH, Elateeq AA, Khan Y, Khan J et al (2023) Plant metabolomics: an overview of the role of primary and secondary metabolites against different environmental stress factors. Life 13(3):706

Salem MA, Perez de Souza L, Serag A, Fernie AR, Farag MA, Ezzat SM, Alseekh S (2020) Metabolomics in the context of plant natural products research: from sample preparation to metabolite analysis. Metabolites 10(1):37

Sangwan NS, Mishra LN, Tripathi S, Kushwaha AK (2014) Omics of secondary metabolic pathways in *Withania somnifera* Dunal (ashwagandha). In: Omics applications in crop science. CRC Press/Taylor & Francis, pp 385–408

Sangwan NS, Tiwari P, Mishra SK, Yadav RK, Tripathi S et al (2015) Plant metabolomics: an overview of technology platforms for applications in metabolism. In: Barh D, Khan MS, Davies E (eds) The omics of plant. Springer, p 257298. https://doi.org/10.1007/978-81-322-2172-2_9. ISBN:978-81-322-2171-5

Satheesh S, Kathayat K, Girisha R, Srivastava P (2024) Plant-microbe interactions and their potential application in improving disease resistance in vegetable crops. J Adv Biol Biotechnol 27(7):1011–1027

Saxena S (2014) Microbial metabolites for development of ecofriendly agrochemicals. Allelopath J 33(1):1–24

Schirawski J, Perlin MH (2018) Plant–microbe interaction 2017—the good, the bad and the diverse. Int J Mol Sci 19(5):1374

Scognamiglio M, D'Abrosca B, Esposito A, Fiorentino A (2015) Chemical composition and seasonality of aromatic mediterranean plant species by NMR-based metabolomics. J Anal Methods Chem 2015(1):258570

Shahbazy M, Moradi P, Ertaylan G, Zahraei A, Kompany-Zareh M (2020) FTICR mass spectrometry-based multivariate analysis to explore distinctive metabolites and metabolic pathways: a comprehensive bioanalytical strategy toward time-course metabolic profiling of *Thymus vulgaris* plants responding to drought stress. Plant Sci 290:110257

Shulaev V, Cortes D, Miller G, Mittler R (2008) Metabolomics for plant stress response. Physiol Plant 132(2):199–208

Siciliano I, Amaral Carneiro G, Spadaro D, Garibaldi A, Gullino ML (2015) Jasmonic acid, abscisic acid, and salicylic acid are involved in the phytoalexin responses of rice to Fusarium fujikuroi, a high gibberellin producer pathogen. J Agric Food Chem 63(37):8134–8142

Singh M, Kumar J, Singh S, Singh VP, Prasad SM (2015) Roles of osmoprotectants in improving salinity and drought tolerance in plants: a review. Rev Environ Sci Biotechnol 14:407–426

Singh N, Mansoori A, Dey D, Kumar R, Kumar A (2021) Potential of metabolomics in plant abiotic stress management. In: Omics technologies for sustainable agriculture and global food security, vol II. Springer, pp 193–214

Singh S, Karadagi AV, Doggalli G, Singh R (2024) Metabolomics and metabolic engineering for crop improvement: current trends and future prospects. Plant Cell Biotechnol Mol Biol 25(1–2):77–99

Smms H, Sivasubramaniam N, Smms A (2021) A review on role of mycorrhizal fungi in plant disease management. Sri Lankan J Technol Res 1(2):41–50

Susniak K, Krysa M, Gieroba B, Komaniecka I, Sroka-Bartnicka A (2020) Recent developments in MALDI MSI application in plant tissues analysis. Acta Biochim Pol 67(3):277–281

Terentev A, Dolzhenko V (2023) Can metabolomic approaches become a tool for improving early plant disease detection and diagnosis with modern remote sensing methods? A review. Sensors 23(12):5366

Timmusk S, Pall T, Raz S, Fetsiukh A, Nevo E (2023) The potential for plant growth-promoting bacteria to impact crop productivity in future agricultural systems is linked to understanding the principles of microbial ecology. Front Microbiol 14:1141862

Tugizimana F, Piater L, Dubery I (2013) Plant metabolomics: a new frontier in phytochemical analysis. S Afr J Sci 109(5–6):01–11

Uchimiya M, Olofsson M, Powers MA, Hopkinson BM, Moran MA, Edison AS (2023) 13C NMR metabolomics: J-resolved STOCSY meets INADEQUATE. J Magn Reson 347:107365

Uhlig S, Rangel-Huerta OD, Divon HH, Rolén E, Pauchon K, Sumarah MW et al (2021) Unraveling the ergot alkaloid and indole diterpenoid metabolome in the Claviceps purpurea species complex using LC–HRMS/MS diagnostic fragmentation filtering. J Agric Food Chem 69(25):7137–7148

Ullah A, Bano A, Janjua HT (2020) Microbial secondary metabolites and defense of plant stress. In: Microbial services in restoration ecology. Elsevier, pp 37–46

van der Heijden MGA, Martin FM, Selosse MA, Sanders IR (2015) Mycorrhizal ecology and evolution: the past, the present, and the future. New Phytol 205(4):1406–1423. https://doi.org/10.1111/nph.13288

Venil CK, Lakshmanaperumalsamy P (2009) An insight into microbial pigments: production and applications. Chil J Agric Res 69(3):396–407. https://doi.org/10.4067/S0718-58392009000300013

Vinaixa M, Samino S, Saez I, Duran J, Guinovart JJ, Yanes O (2012) A guideline to univariate statistical analysis for LC/MS-based untargeted metabolomics-derived data. Metabolites 2(4):775–795

Vinay CM, Udayamanoharan SK, Prabhu Basrur N, Paul B, Rai PS (2021) Current analytical technologies and bioinformatic resources for plant metabolomics data. Plant Biotechnol Rep 15(5):561–572

Wahid A, Gelani S, Ashraf M, Foolad MR (2007) Heat tolerance in plants: an overview. Environ Exp Bot 61(3):199–223

Walder F, Niemann H, Natarajan M, Lehmann MF, Boller T, Wiemken A (2012) Mycorrhizal networks: common goods of plants shared under unequal terms of trade. Plant Physiol 159(2):789–797. https://doi.org/10.1104/pp.112.195727

Wang W, Shi J, Xie Q, Jiang Y, Yu N, Wang E (2017) Nutrient exchange and regulation in arbuscular mycorrhizal symbiosis. Mol Plant 10(9):1147–1158. https://doi.org/10.1016/j.molp.2017.07.012

Wang L, Lui AC, Lam PY, Liu G, Godwin ID, Lo C (2020) Transgenic expression of flavanone 3-hydroxylase redirects flavonoid biosynthesis and alleviates anthracnose susceptibility in sorghum. Plant Biotechnol J 18(11):2170

Wang Y, Zhao S, Li X (2023) New derivatization reagents for improved plant metabolite analysis. Metabolomics 19(2):15–24

Waris M, Kocak E, Gonulalan EM, Demirezer LO, Kır S, Nemutlu E (2022) Metabolomics analysis insight into medicinal plant science. TrAC Trends Anal Chem 157:116795

Xia J, Wishart DS (2010) MetPA: a web-based metabolomics tool for pathway analysis and visualization. Bioinformatics 26(18):2342–2344

Xie K, Ren Y, Chen A, Yang C, Zheng Q, Chen J, Wang D, Li Y, Hu S, Xu G (2022) Plant nitrogen nutrition: the roles of arbuscular mycorrhizal fungi. J Plant Physiol 269:153591

Xu Y, Fan KT (2024) Advancements in metabolomics: illuminating plant-microbe interactions for enhanced ecosystems and agricultural productivity. Preprints.org

Yang Z, Qiao Y, Konakalla NC, Strøbech E, Harris P, Peschel G et al (2023) Streptomyces alleviate abiotic stress in plant by producing pteridic acids. Nat Commun 14(1):7398

Yi Y, Sun Y, Wang D, Li X, Wu X, Pan Y, Zhang L (2022) Advances in liquid chromatography-mass spectrometry analysis of several important secondary metabolites in plant metabolomics. Sheng wu Gong Cheng xue bao/Chin J Biotechnol 38(10):3674–3681

Zaynab M, Fatima M, Sharif Y, Zafar MH, Ali H, Khan KA (2019) Role of primary metabolites in plant defense against pathogens. Microb Pathog 137:103728

Ziedan ESH, Sadek Elewa I, Mostafa HM, Sahab AF (2011) Application of mycorrhizae for controlling root diseases of sesame. J Plant Protect Res 51(4). https://doi.org/10.2478/v10045-011-0058-0

Rhizosphere Engineering for Agriculture and Ecological Sustainability

12

Sunil Kumar Verma, Anand Dave, Juveriya Israr, Devendra Singh, and Sankalp Misra

Abstract

The rhizosphere is considered the most intricate microbial environment. Interactions between plants and microbes and various processes in the rhizosphere are vital for improving plant resilience to environmental disturbances. These interactions contribute to sustaining or restoring ecosystem productivity and help address climate change by promoting ecosystems that retain soil carbon for extended durations. This can be achieved by understanding, predicting, and controlling the structure and function of the rhizosphere. This chapter examines methods for enhancing plant productivity and health by modifying plants and the microbes connected to their roots. Rhizosphere engineering can help alter the pH of the rhizosphere, improve nutrient availability, provide protection against biotic and abiotic stresses, and promote the growth of plants. To inhibit the synthesis of stress-induced hormones like ethylene, which slows down root growth, rhizobacteria that support plant growth have been modified to generate lytic enzymes and antibiotics that are effective against soilborne root infections. Plants may also choose advantageous bacteria populations through rhizosphere engineering. Certain crop species or cultivars, for instance, actively seek out and harbor populations of antibiotic-producing strains, which are important in maintaining soils

S. K. Verma · D. Singh
Faculty of Biotechnology, Institute of Biosciences and Technology, Shri Ramswaroop Memorial University, Barabanki, Uttar Pradesh, India

A. Dave
Department of Microbiology, Seventh Day Adventist Arts and Science College, Ahmedabad, India

J. Israr · S. Misra (✉)
Faculty of Biosciences, Institute of Biosciences and Technology, Shri Ramswaroop Memorial University, Barabanki, Uttar Pradesh, India
e-mail: sankalpmisra.ibst@srmu.ac.in

U. Pankaj et al. (eds.), *Plant-microbiome Interactions for Climate-resilient Agriculture*, https://doi.org/10.1007/978-981-96-3534-4_12

that are inherently resistant to soil-borne fungal diseases. Soil amendments can also improve the fitness of root-associated bacterial communities, which has led to the emergence of bacterial consortia that are resistant to bacterial diseases. In addition, plants can be deliberately modified to affect the bacteria that are connected with them. The intricate chemical and biological interactions that take place in the rhizosphere will continue to be better understood through powerful biotechnological advancements and new molecular tools. Therefore, the strategies used to engineer the rhizosphere are extensive and productive, significantly increasing the sustainability of agricultural systems. This chapter highlights significant knowledge gaps in rhizosphere science and discusses how a mechanistic understanding of rhizosphere interactions can be applied in rhizosphere engineering efforts.

Keywords

Rhizosphere engineering · Abiotic stress · Plant growth-promoting rhizobacteria (PGPR) · Plant-microbe interactions · Rhizosphere

1 Introduction

Soil microorganisms first encounter plant-influenced environments in the rhizosphere, the narrow zone where soil particles and plant roots interact (Kar et al. 2021; Misra et al. 2017a). The rhizosphere is a highly active and densely populated soil region within this interaction zone, marked by intricate relationships among various species and food webs that play a crucial role in the movement and transformation of carbon (Dai et al. 2024). The rhizosphere is divided into three distinct zones: the endorhizosphere, which encompasses sections of the root cortex and endodermis where microorganisms and mineral ions exist in the intercellular spaces; the rhizoplane, the intermediate zone next to the root epidermal cells and mucilage; and the ectorhizosphere, the outermost zone that extends from the rhizoplane into the surrounding bulk soil (Rajguru et al. 2024). Rather than being viewed as a zone with a specific size or shape, the rhizosphere should be considered a gradient of all physicochemical and biological properties along the root (Gupta et al. 2020). Plant metabolism plays a key role in influencing the rhizosphere by releasing carbon dioxide (CO_2) and secreting photosynthate through various root exudates, predominantly from the ectorhizosphere and rhizoplane (Cai et al. 2024). These root exudates, rich in phytohormones, enhance interactions in the rhizosphere by supplying energy to microorganisms and serving as chemical attractants or repellents (Agarwal et al. 2024). By affecting the physicochemical properties of the soil and influencing the soil microbial community, these exudates act as signaling molecules that trigger biological and physiological interactions between plant roots and the soil microbial community (Mishra et al. 2023). They also play a role in promoting beneficial symbioses, such as nitrogen-fixing bacteria with fungi and epiphytes, while inhibiting the growth of phytopathogens (Agarwal et al. 2024).

The rhizosphere plays a vital role in ecosystem services such as crop production, nutrient retention, carbon sequestration, and water and carbon cycling (Misra and Chauhan 2020). The rising temperatures and changing weather patterns due to increased atmospheric CO_2 concentrations will have direct and indirect effects on rhizosphere ecology, affecting ecosystem functioning. For example, global warming between 1981 and 2002 is estimated to have reduced cereal crop yields by over $5 billion annually (Imtiaz et al. 2024). Abiotic stresses like drought, extreme heat, and salinity can lead to declines in agricultural productivity, the loss of native plant species, and reduced capacity for CO_2 absorption (Umrao et al. 2024). Drought stress, in particular, inhibits photosynthesis and root growth (Gupta et al. 2023; Mishra et al. 2020). Salinity stress, caused by high levels of Na^+ and Cl^-, results in ion toxicity that negatively affects plant growth and development (Dixit et al. 2023; Misra et al. 2019a, b). Both salinity/alkalinity and drought stress can increase ethylene levels, inhibiting root growth and impacting numerous plant physiological pathways (Kar et al. 2024; Gupta et al. 2020). Additional environmental stresses can disrupt hormone balance, increase susceptibility to diseases, and impair plant development (Singh et al. 2024).

Plants undergo significant physiological changes to survive under abiotic stress conditions, and plant hormones play a critical role in root development and in managing the plant's response to stress (Misra et al. 2023). Abiotic stressors can disrupt the intricate hormonal networks, altering root development in the rhizosphere. Auxin, a phytohormone, is a key regulator of most aspects of plant growth and development, with effects on root architecture (Etesami and Glick 2024). Studies showed that lowering endogenous cytokinin levels can enhance primary root elongation, whereas exogenous application of cytokinins suppresses it (Wang et al. 2023). Abscisic acid (ABA) ameliorates plant responses to environmental stresses like drought. During drought, roots detect reduced soil moisture, leading to increased ABA production in root tips, which is then transported to the leaves, inducing stomatal closure. Research also shows that ABA restricts primary root growth under drought conditions (Luo et al. 2023). Ethylene, a gaseous plant hormone, significantly affects plant growth by inhibiting root elongation and promoting root hair growth (Misra and Chauhan 2020). Strigolactones, newly recognized phytohormones, inhibit lateral root formation while promoting the elongation of adventitious and primary roots (Li et al. 2023a, b). Hormonal interactions, or crosstalk, are crucial to plant growth; for instance, ethylene influences auxin transport in root tips, which regulates root formation (Jacobsen et al. 2021). It also enhances basipetal auxin transport by promoting the expression of auxin biosynthesis-related genes, including AUX1 and PIN2 (Tao et al. 2023). Plant phytohormones also play an important role in beneficial plant-microbe interactions. Therefore, plant hormones should be considered essential elements in any rhizosphere engineering strategy.

While phenotypic plasticity aids plants in adapting to abiotic stress, interactions with natural microbial communities provide another strategy for enhancing resilience to various stresses (Schneider 2022). Like humans and other eukaryotes, plants host diverse microorganisms both inside and outside their tissues—in the ectosphere and endosphere, respectively—comprising their microbiota or microbiomes (Dubey and Misra 2024; Semwal et al. 2023b). Due to their close association with microorganisms, plants can be viewed as holobionts, or meta-organisms, with their microbiota (Snelders et al. 2022). The plant microbiome's genome is often referred to as the plant's second genome. As noted by Diwan et al. (2022), the plant-microbe interactome encompasses a range of relationships, from parasitic to mutualistic, reflecting an ancient co-evolution. One example of an ancient symbiosis is arbuscular mycorrhizal (AM) symbiosis, which has existed for over 400 million years (Krüger et al. 2021). The chemical interplay between the host plant and rhizospheric microbial partners is vital to plant fitness (Gupta et al. 2020). This complex microbial community is essential for plant growth. Plant growth-promoting bacteria (PGPB) enhance plant growth through various mechanisms, both direct and indirect, including nutrient solubilization, nitrogen fixation, systemic resistance induction, production of organic acids, volatile organic compounds (VOCs), plant growth regulators, and enzymes like chitinase, glucanase, and ACC deaminase (Dai et al. 2024; Misra et al. 2017a).

PGPBs promote plant growth through the secretion of microbial exudates such as metabolites and small peptides/lipids near the rhizosphere. A specific group, known as plant growth-promoting rhizobacteria (PGPR), resides in the vicinity or on the surface of roots, aiding in plant growth and development (Dai et al. 2024; Misra et al. 2017a). While Kloepper et al. (1978) introduced the term "PGPR," the influence of soil microbes on plant growth has been recognized for over a century. Research has demonstrated that PGPRs can boost productivity in various crops under normal and stressful conditions, including tomato, lettuce, wheat, soybean, rice, and apples (Mishra et al. 2023; Umrao et al. 2024; Misra et al. 2019a, b). Their modes of action include broad-spectrum biocontrol of soil-borne pathogens, nutrient immobilization (e.g., phosphorus), plant hormone release, and nitrogen fixation (e.g., by *Bradyrhizobium* and *Rhizobium* species) (Misra et al. 2019a, b).

The physicochemical characteristics of the rhizosphere are shaped by factors like plant physiology, microbial community composition, soil type, and moisture content (Li et al. 2021). By manipulating plants, soil, and microbes, rhizosphere conditions can be optimized for enhanced plant productivity. Plants can be genetically modified to possess desirable traits, microbiomes can be selected for characteristics that support plant growth and root development, and soil quality can be improved through amendments that alter its physical and chemical properties. Natural and engineered plant-microbe interactions can be leveraged to enhance nutrient bioavailability, while genetic engineering can further target the chemical signals involved in plant root-microbe communications. This book chapter focuses on the roles of plants and microbes (particularly PGPRs) in rhizosphere engineering.

2 Utilizing the Rhizosphere Region's Naturally Occurring Plant-Microbe Interaction

To mitigate the adverse environmental stresses on plants various approaches have been reported. One common method for enhancing abiotic stress tolerance in plants is the constitutive overexpression of single genes responsible for encoding proteins that function as ion transporters, reactive oxygen species (ROS) scavengers, or enzymes linked to osmolyte accumulation (Umrao et al. 2024). However, transcriptome engineering shows potential as a method to develop crops that are more tolerant to abiotic stresses. Despite this, the approach has often been unsuccessful, as plant acclimatization to stress involves numerous pathways, and there can be pleiotropic effects that negatively influence plant growth (Skalak et al. 2021). Another strategy for increasing crop yield involves the use of agrochemicals or pesticides, but this remains controversial due to their cost and potential long-term environmental impacts when applied to soil.

An alternative method to mitigate the effects of climate change on plant fitness is to leverage beneficial mutualistic interactions between plants and microorganisms in the rhizosphere. Such mutualism can enhance biomass production, nutrient uptake in roots, and overall plant adaptation to abiotic stress (Kar et al. 2024; Dixit et al. 2020). Identifying rhizospheric microbes that confer stress tolerance to plants and using symbiont-based approaches to improve root biomass, overall plant productivity, and soil carbon storage could significantly reduce the negative effects of abiotic stress on plant ecosystems. This approach offers several advantages, such as the ability of plant growth-promoting rhizobacteria (PGPRs) to confer various types of abiotic stress tolerance, particularly by modifying plant hormone pathways, and its versatility across many plant species (Majhi et al. 2023) (Fig. 12.1). Recently, PGPRs have gained considerable attention for their ability to enhance crop productivity and stress tolerance, similar to the outcomes of labor-intensive plant breeding programs (Misra et al. 2019a, b). For example, growth-promoting bacterium such as *Bacillus* spp., *Pseudomonas* sp., *Ochrobactrum* sp., etc. has been shown to colonize and promote the growth of numerous host plants under various conditions, especially during early growth stages (Semwal et al. 2023b; Mishra et al. 2020, 2023), making them a promising candidate for improving crop production.

2.1 Engineering Rhizosphere for Drought Stress Mitigation

Drought is widely recognized as the most significant threat to agricultural and plant productivity, impeding growth and development more than any other environmental condition (Dietz et al. 2021). Several factors, including climate, soil properties, and farming practices, influence drought stress. It is anticipated that the effects of drought will worsen due to climate change and the ongoing depletion of the world's freshwater resources (Gupta et al. 2023; Dietz et al. 2021). Projected changes in global temperatures and precipitation patterns will likely escalate the risk posed by drought to the production of food, feed, and biomass (Gupta et al. 2020). Evidence

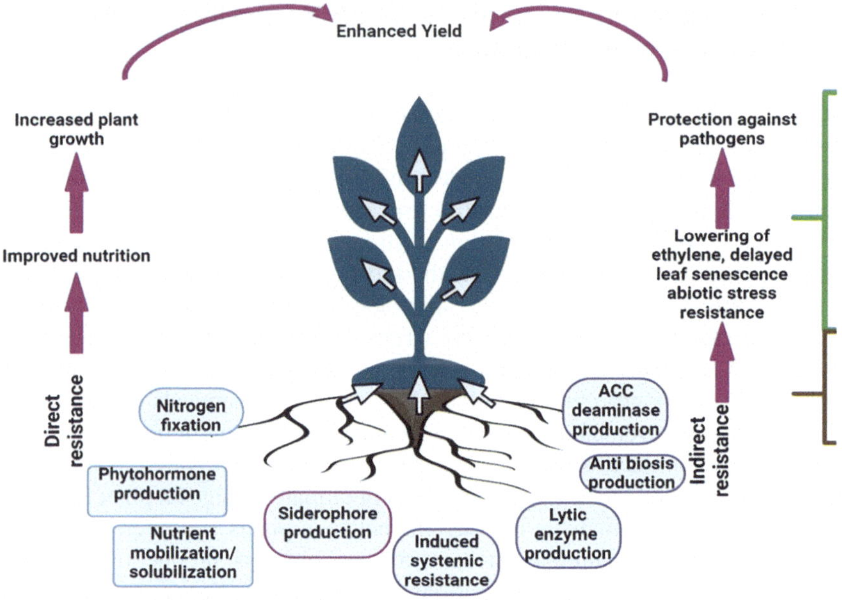

Fig. 12.1 Ecosystem services by plant microbial communities

of poor water-use efficiency (WUE) has been observed in bioenergy crops like Miscanthus and poplar, indicating that many of these crops are somewhat vulnerable to drought (Hoover et al. 2023). Therefore, improving drought tolerance and WUE in bioenergy crops is essential for enhanced production in arid and semi-arid regions, alongside the need to develop drought-resistant crops for food security (Hoover et al. 2023).

Despite the vast amount of research conducted on improving plant drought tolerance, progress has been slow using genetic engineering approaches. This gradual advancement underscores the complexity of the trait and the numerous genes that are involved. The rhizosphere and its associated microbial partners generally govern the plant ecosystems to tolerate drought stress. Plant growth-promoting rhizobacteria (PGPR) commonly resides in the rhizosphere and endorhizosphere, enhancing drought tolerance through various direct and indirect mechanisms (Dixit et al. 2021). These include (i) exopolysaccharides (EPS), 1-aminocyclopropane-1-carboxylate (ACC) deaminase, volatile organic compounds (VOCs), and phytohormones production; (ii) antioxidant and osmolyte accumulation; and (iii) regulating stress-responsive genes and altering root morphology (Dixit et al. 2023; Misra et al. 2017a). For example, *Azospirillum* species that produce IAA were found to increase wheat's resilience to drought by enhancing root growth and lateral root formation (Lastochkina and Allagulova 2023). Similarly, *Bacillus* spp. has improved the nutritional and physiological functions of multiple host crops by producing IAA and other growth-promoting traits, which supported growth under physiological drought conditions (Misra and Chauhan 2020). Additionally, *Bacillus* isolates have been also shown to enhance drought tolerance in Arabidopsis through water stress-induced mechanisms (Gupta et al. 2023). In another study, *Ochrobactrum* sp. treated maize plants showed improved vegetative growth and modulated defense enzyme levels, which conferred drought tolerance (Mishra et al. 2023). Additionally, the *Phyllobacterium brassicacearum* strain, which was isolated from the rhizosphere of oilseed rape (*Brassica napus*), improved osmotic stress tolerance in Arabidopsis by raising ABA levels and decreasing leaf transpiration, thereby assisting in adaptation to drought stress (Kumar et al. 2024). Additionally, maize plants receiving *Alkaligens* strain treatment exhibited increased shoot length, fresh weight, and higher levels of chlorophyll under drought stress compared to control plants (Dixit et al. 2020). Misra et al. (2019a, b) present a promising strategy where microbial strains isolated from extreme environments were used to prime wheat plants. Among these, *Bacillus* strains showed the best results in improving host plant abiotic stress resistance (Misra et al. 2019a, b).

2.2 Engineering Rhizosphere for Salinity Stress Amelioration

Salinity has imposed negative effects on plant productivity worldwide. Excess salt leads to ion toxicity and imbalance in plants, resulting in metabolic disturbances and water deficiency caused by hyperosmotic stress. To counter these effects, plants produce osmolytes and polyamines, activate antioxidants, and regulate ion transport, and compartmentalization. Research by Misra and Chauhan (2020) revealed that inoculating maize with EPS-producing PGPRs, such as *Bacillus*, species, significantly reduced Na^+ uptake and enhanced growth under high salinity conditions. Similarly, Dixit et al. (2023) demonstrated that inoculating Arabidopsis plants with specific PGPRs helped maintain growth in high-salinity and water-limited environments by reducing the negative effects of stress-induced ethylene on root growth through ACC-deaminase activity. Misra et al. (2019a, b) showed that using the halotolerant, carotenoid-producing PGPR strain *Jeotgalicoccu huakuii* conferred salinity tolerance in the host plant by increasing proline levels and other antioxidants, allowing plants to withstand salt stress. This inoculation also activated ABA signaling, SOS pathways, and iron transport. Likewise, different host plants treated with bacterial isolates from *Bacillus* and *Jeotgalicoccus* genera exhibited significant growth under salt stress, maintaining ion balance and reducing ROS accumulation compared to uninoculated seedlings (Misra et al. 2019a, b; Mishra et al. 2020). Misra et al. (2023) found that spinach plant showed improved salt tolerance when inoculated with *Enterobacter* sp. Moreover, when *Arabidopsis* seedlings were subjected to salinity stress using 100 mM NaCl, strain of *Bacillus* significantly boosted plant growth. Inoculated plants had enhanced vegetative parameters as compared to untreated plants (Dixit et al. 2023).

2.3 Engineering Rhizosphere for Alleviating Elevated CO_2 Levels

Photosynthesis enables plants to absorb CO_2 from the atmosphere, storing it as organic carbon in plant sinks, microbial biomass, or soil through root allocation (Xue et al. 2024). Elevated atmospheric CO_2 levels can enhance C3 photosynthesis, which increases the transfer of photosynthates to the soil, potentially stimulating rhizospheric microorganisms (Jin et al. 2023). Over the long term, changes in plant diversity and composition due to climate change affect soil and soil microorganisms, altering the quantity and quality of soil organic matter. Both plant-mediated changes and climate change-induced effects on soil properties and soil biota are likely to significantly influence the carbon cycle, methane and nitrous oxide uptake, and terrestrial ecosystem climate (Xue et al. 2024). PGPR use has become a valuable tool for reforestation, ecosystem restoration, and grassland management (Hoover et al. 2023). As suggested by Semwal et al. (2023a), PGPRs may enhance terrestrial carbon storage by increasing plant productivity while reducing carbon loss through microbial respiration in response to rising atmospheric CO_2 levels. Consequently, higher CO_2 concentrations in future climate scenarios could favor PGPR dominance. The mechanisms behind plant-microbe interactions, the influence of a plant's microbiome on its productivity, and how specific microbiomes alleviate plant stress remain largely unknown. To fully harness the benefits of PGPR in improving plant ecosystem performance, it is essential to thoroughly investigate both the processes that promote plant growth and the steps involved in bacterial colonization of plants.

2.4 Engineering Rhizosphere for Conferring Temperature Stress Tolerance

Elevated temperatures, such as heat stress, are significant abiotic challenges for plant and microorganism growth and stability (Aqeel et al. 2023). Identifying PGPR isolates that promote plant growth under higher temperatures could expand the global range of crop cultivation, especially given future climate predictions. However, few studies have focused on inoculating plants with PGPR to enhance heat stress tolerance (Ahmad et al. 2023). Most research has targeted selecting PGPR isolates (e.g., rhizobial isolates) that can independently tolerate temperatures as high as 60 °C, but their ability to confer heat tolerance to host plants remains unexplored (Misra et al. 2017b). Nevertheless, some studies highlight the use of PGPR to alleviate the negative effects of low temperatures on plant growth (Zhang et al. 2023). Physiological adaptations to cold, such as increases in sugar, proline, and anthocyanin levels, are observed in plants during cold acclimation or hardening (Zhang et al. 2023). Tomato plants receiving *Bacillus methylotrophicus* strain treatment under cold stress demonstrated improved photosynthesis rates, starch deposition, and higher carbohydrate, proline, and phenol accumulation compared to control plants (Li et al. 2023a, b). Additionally, this same PGPR strain helped tomato plants to recover after chilling by reducing biomass loss and electrolyte leakage—an indicator of cell membrane damage—under cold treatment at 4 °C.

2.5 Engineering Rhizosphere for Heavy Metals

When concentrations of heavy metals such as cadmium (Cd), arsenic (As), zinc (Zn), chromium (Cr), mercury (Hg), etc. exceed their tolerance limits or essential needs (with lower levels of Cu, Ni, Cr, and Zn being vital), they present a serious threat to the growth of living forms including plants and microbes (Chauhan et al. 2017). The detrimental effects of these metals in the soil can significantly impact plants and their ability to engage in remediation. However, introducing soil bacteria can enhance this process, referred to as "microbe-assisted phytoremediation" (Dixit et al. 2021).

Several studies have reported heavy metal toxicity alleviation through PGPRs inoculation (Dixit et al. 2021). A wide variety of PGPR genera have been identified as capable of providing this protection, including γ-proteobacteria (*Azotobacter*, *Pseudomonas*, *Mesorhizobium*, *Rhizobium*, and *Bradyrhizobium*), Betaproteobacteria (*Achromobacter*), and Firmicutes (e.g., numerous strains of *Bacillus* spp.) (Kumar et al. 2023). According to Majhi et al. (2023), inoculating rice with *Priestia flexa* strain resulted in improved seed germination and biochemical characteristics under arsenic stress, indicating the strain's potential to protect plants from the adverse effects of heavy metals. Like many microorganisms, plant growth-promoting rhizobacteria (PGPRs) have developed specific strategies for utilizing, transforming, mobilizing, and immobilizing heavy metals to render them inert or less harmful. The five main mechanisms involved in these processes include (1) transport via efflux pumps for extrusion; (2) exclusion by removing metals from targeted sites; (3) inactivation through complexation (such as the formation of thiol-containing compounds); (4) biotransformation to a less toxic redox state (for example, Cr^{+4} compared to Cr^{+6}); and (5) methylation and demethylation (Kumar et al. 2023). While many of the heavy metal resistance strategies are also observed in plants, the relationship between the molecular mechanisms of heavy metal resistance in plants and microbes remains largely unexplored. Further research into the molecular processes responsible for heavy metal tolerance, as well as the interactions between plants and microbes, will enhance our understanding of plant growth in contaminated areas, for agricultural production, and for effective cleanup of spills.

3 Plant-Microbe Symbiotic Interactions: Molecular Mechanisms

The existing literature provides substantial evidence of the ability of PGPRs to promote plant growth in challenging conditions (Misra et al. 2017a, b; Dixit et al. 2021; Majhi et al. 2023). Considerable progress has been made in elucidating the genetic foundations of the beneficial impacts of PGPRs on plants, particularly through whole-genome sequencing initiatives (Gupta et al. 2020, 2023). Nevertheless, there is a scarcity of publications that delve into the molecular mechanisms by which rhizobacteria foster plant growth. Understanding these mechanisms could enhance plant development, particularly under stressful circumstances, through genomic modifications of both microorganisms and plants (Table 12.1).

In this context, Dixit et al. (2023) utilized transcriptome analysis to investigate changes in the physiology and biochemistry of the host plant, Arabidopsis, resulting from the PGPR strains. Their findings indicated that approximately 200 out of 22,810 Arabidopsis genes exhibited altered expression patterns in bacteria-treated plants, with changes exceeding twofold. Most of these genes were associated with functional groups such as stress response, signal transduction, and fundamental metabolism. More precisely, a previous study demonstrated that certain ethylene-related genes were down-expressed, while putative auxin-regulated and nodulin-like genes were up-regulated following PGPR colonization (Gupta et al. 2020). Nong et al. (2023) observed that sugar cane plants treated with the *Gluconacetobacter diazotrophicus* strain triggered ABA-dependent signaling genes that enhance drought resistance, as shown through Illumina RNA-Seq technology. Gupta et al. (2023) emphasized the effects of volatile organic compounds (VOCs) formed by PGPR strain on the gene expression profiles of transgenic Arabidopsis plants. In plants treated with this PGPR strain, there was an upregulation of genes linked to photosynthetic pathways, indicating that VOCs contributed to improved plant growth (Gupta et al. 2023).

In addition to gene expression profiling, proteomic analysis will provide further insights into the proteins and pathways activated during the interaction between host plants and PGPRs. Identifying candidate proteins across various PGPR subtypes will provide valuable information for future studies aimed at promoting growth (Etesami and Glick 2024). Dixit et al. (2018) examined how *Paenibacillus lentimorbus* strain activates extracellular protein elicitors to enhance plant growth and induce systemic resistance in tomato plants. Their investigation employed two-dimensional polyacrylamide gel electrophoresis (2D-PAGE), mass spectrometry (MS), and protein database search to differentiate between the extracellular proteins secreted by the strain, which produces fewer extracellular proteins and lacks growth promotion activity. This research identified twenty extracellular proteins potentially involved in enhancing plant growth and inducing resistance (Dixit et al. 2018). Moreover, Gupta et al. (2020) revealed differentially expressed stress-related proteins in chickpea plants receiving PGPR strain treatment under drought stress. Although proteomic approaches for identifying proteins in PGPR-plant interactions face limitations, such as issues with sample preparation and incomplete protein database information (Gupta et al. 2023), advanced molecular biological techniques like top-down proteomics and MALDI-TOF hold great promise for improving our understanding of the molecular mechanisms underpinning plant-PGPR interactions. Another approach for elucidating the molecular mechanisms of symbiotic relationships is through the analysis of metabolites produced by bacteria and plants. For example, Mishra et al. (2020) employed a GC–MS technique to track emissions of seven stress-related volatile organic compounds (VOCs) from bacterially-primed

Table 12.1 Molecular research employing PGPRs to mitigate abiotic stressors in plants

Plant species	Bacterial species	Mechanism involved	Methodology	References
Arabidopsis	*Pseudomonas fluorescens*	200 differently expressed genes in bacteria inoculated plants	Microarray analysis	Dixit et al. (2023)
Arabidopsis	*Bacillus megaterium*	150 differentially expressed genes in bacteria inoculated plants	Microarray analysis	Vibhuti et al. (2017)
Sugar cane	*Gluconacetobacter diazotrophicus*	Expression of ABA-dependent genes conferring drought tolerance	Illumina sequencing	Vargas et al. (2014)
Arabidopsis	*Burkholderia phytofirmans*	Phytohormone homeostasis	Quantitative gene expression profiling	Poupin et al. (2016)
Chickpea	*Streptomyces* spp.	Increased content of defense enzymes in inoculated plants	Gene expression profiling	Singh and Gaur (2017)
Rice	*Azospirillum brasilense*	Overexpression of ethylene receptors	qRT-PCR analysis	Cook et al. (2022)
Arabidopsis	*Pseudomonas putida*	Upregulation of 520 genes and downregulation of 364 genes	Microarray analysis	Manh et al. (2022)
Pepper	*Bacillus licheniformis*	Differential regulation of stress proteins	PCR and 2D PAGE	Sukkasem et al. (2018)
Wheat	*Dietzia natronolimnaea*	Modulation of salinity tolerance genes	qRT-PCR analysis	Bharti et al. (2016)
Arabidopsis	*Bacillus subtilis*	Regulation of Na^+ transporter and down-regulation of *HKT1* gene	RT-PCR analysis	Venkataraman et al. (2021)
Soybean	*Pseudomonas simiae*	Induced systemic resistance	MALDI-MS/MS analysis	Vaishnav and Choudhary (2019)

(continued)

Table 12.1 (continued)

Plant species	Bacterial species	Mechanism involved	Methodology	References
Okra	*Enterobacter* sp.	Upregulation of antioxidant pathway genes	RT-PCR analysis	Habib et al. (2016)
Cucumber	*Acinetobacter calcoaceticus*, *Burkholdera cepacia*, *Promicromonospora* sp.	Modulation of abiotic stress-related defense enzyme genes	GCMS analysis	Lastochkina and Allagulova (2023)
Arabidopsis	*Pseudomonas fluorescens*	Upregulation of 1910 genes	Microarray analysis and untargeted metabolomics using LC-QTOF-MS	Mavrodi et al. (2021)

drought-stressed wheat seedlings in the rhizosphere. They identified three of these volatiles—benzaldehyde, β-pinene, and geranyl acetone—as promising candidates for a rapid and non-invasive method to assess crop drought stress. By gaining insights into the genes, proteins, and metabolites involved in PGPR-plant interactions that contribute to abiotic stress resistance, it may become possible to develop plants with stress-resilient genes or employ microorganisms that help mitigate stress.

4 Rhizosphere Engineering for Environmental Sustainability

Plant ecosystems are vital for numerous reasons: They create habitats for various species, regulate climate, sequester carbon, cycle carbon dioxide and water, and produce food, feed, and fuel, all of which support human livelihoods and income. Additionally, they offer recreational opportunities. Genetic diversity significantly contributes to maintaining or enhancing the value of plant ecosystems by managing stability and resilience to stress in both native and cultivated environments, boosting productivity in agricultural systems, and supporting ecosystem functions. This is largely due to the wide variety of genotypes available within different plant species. To harness or alleviate the potential adverse effects of climate change and mitigate negative impacts on local or global climates, it is essential to consider the selection of genotypes and species in the design of ecosystems and breeding programs.

To maximize carbon storage in plant ecosystems, it is essential to channel more carbon into above-ground and below-ground biomass. This process allows for the carbon to be allocated to structural components or transferred to the soil, where it can be transformed into stable minerals such as calcite or soil aggregates. Dixit et al. (2021) discussed "Rhizoremediation" while discussing the potential for designing plants with improved carbon storage capabilities. Hoover and colleagues have extensively explored the broader concept of utilizing terrestrial ecosystems for carbon storage (Hoover et al. 2023). Understanding the complex metabolic activities occurring within the rhizosphere, their interactions with plant hosts, and the mechanisms involved in soil carbon deposition is essential for ensuring long-term carbon storage in the soil.

4.1 Engineering the Rhizosphere by Redesigning Plant Metabolism

The enhancement of beneficial genes associated with stress tolerance in a plant's genome can be achieved through advanced plant genome editing methods and traditional breeding techniques. However, efforts to engineer the rhizosphere by modifying the efflux of H^+ ions and organic anions from the roots in transgenic plants have been limited. One primary method by which plants influence the rhizosphere is through root exudation (Agarwal et al. 2024). Given the numerous genes identified that regulate exudate production, it should be feasible to manipulate the expression levels of these genes in plants to improve rhizosphere characteristics. For instance, transgenic rice and tomato plants modified with the Arabidopsis vacuolar H^+-pyrophosphatase gene AVP1 exhibited nearly a 50% increase in citrate and malate efflux when treated with $AlPO_4$ compared to wild-type plants (Cai et al. 2024). This modification was aimed at enhancing the plants' ability to utilize insoluble phosphorus and improve resistance to Al^{3+} stress (Guo et al. 2024). However, several factors complicate the process of engineering plants to impact the rhizosphere effectively, including the degradation or inactivation of engineered compounds in the soil, the slow rates of exudation, the limited understanding of root exudate composition, and the variability in exudate release timing and levels throughout plant development and in response to external stimuli.

To modify how biomass is distributed between aboveground and belowground parts, controlling the allocation of photosynthate in plants is crucial (Mishra et al. 2020). Another strategy involves exploring genotypic diversity when selecting crops or plants from native ecosystems with desirable traits for carbon partitioning and allocation. While one could argue that increased allocation of photosynthates to the soil and roots in the rhizosphere could lead to reduced carbon being directed to harvestable compounds such as grains or lignocellulosics (i.e., harvest index or partitioning efficiency), extensive research indicates that photosynthesis is often limited by sink demand. Insufficient sink demand can inhibit the feedback necessary for effective photosynthesis (Misra et al. 2019a, b; Dixit et al. 2021). Thus, there is potential to channel more carbon underground for long-term storage without negatively impacting overall productivity.

4.2 Rhizosphere Engineering and Microbial Association

The bioengineering of microbial communities presents a significant opportunity for enhancing disease resistance, and stress tolerance and promoting plant or crop growth. Although numerous bacterial strains have demonstrated various beneficial traits, creating a viable synthetic microbial community remains a major challenge. For example, in a simple co-culture of two strains, six ecological interaction parameters must be considered (Kumar et al. 2015): (1) Commensalism, where one strain benefits at the expense of another without causing harm; for instance, one strain's by-products may serve as substrates for another; (2) Competition, which involves competing for substrates; (3) Predation, in which the predator benefits while the prey is harmed; (4) No contact, where there are no interactions such as competition, syntrophy, or predator-prey relationships, resulting in a net-zero effect; (5) Cooperation, where both strains benefit from the interaction; and (6) Amensalism, where one strain suffers while others remain unaffected. As more strains are introduced, the complexity of these ecological interactions increases exponentially (Misra et al. 2017a, b). The primary challenges are to reduce competition and parasitism while enhancing beneficial interactions and cooperation. This is particularly challenging since competition often dominates even in co-cultures of just two strains (Djaya et al. 2019). Moreover, environmental factors such as pH, temperature, host plant exudates, and nutrient availability will influence the stability, growth, seeding rates, and susceptibility to infections of the synthetic microbial community after application.

Despite these challenges, insights from naturally occurring microbial communities that include plant growth-promoting rhizobacteria (PGPR) can aid in developing synthetic microbial communities. Many microbial genera inhabit the rhizosphere, and their genome sequences are publicly available for genetic modification. Notable genera include *Bacillus*, *Ochrobactrum*, *Paenibacillus*, and *Pseudomonas* (Dixit et al. 2021; Misra et al. 2017a, b). Although Streptomyces species are valuable sources of PGPRs due to their well-established genetic systems and the availability of fully sequenced genomes, their large genomes (averaging ~7 MB) and many mobile elements can complicate engineering efforts and cooperative growth in synthetic microbial communities (Khan et al. 2023). *Bacillus* species are preferred as a foundation for a synthetic community, as they are already employed in bio-control applications, are relatively easy to genetically engineer, have a substantial database of complete genome sequences (Djaya et al. 2019), and include many isolates known for their plant growth-promoting properties (Dixit et al. 2018). For example, a *Bacillus* species could be engineered to produce elevated levels of plant hormones, integrate pathways from other *Bacillus* species to combat pathogens, or incorporate nitrogen-fixing mechanisms (e.g., NifH from *Paenibacillus*) (Kumar et al. 2015). Additional genera such as *Bradyrhizobium*, *Rhizobium*, and/or *Pseudomonas* can be added to enhance nitrogen fixation (Ejaz et al. 2020) (Table 12.2). A simple consortium of two strains, consisting of an engineered *Bacillus* species and two naturally occurring or artificially induced nitrogen fixers, could yield numerous benefits typical of a more complex natural rhizosphere community (Kumar et al. 2015). The

Table 12.2 Plant development and the function of rhizosphere engineering using PGPR

Plant host	Microbes used	Effect on plant/soil	References
Growth stimulation			
Wheat	*Enterobacter* sp., *A. chlorophenolicus, S. marcescens, B. megaterium*	P-solubilization, IAA, HCN, N-fixation, Gibberellin, Siderophores	Kumar et al. (2015)
Potato	*Brevundimonas* spp.	P-solubilization, N-fixation	Naqqash et al. (2020)
Pea	*Azospirillum, Agrobacterium tumefaciens*	N-fixation, P-solubilization	Ejaz et al. (2020)
Cabbage	*B. Subtilis*	Gibberellins production	Kang et al. (2019)
Disease infested soils			
Bacterial pathogen			
Rice	*Pseudomonas* sp., *Xanthomonas oryzae*	Peroxidase activity, polyphenol-oxidase, and phenylalanine-ammonia lyase	Yasmin et al. (2016)
Potato	*Lysinibacillus* sp., *Pseudomonas fluorescens, Bacillus subtilis, Ralstonia solanacearum*	Antibiosis	Djaya et al. (2019)
Fungal pathogens			
Tomato	*Paenibacillus lentimorbus, Sclerotium rolfsii*	Expression of defense-related and autophagy-related genes	Dixit et al. (2018)
Chili	*Talaromyces* funiculosus, *Colletotrichum* capsica	ISR, Callose & Lignin deposition	Naziya et al. (2020)

removal of genes that require all strains' biosynthetic machinery to synthesize a critical co-factor could promote cooperation over competition. A two-strain consortium could yield more than 729 potential interactions, whereas a four-member consortium could have up to 531,441 predicted interactions, highlighting how introducing additional strains increases the potential ecological relationships (Kumar et al. 2015). To effectively manage these interactions, it is recommended to limit the number of strains in a synthetic microbial community to no more than three.

When engineering a microbial community for the rhizosphere, several important factors related to microbial competence must be addressed (Misra and Chauhan 2020). Before selecting microbes for an engineered consortium, several trait-related questions need consideration: (1) How well does the microbe colonize the root surfaces of the target plant in the rhizosphere? (2) Can this microorganism successfully colonize the intended host plant? (3) Is the bacterium capable of surviving, proliferating, and competing with other bacteria within the consortium? (4) Does the microbe adhere effectively to the host root surfaces? (5) Does the microorganism promote plant growth or assist consortium members in stimulating plant growth? (6) How does the strain respond to abiotic stressors such as drought? and (7) Does the microbe achieve the necessary density for effective function? Strong attachment to the host plant, along with a motile and chemotactic response to root exudates, facilitates better colonization and potentially enhances beneficial effects on the plant (Misra et al. 2017a, b). A key determinant of a microorganism's beneficial nature is its growth density; for instance, *Paenibacullus* spp. need to reach a minimum density of 10^5–10^6 CFU g^{-1} of root to effectively protect plants from diseases caused by *Sclerotium rolfsii* (Dixit et al. 2018). The microbial consortia must also be resilient against pesticides, fertilizers, and herbicides without losing their beneficial properties if conventional farming practices are employed. Once suitable bacteria are selected, questions about the timing, location, and method of inoculation arise. Should the inoculum be applied to the roots, leaves, or both? Is it necessary for the inoculum to be live, or can it be freeze-dried? Does inoculation have greater efficacy during the day or night? The most common types of inoculants directly applied to seeds to enhance crop performance include pelleted, slurry-coated, film-coated, and bioprimed inoculants.

5 Conclusion

The challenge of increasing crop yield with environmental sustainability, while also understanding the interplay between plant and environmental processes affects both our nation and the world as a whole. To achieve these ambitious objectives, it is essential to maximize the productivity of plant ecosystems across diverse and challenging environments. Moreover, we must build a robust knowledge base that allows us to anticipate the impacts of climate change on plant ecosystems, including the rhizosphere, as well as understand how the biogeochemical cycles within these ecosystems influence climate. Within this framework, rhizosphere and ecosystem engineering offer a promising avenue to address significant knowledge gaps and provide solutions. For example, there is still much to learn about the interactions between the entire ectophytic and endophytic microbial communities and their plant hosts, including both beneficial and detrimental activities, as well as the interactions among these communities themselves, particularly concerning fungal-rhizospheric microbiome interactions. To fully exploit a cropping system's yield potential, understand the rationale behind priming the rhizosphere, and manage the soil carbon cycle in both current and future climatic scenarios, it will be necessary to decode the plant-microbe interactome. Future research initiatives will leverage synthetic biology techniques, harness advantageous root-microbe interactions, and integrate these approaches to enhance the productivity of key food and bioenergy crops under various biotic and abiotic stresses. Simultaneously, this will facilitate increased atmospheric CO_2 absorption, helping to stabilize carbon pools in the soil.

Acknowledgments The author acknowledges the financial assistance from the Council of Science & Technology, Uttar Pradesh (UPCST) project (Project ID 2707). The authors also thank the in-house seed grant project (File No. FY-2023-24/SMG/001) funded by Shri Ramswaroop Memorial University, Lucknow- Deva Road, Barabanki-225003, U.P., India.

References

Agarwal P, Vibhandik R, Agrahari R, Daverey A, Rani R (2024) Role of root exudates on the soil microbial diversity and biogeochemistry of heavy metals. Appl Biochem Biotechnol 196(5):2673–2693

Ahmad M, Imtiaz M, Nawaz MS, Mubeen F, Sarwar Y, Hayat M, Asif M, Naqvi RZ, Ahmad M, Imran A (2023) Thermotolerant PGPR consortium B3P modulates physio-biochemical and molecular machinery for enhanced heat tolerance in maize during early vegetative growth. Ann Microbiol 73(1):34

Aqeel M, Ran J, Hu W, Irshad MK, Dong L, Akram MA, Eldesoky GE, Aljuwayid AM, Chuah LF, Deng J (2023) Plant-soil-microbe interactions in maintaining ecosystem stability and coordinated turnover under changing environmental conditions. Chemosphere 318:137924

Bharti N, Pandey SS, Barnawal D, Patel VK, Kalra A (2016) Plant growth promoting rhizobacteria *Dietzia natronolimnaea* modulates the expression of stress responsive genes providing protection of wheat from salinity stress. Sci Rep 6(1):1–6

Cai Y, Yu X, Zou Y, Ding S, Min Y (2024) Shrubification of herbaceous peatlands modulates root exudates, increasing rhizosphere soil CO_2 emissions while decreasing CH_4 emissions. Catena 245:108282

Chauhan PS, Mishra SK, Misra S, Dixit VK, Pandey S, Khare P, Khan MH, Dwivedi S, Lehri A (2017) Evaluation of fertility indicators associated with arsenic contaminated paddy fields soil. Int J Environ Sci Technol 15:2447–2458

Cook J, Degon Z, Ruiz D, Pope J, Rahmatallah Y, Mukherjee A (2022) The plant growth-promoting bacteria, *Azospirillum brasilense*, induce a diverse array of genes in rice shoots and promote their growth. J Plant Growth Regul 97(1):143–155

Dai H, Wu B, Zhuang Y, Ren H, Chen Y, Zhang F, Chu C, Lv X, Xu J, Ma B (2024) Dynamic in situ detection in iRhizo-Chip reveals diurnal fluctuations of *Bacillus subtilis* in the rhizosphere. Proc Natl Acad Sci 121(40):e2408711121

Dietz KJ, Zörb C, Geilfus CM (2021) Drought and crop yield. Plant Biol 23(6):881–893

Diwan D, Rashid MM, Vaishnav A (2022) Current understanding of plant-microbe interaction through the lenses of multi-omics approaches and their benefits in sustainable agriculture. Microbiol Res 265:127180

Dixit R, Agrawal L, Singh SP, Singh PC, Prasad V, Chauhan PS (2018) *Paenibacillus lentimorbus* induces autophagy for protecting tomato from *Sclerotium rolfsii* infection. Microbiol Res 215:164–174

Dixit VK, Misra S, Mishra SK, Tewari SK, Joshi N, Chauhan PS (2020) Characterization of plant growth-promoting alkalotolerant *Alcaligenes* and *Bacillus* strains for mitigating the alkaline stress in *Zea mays*. Antonie Van Leeuwenhoek 113:889–905

Dixit VK, Misra S, Mishra SK, Joshi N, Chauhan PS (2021) Rhizobacteria-mediated bioremediation: insights and future perspectives. In: Parray JA, Mahmoud AHAE, Sayyed R (eds) Soil bioremediation: an approach towards sustainable technology. Wiley, pp 193–211

Dixit R, Bisht N, Misra S, Gupta SC, Chauhan PS (2023) Bacillus consortia modulate transcriptional and metabolic machinery of arabidopsis plants for salt tolerance. Curr Microbiol 80:77

Djaya L, Istifadah N, Hartati S, Joni IM (2019) *In vitro* study of plant growth promoting rhizobacteria (PGPR) and endophytic bacteria antagonistic to *Ralstonia solanacearum* formulated with graphite and silica nano particles as a biocontrol delivery system (BDS). Biocatal Agric Biotechnol 19:101153

Dubey R, Misra S (2024) Biostimulants: an eco-friendly regulator of plant stress tolerance and sustainable solution to future agriculture. Proc Indian Natl Sci Acad 20:1–8

Ejaz S, Batool S, Anjum MA, Naz S, Qayyum MF, Naqqash T, Shah KH, Ali S (2020) Effects of inoculation of root-associative *Azospirillum* and *Agrobacterium* strains on growth, yield and quality of pea (*Pisum sativum* L.) grown under different nitrogen and phosphorus regimes. Sci Hortic 270:109401

Etesami H, Glick BR (2024) Bacterial indole-3-acetic acid: a key regulator for plant growth, plant-microbe interactions, and agricultural adaptive resilience. Microbiol Res 11:127602

Guo C, Shabala S, Chen ZH, Zhou M, Zhao C (2024) Aluminium tolerance and stomata operation: towards optimising crop performance in acid soil. Plant Physiol Biochem 10:108626

Gupta S, Mishra SK, Misra S, Pandey V, Agrawal L, Nautiyal CS, Chauhan PS (2020) Revealing the complexity of protein abundance in chickpea root under drought-stress using a comparative proteomics approach. Plant Physiol Biochem 151:88–102

Gupta S, Misra S, Kumar M, Mishra SK, Tiwari S, Narayan S, Agrawal L, Chauhan PS (2023) Enhancement of drought tolerance in transgenic *Arabidopsis thaliana* plants overexpressing chickpea Ca14-3-3 gene. J Plant Growth Regul 42:1544–1557

Habib SH, Kausar H, Saud HM (2016) Plant growth-promoting rhizobacteria enhance salinity stress tolerance in okra through ROS-scavenging enzymes. Biomed Res Int 2016(1):6284547

Hoover DL, Abendroth LJ, Browning DM, Saha A, Snyder K, Wagle P, Witthaus L, Baffaut C, Biederman JA, Bosch DD, Bracho R (2023) Indicators of water use efficiency across diverse agroecosystems and spatiotemporal scales. Sci Total Environ 15(864):160992

Imtiaz S, Shahid S, Ishfaq T, Ilyas M, Nawaz AF, Shamshad J, Fiaz S, Arif M (2024) Impact of climate change on agriculture. In: Fahad S, Saud S, Nawaz T, Gu L, Ahmad M, Zhou R (eds) Environment, climate, plant and vegetation growth. Springer Nature Switzerland, pp 285–305

Jacobsen AG, Jervis G, Xu J, Topping JF, Lindsey K (2021) Root growth responses to mechanical impedance are regulated by a network of ROS, ethylene and auxin signalling in Arabidopsis. New Phytol 231(1):225–242

Jin K, Chen G, Yang Y, Zhang Z, Lu T (2023) Strategies for manipulating rubisco and creating photorespiratory bypass to boost C3 photosynthesis: prospects on modern crop improvement. Plant Cell Environ 46(2):363–378

Kang SM, Hamayun M, Khan MA, Iqbal A, Lee IJ (2019) *Bacillus subtilis* JW1 enhances plant growth and nutrient uptake of Chinese cabbage through gibberellins secretion. J Appl Bot Food Qual 1:92

Kar S, Jaiswal P, Misra S, Chauhan PS (2021) Endospheric microbiome-assisted alteration in the metabolomic profiling of host towards abiotic stress mitigation. In: Pudake RN, Sahu BB, Kumari M, Sharma AK (eds) Omics science for rhizosphere biology. Springer Nature Switzerland, pp 263–279

Kar S, Mishra SK, Misra S, Agarwal R, Kumar S, Chauhan PS (2024) Endophytic alkalotolerant plant growth-promoting bacteria render maize (*Zea mays* L.) growth under alkaline stress. Curr Microbiol 81(1):43

Khan S, Srivastava S, Karnwal A, Malik T (2023) *Streptomyces* as a promising biological control agents for plant pathogens. Front Microbiol 14:1285543

Kloepper JW, Schroth MN, Miller TD (1978) Effects of rhizosphere colonization by plant growth-promoting rhizobacteria on potato plant development and yield. Phytopathology 70(11):1078–1082

Krüger M, Krüger C, Wang Y, Stajich JE, Keller J, Chen EC, Yildirir G, Villeneuve-Laroche M, Roux C, Delaux PM, Corradi N (2021) The genome of *Geosiphon pyriformis* reveals ancestral traits linked to the emergence of the arbuscular mycorrhizal symbiosis. Curr Biol 31(7):1570–1577

Kumar M, Mishra S, Dixit V, Kumar M, Agarwal L, Chauhan PS, Nautiyal CS (2015) Synergistic effect of *Pseudomonas putida* and *Bacillus amyloliquefaciens* ameliorates drought stress in chickpea (*Cicer arietinum* L.). Plant Signal Behav 11:e1071004

Kumar RS, Koner S, Tsai HC, Chen JS, Huang SW, Hsu BM (2023) Deciphering endemic rhizosphere microbiome community's structure towards the host-derived heavy metals tolerance and plant growth promotion functions in serpentine geo-ecosystem. J Hazard Mater 452:131359

Kumar A, Naroju SP, Kumari N, Arsey S, Kumar D, Gubre DF, Roychowdhury A, Tyagi S, Saini P (2024) The role of drought response genes and plant growth promoting bacteria on plant growth promotion under sustainable agriculture: a review. Microbiol Res 286:127827

Lastochkina OV, Allagulova CR (2023) The mechanisms of the growth promotion and protective effects of endophytic PGP bacteria in wheat plants under the impact of drought. Appl Biochem Microbiol 59(1):14–32

Li M, Wang J, Yao T, Zhang T, Zhou Q (2021) Bacterial diversity and community structure in the rhizosphere of four halophytes. Curr Microbiol 78(7):2720–2732

Li J, Fan M, Zhang Q, Lü G, Wu X, Gong B, Wang Y, Zhang Y, Gao H (2023a) Transcriptome analysis reveals that auxin promotes strigolactone-induced adventitious root growth in the hypocotyl of melon seedlings. Front Plant Sci 14:1192340

Li G, Peng T, Qu F, Wang J, Long Y, Hu X (2023b) *Bacillus methylotrophicus* could improve the tolerance and recovery ability of the tomato to low-temperature stress and improve fruit quality. Agronomy 13(7):1902

Luo X, Xu J, Zheng C, Yang Y, Wang L, Zhang R, Ren X, Wei S, Aziz U, Du J, Liu W (2023) Abscisic acid inhibits primary root growth by impairing ABI4-mediated cell cycle and auxin biosynthesis. Plant Physiol 191(1):265–279

Majhi B, Semwal P, Mishra SK, Misra S, Chauhan PS (2023) Arsenic stress management through arsenite and arsenate-tolerant growth-promoting bacteria in rice. Int Microbiol 18:1–5

Manh TH, Garcia MS, Vandecasteele M, Willems A, Luo D, Beirinckx S, Goormachtig S (2022) *Stenotrophomonas* sp. SRS1 promotes growth of Arabidopsis and tomato plants under salt stress conditions. Plant Soil 473(1):547–571

Mavrodi OV, McWilliams JR, Peter JO, Berim A, Hassan KA, Elbourne LD, LeTourneau MK, Gang DR, Paulsen IT, Weller DM, Thomashow LS (2021) Root exudates alter the expression of diverse metabolic, transport, regulatory, and stress response genes in rhizosphere Pseudomonas. Front Microbiol 12:651282

Mishra SK, Khan MH, Misra S, Dixit VK, Gupta S, Tiwari S, Gupta SC, Chauhan PS (2020) Drought tolerant *Ochrobactrum* sp. inoculation performs multiple roles in maintaining the homeostasis in *Zea mays* L. subjected to deficit water stress. Plant Physiol Biochem 150:1–14

Mishra SK, Misra S, Dixit VK, Kar S, Chauhan PS (2023) *Ochrobactrum* sp. NBRISH6 inoculation enhances *Zea mays* productivity, mitigating soil alkalinity and plant immune response. Curr Microbiol 80:328

Misra S, Chauhan PS (2020) ACC deaminase-producing rhizosphere competent *Bacillus* spp. mitigate salt stress and promote *Zea mays* growth by modulating ethylene metabolism. 3 Biotech 10:1–4

Misra S, Pandey S, Dixit V, Mishra SK, Khan MH, Agarwal L, Chauhan PS (2017a) Soil microbiome for enhanced crop productivity. In: Kalia VC, Shouche Y, Purohit HJ, Rahi P (eds) Mining of microbial wealth and metagenomics. Springer Nature Singapore, pp 227–247

Misra S, Dixit VK, Khan MH, Mishra SK, Dviwedi G, Yadav S, Lehri A, Chauhan PS (2017b) Exploitation of agro-climatic environment for selection of 1-aminocyclopropane-1-carboxylic acid (ACC) deaminase producing salt tolerant indigenous plant growth promoting rhizobacteria. Microbiol Res 205:25–34

Misra S, Dixit VK, Mishra SK, Chauhan PS (2019a) Demonstrating the potential of abiotic stress-tolerant *Jeotgalicoccus huakuii* NBRI 13E for plant growth promotion and salt stress amelioration. Ann Microbiol 69:419–434

Misra S, Dixit V, Pandey S, Mishra SK, Bisht N, Chauhan PS (2019b) Exploration of soil resistome through a metagenomic approach. In: Ahmad I, Ahmad S, Rumbaugh K (eds) Antibacterial drug discovery to combat MDR. Springer Nature Singapore, pp 313–325

Misra S, Semwal P, Pandey DD, Mishra SK, Chauhan PS (2023) Siderophore-producing *Spinacia oleracea* bacterial endophytes enhance nutrient status and vegetative growth under iron deficit conditions. J Plant Growth Regul 43(5):1317–1330

Naqqash T, Imran A, Hameed S, Shahid M, Majeed A, Iqbal J, Hanif MK, Ejaz S, Malik KA (2020) First report of diazotrophic *Brevundimonas* spp. as growth enhancer and root colonizer of potato. Sci Rep 10:1–14

Naziya B, Murali M, Amruthesh KN (2020) Plant growth-promoting fungi (PGPF) instigate plant growth and induce disease resistance in *Capsicum annuum* L. upon infection with *Colletotrichum capsici* (Syd.) Butler & Bisby. Biomol Ther 10:41

Nong Q, Lin L, Xie J, Mo Z, Malviya MK, Solankim MK, Wang Z, Song X, Li Y, Li C (2023) Regulation of an endophytic nitrogen-fixing bacteria GXS16 promoting drought tolerance in sugarcane. BMC Plant Biol 23(1):573

Poupin MJ, Greve M, Carmona V, Pinedo I (2016) A complex molecular interplay of auxin and ethylene signaling pathways is involved in Arabidopsis growth promotion by *Burkholderia phytofirmans* PsJN. Front Plant Sci 7:492

Rajguru B, Shri M, Bhatt VD (2024) Exploring microbial diversity in the rhizosphere: a comprehensive review of metagenomic approaches and their applications. 3 Biotech 14(10):224

Schneider HM (2022) Characterization, costs, cues and future perspectives of phenotypic plasticity. Ann Bot 130(2):131–148

Semwal P, Majhi B, Pandey DD, Misra S, Mishra SK, Chauhan PS (2023a) Agriwaste burning management through microbial intervention. In: Ravindran B, Gupta SK, Bhat SA, Chauhan PS, Tyagi N (eds) Recent trends in solid waste management. Woodhead Publishing (Elsevier), pp 333–347

Semwal P, Misra S, Misra A, Kar S, Majhi B, Mishra SK, Srivastava S, Chauhan PS (2023b) Endophytic *Bacillus* strains enhance biomass and bioactive metabolites of *Gloriosa superba*. Ind Crop Prod 204:117296

Singh SP, Gaur R (2017) Endophytic *Streptomyces* spp. underscore induction of defense regulatory genes and confers resistance against *Sclerotium rolfsii* in chickpea. Biol Control 104:44–56

Singh K, Gupta R, Shokat S, Iqbal N, Kocsy G, Pérez-Pérez JM, Riyazuddin R (2024) Ascorbate, plant hormones and their interactions during plant responses to biotic stress. Physiol Plant 176(4):e14388

Skalak J, Nicolas KL, Vankova R, Hejatko J (2021) Signal integration in plant abiotic stress responses via multistep phosphorelay signaling. Front Plant Sci 12:644823

Snelders NC, Rovenich H, Thomma BP (2022) Microbiota manipulation through the secretion of effector proteins is fundamental to the wealth of lifestyles in the fungal kingdom. FEMS Microbiol Rev 46(5):fuac022

Sukkasem P, Kurniawan A, Kao TC, Chuang HW (2018) A multifaceted rhizobacterium *Bacillus licheniformis* functions as a fungal antagonist and a promoter of plant growth and abiotic stress tolerance. Environ Exp Bot 155:541–551

Tao L, Zhu H, Huang Q, Xiao X, Luo Y, Wang H, Li Y, Li X, Liu J, Jásik J, Chen Y (2023) PIN2/3/4 auxin carriers mediate root growth inhibition under conditions of boron deprivation in Arabidopsis. Plant J 115(5):1357–1376

Umrao V, Yadav S, Semwal P, Misra S, Mishra SK, Chauhan PS, Shirke PA (2024) Endophytic bacilli from *Cyamopsis tetragonoloba* (L.) Taub. induces plant growth and drought tolerance. Int Microbiol 12:1–6

Vaishnav A, Choudhary DK (2019) Regulation of drought-responsive gene expression in *Glycine max* l. Merrill is mediated through *Pseudomonas simiae* strain AU. J Plant Growth Regul 38:333–342

Vargas L, Santa Brigida AB, Mota Filho JP, de Carvalho TG, Rojas CA, Vaneechoutte D, Van Bel M, Farrinelli L, Ferreira PC, Vandepoele K, Hemerly AS (2014) Drought tolerance conferred to sugarcane by association with *Gluconacetobacter diazotrophicus*: a transcriptomic view of hormone pathways. PLoS One 9(12):e114744

Venkataraman G, Shabala S, Véry AA, Hariharan GN, Somasundaram S, Pulipati S, Sellamuthu G, Harikrishnan M, Kumari K, Shabala L, Zhou M (2021) To exclude or to accumulate? Revealing the role of the sodium HKT1; 5 transporter in plant adaptive responses to varying soil salinity. Plant Physiol Biochem 169:333–342

Vibhuti M, Kumar A, Sheoran N, Nadakkakath AV, Eapen SJ (2017) Molecular basis of endophytic *Bacillus megaterium*-induced growth promotion in *Arabidopsis thaliana*: revelation by microarray-based gene expression analysis. J Plant Growth Regul 36:118–130

Wang Y, Li J, Yang L, Chan Z (2023) Melatonin antagonizes cytokinin responses to stimulate root growth in Arabidopsis. J Plant Growth Regul 42(3):1833–1845

Xue T, Zhang L, Yang F, Cheng C, Wang J, Sang Q, Yang S, Zhao X, Zhang L (2024) Carbon sink and soil organic carbon sequestration mechanisms in vineyards. J Clean Prod 21:143217

Yasmin S, Zaka A, Imran A, Zahid MA, Yousaf S, Rasul G, Arif M, Mirza MS (2016) Plant growth promotion and suppression of bacterial leaf blight in rice by inoculated bacteria. PLoS One 11(8):e0160688

Zhang X, Yang Z, Wang L, Yue Y, Wang L, Yang X (2023) The effects of plant growth-promoting rhizobacteria on plants under temperature stress: a meta-analysis. Rhizosphere 26:100788

CRISPR/Cas System-Based Genome Editing of Crops and Microbiome to Achieve Food and Nutrition Security

13

Anu Chaudhary, Rohit Kumar Singh, Tanvi Bisht, Aayushi Verma, Aditya Sharma, Pushpak Dahiya, Neelam Kaushik, and Ravi Kant Bhatia

Abstract

Genome editing has been transformed by the CRISPR/Cas system, which provides unmatched accuracy and adaptability for tackling issues with global food and nutrition security. In order to fulfil the demands of a growing population under the challenges of climate change, this cutting-edge technology allows targeted genetic alterations in crops to improve yield, nutritional content, and tolerance to biotic and abiotic stresses. Beyond crop development, CRISPR plays a key role in improving soil health, boosting nutrient availability, creating plant-associated microbiomes, and building resilience to environmental stressors. By combining CRISPR/Cas with systems biology and sophisticated breeding methods, sustainable agricultural systems can be developed more quickly, lowering dependency on chemical inputs while boosting output. By improving their metabolic pathways for greater nutritional value and biofortifying crops with vital nutrients, CRISPR/Cas-based therapies also hold promise for reducing malnutrition. To ensure safe, egalitarian, and sustainable deployment, however, ethical, legal, and societal issues must be addressed before this technology may be widely adopted. CRISPR/Cas is a revolutionary technique with great potential for developing robust food systems that support international sustainability objectives. By utilizing its potential, we can create a future in which everyone can achieve food and nutrition security, guaranteeing fair access to wholesome and sustainable food supplies.

A. Chaudhary · R. K. Singh · T. Bisht · A. Verma · A. Sharma · P. Dahiya · N. Kaushik
R. K. Bhatia (✉)
Department of Biotechnology, Himachal Pradesh University, Summer Hill, Shimla, Himachal Pradesh, India

U. Pankaj et al. (eds.), *Plant-microbiome Interactions for Climate-resilient Agriculture*, https://doi.org/10.1007/978-981-96-3534-4_13

Keywords

Genetic engineering · CRISPR/Cas System · Crops · Microbiome · Food · Nutrition

1 Introduction

CRISPR/Cas technology offers innovative strategies for manipulating soil microbiomes, crucial to plant health and productivity. Farmers can enhance soil quality and crop yields by fostering beneficial microbial communities, ultimately promoting more sustainable agricultural practices. This integration of CRISPR/Cas technology with traditional farming methods represents a new paradigm in sustainable agriculture (Lino et al. 2018). This approach focuses on improving crop resilience through targeted genetic modifications while optimizing beneficial microorganisms within the soil ecosystem. Engineering plants to promote specific beneficial bacteria or fungi can lead to healthier root systems, enhancing nutrient absorption and supporting plant growth even in adverse conditions (Lio 2023). The synergistic relationship between genetically modified crops and their microbiomes holds significant potential for addressing the pressing food security challenges intensified by climate change and the increasing global population.

CRISPR/Cas technology offers transformative benefits for agricultural enhancement and food security by enabling precise genetic modifications that improve crop resilience, productivity, and nutritional value (Huang et al. 2021). This ground breaking gene-editing tool addresses critical challenges such as climate change, population growth, and resource scarcity, making it a crucial solution for sustainable agricultural practices. CRISPR/Cas allows for targeted modifications of genes associated with disease susceptibility, leading to crops with increased resistance to pathogens (Ajayi et al. 2024). This advancement reduces reliance on chemical pesticides, fostering environmentally sustainable agricultural methods. Additionally, this technology enhances traits such as photosynthetic efficiency and nutrient uptake, which are essential for boosting crop yields (Chen et al. 2024). Furthermore, CRISPR/Cas has the potential to improve the nutritional profile of crops, aiding food security by addressing malnutrition (Mishra et al. 2024). With CRISPR/Cas technology, farmers can grow crops that withstand abiotic stresses like drought and salinity, challenges that are exacerbated by climate change (Doggalli et al. 2024). This resilience is vital for maintaining agricultural productivity despite changing environmental conditions. While CRISPR/Cas technology presents significant advantages, concerns about off-target effects and regulatory barriers remain. To effectively implement this technology in agriculture, it is essential to address these issues, ensuring that its full potential is realized in the pursuit of global food security (Chen et al. 2024; Mishra et al. 2024).

2 CRISPR/Cas System: Mechanism and Function

The CRISPR/Cas system is an adaptive immune system implicated in bacteria and archaea and has revolutionized molecular biology and applications such as genetic engineering in crops or medicinal plants. As prokaryotic organisms, it protects them naturally from foreign elements like bacteriophages and plasmids by targeting and cleaving foreign DNA (Zheng et al. 2020a, b). The CRISPR locus contains short, repeated DNA sequences interspersed with unique spacer sequences sourced from previous invaders, which together provide a molecular "memory" of past infections (Muzyukina 2023). When the organism encounters the same invader, it transcribes the CRISPR array into precursor RNA, which is processed into guide RNAs (gRNAs). These gRNAs associate with Cas proteins, forming a complex that recognizes and cleaves matching sequences in the invader's genome (Beljouw et al. 2023). This targeted cleavage neutralizes the threat and showcases the system's precision and adaptability. Scientists have harnessed this mechanism to create CRISPR/Cas genome-editing tools, with the most widely used system being CRISPR/Cas9 (Akram et al. 2023). This technology uses two main components: a synthetic guide RNA designed to target specific DNA sequences and the Cas9 nuclease, which introduces double-stranded breaks (DSBs) at the target site. These DSBs are repaired by cellular mechanisms either non-homologous end joining (NHEJ), which can cause small insertions or deletions disrupting the target gene, or homology-directed repair (HDR), which uses a donor template to introduce precise modifications. The simplicity, accuracy, and flexibility of CRISPR/Cas9 have made it a preferred tool for genetic modifications in organisms ranging from microbes to plants by various means as shown in Fig. 13.1 (Lio 2023).

In agriculture, CRISPR/Cas technology offers transformative potential for crop improvement and sustainable farming practices. One of its most significant

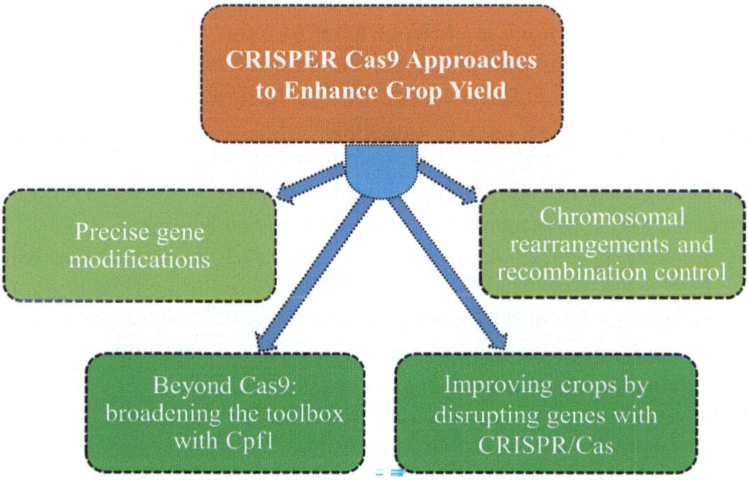

Fig. 13.1 Various approaches to enhance crop yield using CRISPR/Cas

contributions is the enhancement of agronomic traits in crops (Shan et al. 2013). It has been used to develop disease-resistant varieties by knocking out susceptibility genes or incorporating resistance-conferring mutations. Various CRISPR/Cas modified crops including wheat resistant to powdery mildew, rice resistant to bacterial blight, and tomatoes resistant to viral infections have been developed (Erdogan et al. 2023). Similarly, CRISPR enables the development of crops with improved tolerance to abiotic stresses such as drought, salinity, and heat, which are increasingly critical as climate change intensifies. Editing stress-responsive genes in maize and rice has resulted in varieties with enhanced drought resistance (KhokharVoytas et al. 2023). Beyond resistance traits, CRISPR is being used to boost crop yield and nutritional quality. Genes regulating growth, grain size, and productivity have been modified in staple crops like rice and wheat to enhance yield. Biofortification, another vital application, has been achieved by increasing nutrient levels in crops such as rice enriched with iron and zinc, addressing malnutrition in vulnerable populations (Bouis et al. 2011). Additionally, harmful compounds like allergens or antinutritional factors can be removed, as demonstrated by the development of gluten-reduced wheat for people with sensitivities. CRISPR/Cas technology also plays a pivotal role in engineering microbial communities associated with crops. These microbes significantly influence plant health and productivity by facilitating nutrient uptake, promoting growth, and suppressing pathogens (Lio 2023). CRISPR can also be applied to enhance the efficiency of nitrogen-fixing bacteria, which helps in reducing the need for synthetic fertilizers and mitigating environmental pollution (Huang 2024). Microbial engineering also includes developing plant-associated microbes that enhance tolerance to biotic and abiotic stresses or promote sustainable farming by improving soil health and fertility. Beyond crops and microbes, CRISPR offers solutions for addressing climate change-related challenges in agriculture. By creating salt-tolerant rice or heat-resistant wheat, CRISPR enables farming in marginal lands affected by rising sea levels and extreme weather conditions (Cordones et al. 2017). Furthermore, engineering crops with improved carbon sequestration abilities could contribute to reducing greenhouse gas emissions, aligning agriculture with global sustainability goals. Despite its transformative potential, the application of CRISPR/Cas technology in agriculture faces challenges. Off-target effects, where unintended edits occur, can compromise safety and effectiveness (Shan et al. 2013). Researchers are addressing this by improving guide RNA design and developing high-fidelity Cas proteins. Regulatory and public acceptance also plays a crucial role in determining the technology's success (Cossart 2020). While some countries have adopted regulatory frameworks that distinguish genome-edited crops from traditional GMOs, others impose stringent regulations, potentially hindering adoption (Bhattacharjee et al. 2023). Public concerns about safety, ethics, and long-term impacts require transparent communication and stakeholder engagement to foster acceptance (Lynch and Vogel 2001). Additionally, technical challenges such as efficient delivery of CRISPR components into plant cells, especially in recalcitrant species, need innovative solutions to expand the technology's reach. The CRISPR/Cas system, originally part of the adaptive immune response in bacteria, has been repurposed into a powerful genome-editing tool known for its unmatched precision and versatility (Abdi et al. 2024). Its application in agriculture offers significant potential to address global challenges such as food security, climate change, and environmental sustainability. By enabling precise genetic modifications, CRISPR facilitates the development of crops with improved traits, biofortification, and enhanced stress tolerance while also advancing the engineering of beneficial microbial communities (Ajeng et al. 2024). Although there are still challenges to overcome, on-going research and technological advancements are expected to address these issues, ensuring CRISPR's role as a transformative tool in shaping the future of sustainable agriculture.

3 CRISPR/Cas for Crop Improvement

The world's population is rising alarmingly, increasing demand for crops more resilient to biotic and abiotic stresses. There is also a need for crops with improved morphological and nutritional characteristics that can produce high-yield varieties. This situation has encouraged the development of new plant breeding technologies (Shen et al. 2018). With genome sequences for many crops now readily available and advancements in genome editing technologies, such as zinc finger nucleases (ZFNs) and transcription activator-like effector nucleases (TALENs), it is becoming possible to create new crop plant varieties with enhanced traits (Mishra et al. 2024). However, these methods can be expensive, time-consuming, and involve complicated processes. The CRISPR/Cas genome editing system has attracted significant research attention due to its ability to modify the crop genome in a flexible and targeted manner, contributing to the development of new kinds (Zhang et al. 2015). This revolutionary and exciting technology holds the potential to help end world hunger and support the sustainable growth of the world's expanding population. The application of CRISPR/Cas genome editing in crops has the potential to transform the agricultural sector and address global food security challenges. Utilizing CRISPR/Cas technology, researchers have also looked at the possibility of altering plants' secondary metabolism to raise crop quality (Sánchez 2020). Carbohydrates, proteins, oil, and bioactive compounds are the primary functional components of food, and the amount of each affects the quality of an agricultural product and protein is one of the important one (Shukla et al. 2018). Nonetheless, the protein composition dictates the quality of the protein combination. Gluten, which is present in wheat and other foods, causes sensitivity in 1–2% of people. Wheat that is low in gluten and transgene-free was created using CRISPR/Cas9 (Sánchez-León et al. 2017). Another crucial feature that scientists hope to improve using genome editing is oil. Karunarathna and colleagues employed CRISPR/Cas9 in rapeseed to stop the BnSFAR4 and BnSFAR5 genes from negatively regulating the amount of oil in the seed. The outcome demonstrated that increasing the seed oil content had no effect on oil mobilization, vigor, or germination (Karunarathna et al. 2020). Likewise, *B. napus L.* CRISPR/Cas9-edited BnTT8 mutants exhibit increased protein and seed oil content with changed fatty acid composition and without adverse impacts on yield-related characteristics (Zhai et al. 2019). CRISPR/Cas9 has also been shown improvement in disease resistance of plants against pathogens. A devastating fungal disease called rice blast reduces yield in nations that farm rice by more than 20%. Using CRISPR-Cas technology to knock out the ethylene responsive factor transcription factor gene OsERF922 gives rice resistance to blast and salinity (Wang et al. 2016a, b, c). The bacteria *Xanthomonas oryzae* pv. *oryzae* is the cause of bacterial blight in rice. Rice plants that are resistant to bacterial blight disease were developed because of the OsSWEET13 promoter being deleted (Zhou et al. 2015). Two detrimental rice diseases are plant hoppers and stem borers. OsCYP71A1 gene disruption resulted in inhibited serotonin biosynthesis and significantly elevated salicylic acid levels in rice plants (Lu et al. 2018). These rice plants with increased salicylic content demonstrated strong resistance to stem borers and plant hoppers. CRISPR/Cas9 showed improvement of crops against bacterial and viral diseases. Despite its great potential, there is still a problem with accurately forecasting the results of CRISPR-Cas9 editing (Zhang et al. 2014). Because it guides the creation of guide RNAs (gRNAs), which lead Cas9 to specific genomic regions, the prediction of these results is essential to genome editing (Gasiunas et al. 2012). This has led to an abundance of research using deep learning techniques to forecast the impact of gRNAs in the CRISPR-Cas9 system, both on and off target.

Unlike standard linear models, deep learning can identify complicated, non-linear relationships in data, which is what makes it so appealing. Deep learning networks, which are made up of several interconnected layers, allow the input data to undergo a few non-linear changes in order to provide a prediction. Given the complexity of sequence-activity correlations, this technique reveals complex patterns in the input data, a feature that is especially relevant in gRNA activity prediction (LeCun et al. 2015). The field of deep learning for CRISPR-Cas9 editing result prediction is developing quickly, and there is increasing interest in using attention processes to improve interpretability. Attention mechanisms, which have their roots in natural language processing, guide the model to concentrate on important segments of the input sequence, providing information on the impact of certain nucleotides in gRNA sequences (Vaswani et al. 2017). Synthetic biology involves the use of microbes in agriculture and can alter the genetic code and metabolic processes of crops. As a result, it offers promising opportunities for crop breeding, yield enhancement, and maintaining the security of the agricultural production environment (Wang et al. 2022).

3.1 Enhancing Crop Resistance Against Biotic and Abiotic Stresses by CRISPER/Cas9

CRISPR/Cas9 technology has revolutionized crop improvement by enabling precise genetic modifications to enhance resistance against biotic and abiotic stresses. By targeting specific genes, CRISPR/Cas9 can develop crops with improved resistance to pathogens, pests, and diseases, reducing dependency on chemical pesticides (Wang et al. 2016a, b, c). It also facilitates tolerance to abiotic stresses like drought, salinity, and extreme temperatures by modifying stress-responsive genes,

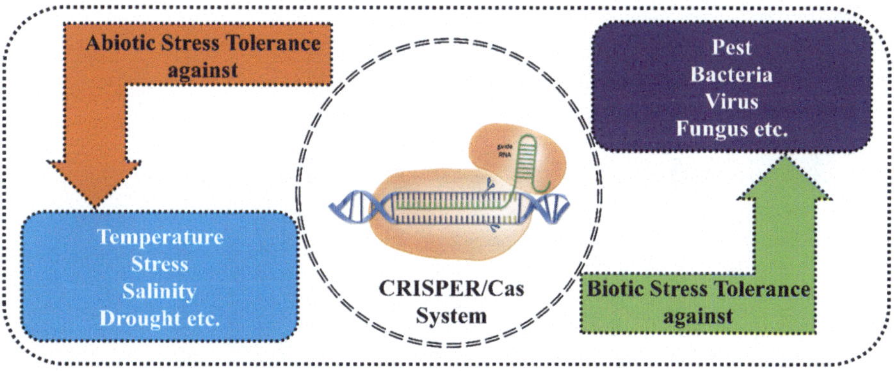

Fig. 13.2 Various potential applications of CRISPR/Cas system in enhancing plant stress tolerance to biotic and abiotic stresses

ensuring stable yields under adverse conditions. This genome-editing approach is faster, more efficient, and more accurate than traditional breeding methods, making it a powerful tool for sustainable agriculture as shown in Fig. 13.2. CRISPR/Cas9 offers a promising solution to global food security challenges amid changing climates.

3.1.1 Enhancing Crop Resilience to Biotic Stresses

CRISPR/Cas9 technology has successfully created multiple disease-resistant plants, as outlined in Table 13.1. Some notable examples include achieving resistance to rice blast disease by targeting the *OsERF922* gene in rice (Wang et al. 2016a, b, c). Segregation was used to select and further investigate transgene-free mutant lines from the T1 and T2 generations. The transgenic lines exhibited a significant reduction in blast lesions caused by pathogen infection. Additionally, Wang et al. (2014) used site-specific endonucleases to introduce mutations in the homeoalleles that encode *Mildew-resistance locus (MLO)* proteins in hexaploid bread wheat. Peng et al. (2017) focused on citrus canker, which is induced by *Xanthomonas citri* subsp. Xcc, in *Citrus sinensis*. Their work involved CRISPR/Cas9-targeted alteration of the susceptibility gene *Lateral organ boundaries1 (CsLOB1)* promoter, which increases disease resistance. Deletion of the entire EBE_{PthA4} sequence from both *CsLOB1* alleles resulted in the greatest resistance to *Wanjincheng orange*. Notably, 42% of the mutant plants included desirable alterations, and 23.5% of the mutants were resistant to citrus canker. The CRISPR/Cas9 system piles numerous nucleases into a single transgene, resulting in targeted cleavage of various viral infections (Iqbal et al. 2016). The long-term effects of the CRISPR/Cas9 modifications are expected to include increased productivity and the potential for developing superior crop varieties through precise genetic alterations (Akhil et al. 2024).

Table 13.1 List of some crops that are made resistant to diseases via CRISPR/Cas9 system

Crop	Disease/symptoms	Causal/target organism	Targeted gene	Significance	References
Triticum aestivum	Powdery mildew disease	*Blumeria graminis f.* sp. *Tritici*	TaMLO-A1 (wheat mildew resistance locus1)	Simultaneous modification in three homoalleles, heritable broad-spectrum resistance to powdery mildew	Wang et al. (2014)
Oryza sativa	Bacterial blight of rice	*Xanthomonas oryzae*	OsSWEET11, OsSWEET14 (rice bacterial blight susceptibility genes)	PEG stimulated Cas9/sgRNA gene uptake in rice protoplast (Agrobacterium independent method), Cas9/sgRNA mutations occur within plant cells, free of bacterial cell involvement	Jiang et al. (2013)
Oryza sativa	Rice blast disease	*Magnaporthe oryzae*	OsERF922 (ethylene responsive factor transcription factor)	42% T0 mutant lines; 6 T2 homozygous mutants showed high blast resistance and have same agronomic traits	Wang et al. (2016a, b, c)
Arabidopsis thaliana	Turnip mosaic virus disease	*Potyvirus (TuMV)*	elF(iso)4E (elF transcription factor)	Mutants show no growth defects, morphologically similar to wild type	Pyott et al. (2016)
Gossypium hirsutum	Cotton leaf curl disease	*Begomovirus*	CLCuD IR and Rep regions	Targeted cleavage of mixed infections by multiple viruses and associated DNA satellites, such as CLCuD-complex Mutants show no growth defects, morphologically similar to wild type	Iqbal et al. (2016)

Cucumis sativus L.	Ring spot disease, vein yellowing disease	*Cucumber vein yellowing virus (Ipomovirus), potyviruses Zucchini yellow mosaic virus and Papaya ring spot mosaic virus-W*	eIF4E (eukaryotic translation initiation factor 4E)	eLF4E disruption generated virus resistant heterozygous non-transgenic mutants	Chandrasekaran et al. (2016)
Nicotiana benthamiana	Leaf thickening, chlorosis, curling	*Bean yellow dwarf virus (BeYDV)*	BeYDV (short intergenic region, trans acting replication initiation protein)	87% reduction in targeted viral load. Study proved that IR targeting via sgRNA confer better resistance	Baltes et al. (2014)
Nicotiana benthamiana	Leaf curl disease	*Tomato yellow leaf curl virus, Beet curly top virus*	TYLCV-IR (intergenic regions), RCA regions	Mutants showed delayed and reduced viral DNA accumulation	Ali et al. (2015)

3.1.2 Enhancing Crop Resilience to Abiotic Stresses

Multiple genes regulate key features such as yield and abiotic stress tolerance. Numerous studies on crop improvement have attempted to map quantitative trait loci (QTL) that influence essential agronomic variables. Once identified, these QTL regions are introduced into elite lines to create improved variety. However, the process of introgression can be time-consuming, especially when QTLs are intricately linked (Wang et al. 2016a, b, c). Additionally, incorporating non-target areas into the elite line can have negative consequences. The CRISPR/Cas9 system offers an effective way to introduce and study rare mutations in crop plants. Shen et al. (2018) used CRISPR-based QTL editing to investigate the impact of grain size (GS3) and grain number QTLs (Gn1a) on rice varieties. Their study revealed that the same QTL can have varying and even opposing effects across different genetic backgrounds. Furthermore, CRISPR/Cas9 technology has been implemented to develop multiple plant varieties that tolerate abiotic stress (Table 13.2).

Table 13.2 List of some crops that are made resistant to abiotic stresses/traits via CRISPR/Cas9 system

Crop	Method	Target gene	Stress/trait	References
Maize	HDR	ARGOS8	Increased grain yield under drought stress	Shi et al. (2017)
Tomato	NHEJ	SlMAPK3	Drought tolerance	Wang et al. (2017)
A. thaliana	NHEJ HDR	UGT79B2, UGT79B3, MIR169a	Susceptibility to cold, salt, and drought stresses Drought tolerance	Zhao et al. (2016)
A. thaliana	NHEJ	OST2(OPENSTOMATA2) (AHA1)	Increased stomatal closure in response to abscisic acid (ABA)	Osakabe et al. (2016)
Oryza sativa	HDR, NHEJ	OsPDS,OsMPK2,OsBADH2	Involved in various abiotic stress tolerance	Shan et al. (2013)
Oryza sativa	NHEJ	OsMPK5	Various abiotic stress tolerance and disease resistance	Xie and Yang (2013)
Oryza sativa	NHEJ, HDR	OsMPK2,OsDEP1	Yield under stress	Shan et al. (2013)
Oryza sativa	NHEJ	OsDERF1,OsPMS3, OsEPSPS,OsMSH1,OsMYB5	Drought tolerance	Zhang et al. (2014)
Oryza sativa	NHEJ	OsAOX1a, OsAOX1b,OsAOX1c,OsBEL	Various abiotic stress tolerance	Xu et al. (2015)
Oryza sativa	NHEJ	OsHAK-1	Low cesium accumulation	Cordones et al. (2017)
Oryza sativa	NHEJ	OsPRX2	Potassium deficiency tolerance	Mao et al. (2018)

3.1.3 Nutritional Quality Enhancement

According to the World Health Organization, malnutrition is one of the threatening problems all around the world that has an impact on around two billion people (WHO 2020a, b). It is defined in the term in stern way: nutrients deficiency or excessive deficiency "under nutrition" (stunting, wasting); "micronutrients deficient" (vit. A, anemia); or "over nutrition" (obesity) (FAO 2020). Citing corroborative evidence from UNICEF (2020), it is estimated that 149 million children less than 5 years are affected by stunting, whereas 49.2 million are affected by wasting. Cause attributive are emergency, poverty, regular devastations of the food, ethnic interactions, changes of the weather (Lancet 2018). These interventions are also some of the interventions which can be perceived as being effective, for they are evidence-based: nutrition education, food supports such as—iodized salt, biofortification such as—Vitamin A rice and infant feeding (WHO 2018). This convergence of needs indorsed serious changes in how malnutrition is perceived, its relationship with health, and far more importantly how it is dealt with.

The global food crisis is mainly caused by several factors, namely, rise in population, reduction in the available land for cultivation, water constraints, and climatological factors. Apart from these factors an estimated 10 billion people will inhabit the earth by 2050 if the current trends in population growth persist. There is an estimated demand for an increase of agricultural production levels by up to 60% to be able to cater for this population by 2050 (Huang et al. 2021). Biofortification, which refers to enhancing the nutrient content in crops, is a solution that is long term. Biofortification is an affordable and environment-friendly strategy in agriculture that improves nutrition in main food plants either by selective breeding or transgenic (Ajeng et al. 2024). This is aimed at improving most of the people in developing nations that are nutritionally vulnerable as these are countries high in the levels of malnutrition and very low of micronutrients (Samtiya et al. 2020). Biofortified foods are those that contain additional important vitamins and minerals such as iron, zinc, vitamin A, and iodine; thus, enhancing the quality of food irrespective of its consumption style. Research has demonstrated the effectiveness of biofortified crops in tackling micronutrient reduction and enhancing health benefits (Bouis et al. 2011). With the introduction of CRISPR technology, the element of biofortification has gained esteem in allowing scientists to introduce better nutrient value crops. This fresh approach inspired the creation of dense crops, like iron cereals (Zheng et al. 2020a, b), beta-carotene "golden" crops as corn and sweet potatoes (Dong et al. 2020) and protein wheat (Ibrahim et al. 2021). These biofortified crops correct micronutrient deficiencies typical in developing nations thus increasing health standards. Vitamin A deficiency is responsible for 250 million people across the globe therefore, β-carotene crops could solve this problem (Bouis et al. 2011). Likewise, the rice that is fortified with a high quantity of iron solves the problem of iron deficiency that affects over 1.6 billion people around the world (WHO 2020a, b). There is sufficient evidence to confirm that biofortification through CRISPR technologies can enhance nutritional quality of crops and thus the world should be free of micronutrient malnutrition.

3.1.4 Reducing Anti-nutritional Factors

Anti-nutritional factors are substances that preclude the absorption of nutrients and could be detrimental to both humans and livestock. Common ANFs include phytic acid, lectins, saponins, oxalates, and tannins (Samtiya et al. 2020). Most commonly, phytate is the cheapest phosphorus reserve located in the seeds and it prevents the bioavailability of minerals, especially that of zinc and iron. Due to the presence of lectins in legumes, there are gastrointestinal disturbances such as nausea and vomiting which also lead to difficulty in taking in nutrients (Raboy 2001). Some beans such as saponins and peas contain protein but the protein digestibility and mineral availability will be diminished. Calcium oxalate can be found in spinach and beets; they can combine with the minerals in the external environment such as magnesium and calcium which the body cannot absorb (Ali et al. 2015). Iron-containing foods, such as tea and coffee, are also accompanied by exacerbating the negative impact of tannins. Awareness of the situations encompassing the strikes of ANFs should strike a balance between minimizing the negative impacts and improving the nutritional optima. CRISPR/Cas technology on the other hand has improved the crop quality as undesirable anti-nutritional factors are removed through it (Haque et al. 2018). One of the practices is the lowering of phytates in grains, which is the single most important contributor to the non-availableness of many minerals. Phytate plants do limit human intake of Zn. Fe, Ca, and Mg which are very important micronutrients (Raboy 2001). It has been possible to deploy CRISPR/Cas technology to produce low-phytate crops by targeting genes responsible for the biosynthetic pathway of phytic acid. It has been shown that bioavailable minerals content in these edited crops is higher than that of the non-edited counterpart crops, thus better palatability and nutritional value. Shi et al. (2017) has been reported that CRISPR/Cas edited low phytic acid maize improving 20–30% bioavailability of zinc and iron. The adverse effects of phytic acid used in crops contain vast potential for addressing the challenge of micronutrient eradication, especially in third-world countries. The deficiency of zinc and iron is a concern to 2 billion people in the world as indicated by the World Health Organization (WHO) as causing stunted growth, low immunity, and high death rates (WHO 2020a, b). Phytate-deficient crops can work towards reducing this deficiency hence improving Global Health outcomes.

4 CRISPR/Cas in Plant and Soil Microbiome Engineering

4.1 Soil Nutrient Cycling

As an important natural resource, soil is a key contact between the hydrosphere, lithosphere, atmosphere, and biosphere. It supports a diverse array of ecosystem services and is made up of a dynamic reservoir of biodiversity that spans large landscapes as well as micro-niches with organisms including plants, annelids, insects, and microbes that exist above or below ground (Igalavithana et al. 2017). One type of living form that is found in large quantities in soil is microorganisms, often known as soil microbiomes. Numerous microorganisms, including bacteria, fungus, archaea, algae, and nematodes, can be either beneficial or harmful while present in soil (Mishra et al. 2016). Numerous species of these microbes can be found in different types of soil, such as alluvial, loamy, black, laterite, red, and other types, as well as in different types of extreme habitats, such as saline, heavy metal, dry, cold-climate, acidic, and alkaline soil, and soil that floods with water. In agriculture, soil microbes can be used as a plant growth promoter and protector (Biswas et al. 2018). A variety of mechanisms are used by oil microbes to enhance plant growth, including nitrogen fixation, the solubilization and mobilization of minerals such as phosphorus, potassium, zinc, and selenium, the production of siderophores for iron chelation, and the release of phytohormones such as auxin, abscisic acid, cytokinin, gibberellins, and ethylene (Backer et al. 2018). The processes by which soil microorganisms protect plants include the creation of siderophores, which chelate soil iron and restrict the uptake of these micronutrients by pathogens; other strategies include hydrogen cyanide (HCN), antibiotics, hydrolytic enzymes, and ACC deaminase, which are displayed by some microbes (Kaur 2020).

4.2 Soil Role in Plant Health

Unquestionably, soil microorganisms play a crucial role in plant health and are essential to the functioning of terrestrial ecosystems. Soil microbial communities are involved in the delivery of 18 of the approximately 29 required components for plants, which are derived from the soil (Brevik et al. 2020). Plants take up a portion of the soil microbiome because they draw microorganisms from the rhizosphere that is vital to their growth and development. Plant-associated microbiota's structure, makeup, and overall activity can be significantly influenced by soil bacteria (Gouda et al. 2017). A variety of functions, such as seed germination, seedling establishment, feeding, water uptake, growth promotion, pathogen suppression, stress tolerance, and hormone control, are made easier by the rhizosphere microbiome, which also helps plants' metabolic repertoire (Carthey et al. 2018). Microbial activity consumes up to 80% of plant nitrogen and 90% of plant phosphorus. Additionally, bacteria in the rhizosphere can indirectly affect important functional characteristics such as shoot, nutrient levels, leaf area, leaf longevity, and root proportion (Berg et al. 2014). Numerous bacteria, including *Rhizobium*, *Arthrobacter*, *Bacillus*, *Alcaligens*, *Rhodococcus*, *Methylobacterium*, *Pseudomonas*, and *Azospirillum*, are well-known for their functions in hormone regulation, stress management, plant feeding, and growth promotion. Mycorrhizal fungi are among the most well-known examples of helpful microorganisms for plants; they coexist in symbiotic relationships with around 90% of land plants, including numerous crops (van der Heijden et al. 2015). After a complex molecular exchange, mycorrhizal fungi are drawn from the soil and settle on plant roots. The host plants receive numerous advantages from mycorrhizae. Plants provide up to one-fourth of their photosynthates to mycorrhizal fungus in exchange for water and vital macro- and micronutrients (Lekberg and Koide 2005). Mycorrhizal fungi can promote the growth of a variety of crops, with reports of growth increases of up to 50% (Hoeksema et al. 2010). Mycorrhizal fungus inoculation in the field can significantly increase plant output, but the results are quite unpredictable (Lekberg and Koide 2005) and many commercial inoculants are of poor quality. Other examples of beneficial bacteria that can increase plant development, stress tolerance, and/or generate local and systemic resistance to diseases are fungi like *Trichoderma* spp. and the endophytic fungus *Piriformospora indica* (Bünemann et al. 2018; Gill et al. 2016). By grazing on harmful bacteria and promoting mineralization, soil protists can also have a significant effect on the availability of nutrients for plants and their overall health (Geisen et al. 2018). However, not all soil microbes promote plant health and there are numerous soil-borne pathogens such as *Ralstonia*, *Rhizoctonia*, *Fusarium*, *Phytophthora*, *Gaeumannomyces* that cause deadly outbreaks in plants (Backer et al. 2018; Philippot et al. 2013). Microbes support the construction and upkeep of urban agriculture and green roofs by encouraging plant growth and seedling survival, which also contributes to the creation of a pleasant and healthy atmosphere (Igalavithana et al. 2017). There are many ways in which plants and these soil bacteria interact, including commensal, neutral, exploitative, competitive, and symbiotic interactions. Plant-specific root exudates can reveal the specific selection of rhizosphere microbial populations (Yuan et al. 2018). The cucumber plant secretes citric acid from its roots, which influences *Bacillus amyloliquefaciens*' attraction to the plant, and the banana plant secretes fumaric acid from its roots, which attracts Bacillus subtilis to the plant and leads to the formation of biofilms (Zhang et al. 2014).

4.3 Role of Soil Microbiome in Plant Growth

Plant growth is aided by soil microbes in both direct and indirect ways. Plant growth can be hampered by a variety of biotic (related to living things) and abiotic (related to non-living causes) stressors. Microorganisms can enhance the synthesis of indole-3-acetic acid (IAA) or trigger the transcription of ACC synthase genes in response to both biotic and abiotic stress conditions (Martínez-Viveros et al. 2010). Low-molecular-weight osmolytes such as proline, glycine betaine, and other amino acids are subsequently produced as a result of this. Furthermore, by aiding in the solubilization of minerals and phosphates, fixing nitrogen, generating organic acids, and synthesizing important enzymes as ACC-deaminase, chitinase, and glucanase, microbes support plant growth in stressful environments. Microbial activities possess the capability to augment plant growth and mitigate the consequences of abiotic stressors (Nanda et al. 2019). By inducing gibberellin production, increasing the growth of different plant parts like stems, leaves, flowers, and fruits, and aiding in seed germination, plant growth-promoting microorganisms (PGPM) are essential in augmenting plant growth. These bacteria can enter plant roots and benefit the plants they live in as hosts (Lopes et al. 2021). The synthesis of cytokines is stimulated by the presence of PGPM, which improves root development, increases vascular cambium activity, facilitates cell differentiation, and decreases apical dominance (Gouda et al. 2017). In response to environmental stress, plants that possess PGPB that contain ACC deaminase function as sinks for ACC

(1-aminocyclopropane-1-carboxylate), which lowers the amount of ethylene the plant produces. Because less ethylene is produced, the plant's stress response is suppressed, making it less vulnerable to stress (Martínez-Viveros et al. 2010).

4.4 Soil Microbiome Against Disease Resistance in Plants

According to numerous research, plant microbiome imparts disease resistance against a variety of pathogens with varying lifestyles. It has been well documented that the root microbiota in plants activates innate immunity, hence offering resistance against a variety of above-ground plant diseases through ISR (induced systemic resistance) (Chialva et al. 2018). The majority of plant-growth-promoting bacteria (PGPR)-mediated disease resistance in plants has been found to be significantly influenced by ISR activation (Shoresh et al. 2010). Prior research has demonstrated that oomycete and bacterial infections in Arabidopsis leaves affect the root exudates to encourage the growth of advantageous ISR-triggering microbes (Yuan et al. 2018). In addition to influencing the beneficial microbiome, root exudates help plants resist disease by either directly preventing the growth of pathogens or stimulating the immune system. Numerous helpful bacteria actively react to root exudates by modifying their transcriptional program to produce characteristics that aid in root colonization, chemotaxis, activating ISR, biofilm formation, and energy metabolism all of which are critical for preventing pathogens and the advancement of disease (Yin et al. 2023). Growing research on *Arabidopsis thaliana*, *Beta vulgaris*, and *Triticum avesticum* has shown that the roots of pathogen-infected plants can draw beneficial bacteria to save or protect subsequent generations. It has also been discovered that members of the endosphere and rhizosphere microbiomes prevent plant diseases such the damping-off and take-all (Liu et al. 2022).

According to Kwak et al. tomato plants' rhizosphere microbiota improves their resistance to *Ralstonia solanacearum*-caused wilt disease (Kwak et al. 2018). Numerous studies have demonstrated that the beneficial microbiome directly inhibits infections through a range of mechanisms, such as competition for nutrients and space, hyper parasitism, and the manufacture of antimicrobial chemicals (Hacquard et al. 2017). Although they may differ from microbe to microbe, all of these tactics lead to pathogen limiting. According to a prior study, *Pseudomonas fluorescens* WCS417 stimulates the production of defence signature genes (pathogenesis-related genes) that limit pathogen entrance and disease development in addition to promoting callose deposition at the pathogen entry site (Van der Ent et al. 2009). Numerous studies have also demonstrated that arbuscular mycorrhiza fungi (AMF) improve plant resistance to a range of illnesses. Mycorrhiza colonization in many agricultural plants confers disease resistance against a variety of Phytopathogens, including *Pseudomonas syringae*, *Phytophthora nicotianae var. parasitica*, *Fusarium oxysporum f.* sp. *Lycopersici*, *Pyrenochaeta terrestris*, and *Phytophthora nicotianae* (Akköprü et al. 2015). Furthermore, AMF triggers systemic acquired resistance (SAR) and ISR, which confer disease resistance against pathogens with diverse lifestyles. During disease development or pathogen attacks, susceptible plants release volatile chemicals (VOCs) that can activate the plant immune system or microbiome assembly in nearby plants, making them vigilant and impervious to targeted pathogen attacks (Hammerbacher et al. 2019).

4.5 Microbes for Nitrogen Fixation

One of the most advantageous interactions between legumes and Rhizobium in the globe is symbiotic nitrogen fixation. It has been possible to identify several genes involved in nitrogen-fixing by using the CRISPR-Cas9 method. CRISPR is a fascinating field of study that blends genetic engineering and microbiology to target beneficial microbes for better nitrogen fixation (Wang et al. 2016a, b, c). In nitrogen-fixing bacteria, researchers can pinpoint genes that control the nitrogenase enzyme, which is essential to the fixation process. It is possible to increase nitrogen fixation efficiency by improving these mechanisms (Huang 2024). By altering plant genomes, CRISPR can strengthen symbiotic connections by making them more compatible with nitrogen-fixing bacteria. The nodulation-specific promoter region of the leghemoglobin genes (LjLb1, LjLb2, and LjLb3) was utilized to effectively express guide RNA in Lotus japonica nodules (Wang et al. 2016a, b, c). Moreover, nodule development was inhibited by the loss-of-function of the cowpea (*Vigna unguiculata*) symbiotic receptor-like kinase (VuSYMRK) gene created by CRISPR-Cas9 (Ji et al. 2019). Through editing of the Rfg1 gene (Fan et al. 2017) and the Rj4 gene (Tang et al. 2015), it was possible to identify the dominant gene in soybeans that limits the production of nodules by *Sinorhizobium fredii*. Furthermore, the conservation of the nodulation process in contrast to that of legumes was identified through deleting four genes (PanHK4, PanEIN2, PanNSP1, and PanNSP2) linked to hormonal regulation of nodulation in the tropical tree *Parasponia andersonii* (Van Zeijl et al. 2018). It was recently discovered that 52 soybean genes were regulated by small tRNA fragments (tRFs) generated in Rhizobia, and these tRFs were implicated in the nodulation process. Further, CRISPR-Cas9 based editing of a few selected genes enhanced nodulation in soybean (Ren et al. 2019). In Medicago, CRISPR/CAS9 based editing of gibberlin oxidase (MtGA2Ox10) inhibited the infection thread development in the first nodulation process (Kim et al. 2019a, b). Furthermore, the CRISPR/CAS9 method in Medicago was used to identify and confirm the MtNFS2 gene, which is responsible for strain specificity (Wang et al. 2016a, b, c). Therefore, CRISPR/CAS9 genome editing helps identify target genes, particularly for improving nodulation in legumes and provides mechanistic insights into nodule formation.

4.6 Phosphorus Solubilization

A possible approach to improving phosphorus solubilization in plants and overcoming the crucial nutrient constraints in agriculture is to engineer the microbiome using CRISPR technology (Prabhu et al. 2019). Researchers may be able to improve the efficacy of phosphate-solubilizing microorganisms (PSMs) in transforming insoluble phosphorus into forms that plants can absorb by modifying particular genes in these organisms. Plants cannot access phosphorus because of its high reactivity, which makes it available in the soil as a variety of organic and inorganic molecules (Liang et al. 2020a, b). By releasing complex substances such as organic acid anions, protons, exopolysaccharides, siderophores, hydroxyl ions, and extracellular enzymes, PSMs convert these insoluble forms of phosphates into soluble ones (Ingle and Padole 2017). Depending on the P substrate, the two main components of the phosphate-solubilizing process of PSM are inorganic phosphate solubilization and organic phosphate solubilization (Prabhu et al. 2019). Using CRISPR, scientists may either increase the synthesis of organic acids and the enzymes that solubilize phosphorus or eliminate genes that prevent phosphorus from being dissolved. The ability of some microbial strains to solubilize phosphorus can be enhanced by engineering (Liang et al. 2020a, b). Altering the phosphate metabolic processes in fungi or bacteria to boost the synthesis of organic acids or phosphatases *Enterobacter, Serratia,* and *Pseudomonas* are important bacterial genera, while *Penicillium, Aspergillus,* and *Bacillus* are important fungal genera for phosphate solubilization (Chen et al. 2023).

5 Engineering Microbiomes to Increase Resilience to Pathogens and Environmental Stress

Plant products are harmed annually by biotic stresses, including viruses, bacteria, and fungal diseases, along with herbivores. Plant defense against pathogens can lessen the impact of disease on plant growth and productivity, which is important given the world's growing population and scarcity of food (Hammerbacher et al. 2019). The new resistant crops have been enhanced by advancements in GE (Genome Editing) technology or new ways, which has decreased yield losses brought on by plant defense. Many pathogen species resistance has been developed through the use of the CRISPR/Cas system. Genome editing technology's future is represented by CRISPR/Cas, which can also be used to a variety of fields in response to biotic and abiotic agricultural stresses. Plants can now undergo gene knockout, knockin, and knockup processes using CRISPR/Cas. Thus, to improve crop growth and development and adapt to different environmental challenges, CRISPR/Cas can be utilized to change critical genes of biotic and abiotic stresses.

5.1 CRISPR/Cas-Mediated Fungal Resistance in Plants

Several fungal pathogens produce lethal diseases in crop plants, like rust, mildew, rot, and smut, which degrade the product's quality in addition to yearly output losses in the biosphere. Based on the knowledge of the genetic pathways underlying crop-pathogen interaction that is currently available, CRISPR/Cas has enhanced mycological resistance in a variety of crop species. Increased plant defense against fungus may be possible with defined candidate genes and gene products (Jeon et al. 2020). RNA-guided Cas9 endonuclease was employed to target MLO loci in three crop varieties: wheat, grapevine (*Vitis vinifera*), and tomato (*Solanum lycopersicum*) (Malnoy et al. 2016) and transgene-free plants were produced. Seven transmembrane domains found in an MLO-encoded protein that is located in the cell membrane are shared by all dicots and monocots. Plants like *A. thaliana*, tomato, and barley that have loss of function alleles (mlo) of the MLO provide long-lasting resistance to powdery mildew (Bai et al. 2008). The MLO-B1 region of the wheat genome is targeted with precision GE to create a 304K deletion Tamlo-R32 mutant that retains wheat growth and yield and offers strong resistance to powdery mildew (Li et al. 2022). One (TaMLO-A1) of the three MLO home alleles has been altered with CRISPR/Cas9 in *Triticum aestivum*, and it has demonstrated strong resistance against infection by *Blumeria graminis f.* sp. *tritici* (Wang et al. 2014). Developed by deleting a 48 bp segment in the SlMlo1 gene (one of 16 key SlMlo genes), the CRISPR/Cas-mediated transgene-free and self-pollinated tomato variety provides resistance against *Oidium neolycopersici*, a powdery mildew (Nekrasov et al. 2017).

5.2 CRISPR/Cas-Mediated Viral Resistance in Plants

CRISPR/Cas systems have been shown to be very successful platforms for engineering virus resistance, and genome-editing techniques have advanced rapidly in recent years. Either altering the host components that allow viral species to replicate or directly destroying and targeting different viral genomes to prevent their reproduction are two ways to develop resistance to viruses (Ali et al. 2015). Enhancing plant virus susceptibility is mostly dependent on the cap-binding protein, also known as the eukaryotic initiation factor 4E (eIF4E) (German-Retana et al. 2008). Several plant species have been shown to develop innate immunity against potyviruses (a family of viruses known as potyviridae) when eIF4E is deactivated. According to a study, eIF4E alteration in cucumber plants using CRISPR/SpCas9 led to the creation of resistance against a number of viruses, such as "cucumber vein yellowing virus (CVYV)," "zucchini yellow mosaic virus (ZYMV)," and "papaya ringspot mosaic virus-W (PRSMV-W)" (Gosavi et al. 2020). The symptoms of turnip mosaic virus (TuMV) infection and cassava "brown streak disease" were lessened as a result of the mutagenesis of eIF4E in A. thaliana and cassava mediated by CRISPR/SpCas9 (Cao et al. 2020). When eIF4G was targeted for alteration using the CRISPR/SpCas9 system, tolerance against the rice "tungro spherical virus (RTSV)" was discovered. Tolerance to the "clover yellow vein virus" was achieved through cytidine base editing in A. thaliana that involved a C > G conversion at N176K in eIF4E1 (wild type) (Bastet et al. 2019). The RNA sequences of RNA viruses can also be targeted by the CRISPR system, which is made up of the Cas13a and FnCas9 proteins. In *N. benthamiana*, CRISPR/Cas13a was used to destroy the TuMV viral RNA (Zhang et al. 2019). Similarly, in genetically modified *A. thaliana and N. benthamiana*, the CRISPR/FnCas9 system was used to target the mutation of "tobacco mosaic virus (TMV)" and "cucumber mosaic virus (CMV)" (Zhang et al.

2018). Additionally, resistance to the "black-streaked draft virus (SRBSDV)," "potato virus Y (PVY)," and "rice stripe mosaic virus (RSMV)" of southern rice has been developed using this method. CRISPR/Cas technology works well in genetically modified crops and is useful enough to inhibit the spread of viruses in plants (Tripathi et al. 2022). Nonetheless, there is a possibility that the virus will evade the system and cause viral infection, which is particularly frequent in strains that are changing quickly.

5.3 Disease Resistance Against Bacteria

During bacterial infections, plants release effectors, which disrupt host cell defenses and trigger susceptibility (S) genes for disease progression. Without affecting plant growth or yield, CRISPR-Cas technology can be used to edit S genes, making plants less vulnerable to bacterial infections (Zhou et al. 2015). It is possible to lessen *Xanthomonas's* capacity to colonize plants and spread illness by carefully identifying and altering particular S genes. Editing crop plants' susceptibility genes (S genes) is one possible strategy. Because they encode proteins necessary for bacterial colonization and disease development, S genes are crucial to the interaction between plants and microbes (Kim et al. 2019a, b). Sugar transporters necessary for bacterial pathogenicity are encoded by the SWEET gene family. In a similar way, Xanthomonas cannot infect some crops unless the LOB1 gene is present, which is involved in controlling plant immunity (Zaidi et al. 2018). Numerous crops, such as rice, tomatoes, and citrus, have previously shown effectiveness with this strategy. A promising strategy for managing bacterial infections in crops is the use of CRISPR-Cas technology to edit S genes.

Plant sensitivity to Xanthomonas and other bacterial diseases can be decreased by focusing on and altering particular genes that are necessary for bacterial colonization and disease development (Kim et al. 2019a, b). This strategy could help farmers develop more sustainable and efficient disease management methods while lowering their need on agrochemicals (Xu et al. 2019). Therefore, to increase plant resistance to microorganisms, the CRISPR/Cas system targets S genes and negative regulators of innate immunity (Zaidi et al. 2018). Knocking down the rice gene Os8N3 or OsSWEET11, which is the target for TALEs, can result in resistance against "*Xanthomonas oryzae* pv. oryzae (Xoo)" (Kim et al. 2019a, b). Targeting the "TALE-binding elements (EBEs)" found in the promoter sequences of the "OsSWEET11" and "OsSWEET14" genes can provide strong resistance against a variety of Xoo strains (Xu et al. 2019). Additionally, when the CRISPR/Cas system targeted the promoter sequence of OsSWEET11/OsSWEET13/OsSWEET14, broad-spectrum resistance was produced (Oliva et al. 2019). It has also been demonstrated that the "CRISPR/Cas12" and "CRISPR/SpCas9" tools are useful for generating resistance against a variety of plant diseases. Targeting the promoter sequence of the CsLOB1 gene in citrus can result in resistance against citrus canker disease, which is caused by *Xanthomonas citri* pv. citri (Peng et al. 2017).

6 CRISPR/Cas-Mediated Abiotic Stress Resistance in Plants

Globally, abiotic stressors such as temperature, salt, drought, and others provide a serious obstacle to crop production and significantly reduce production. Food security and agriculture are at risk due to climate change. The frequent occurrence of high temperatures and drought stress in crop plants is caused by excessive greenhouse gas emissions (Wang et al. 2016a, b, c). Therefore, a sustainable approach to tackling these issues would be to breed climate-smart crops that are able to withstand abiotic stresses.

6.1 CRISPR for Improving Drought Tolerance in Crop Plants

In many regions of the world, drought is becoming a challenge to sustainable agriculture because of climate change's effects, such as irregular rainfall patterns and rising temperatures. Many physiological and pharmacological processes are responsible for the complicated quantitative property of drought stress tolerance (Shendekar et al. 2024). Many attempts were made to modify these genes and use transgenic techniques to add additional genes. Numerous researches have documented using CRISPR to give plants drought resistance. It has been demonstrated that rice's drought tolerance is increased when the expression of the DERF1, PMS3, MSH1, MYB5, and SPP regulatory genes is down regulated using CRISPR/Cas9 (Zhang et al. 2014). Drought tolerance was demonstrated by a mutation in the Arabidopsis OST2 structural gene that was created using CRISPR/Cas9 (Osakabe et al. 2016). Another study found that Arabidopsis's ability to withstand drought was significantly enhanced when the miR169a gene was knocked out using CRISPR/Cas9 (Zhao et al. 2016). Similarly, Arabidopsis's ability to withstand drought improved when the vacuolar H+-pyrophosphate (AVP1) regulating gene was activated using CRISPR/Cas9 (Park et al. 2017). Drought tolerance in rice was enhanced by CRISPR/Cas9-mediated deletion of the SRL1, SRL2, and ERA1 genes (Ogata et al. 2020). Using the CRISPR/Cas9 assay, several genes were also altered in tomato plants to increase their resistance to drought such as LBD40 gene (Liu et al. 2020) and the gibberellin insensitive dwarf1 (GID1) gene (Illouz-Eliaz et al. 2020).

6.2 CRISPR for Improving Salt Tolerance in Crop Plants

According to UN estimates, 70–100% of crop production needs be raised by the end of 2050 to fulfil the growing global demand for food. On the other hand, growing agricultural cultivation has resulted in decreased soil fertility and salinization, which are highly inappropriate for crop cultivation and growth (Yue et al. 2012). Excessive soluble salt buildup in the crop root zone causes soil salinization, which prevents roots from absorbing water. By decreasing chlorophyll concentration and stomatal conductance and affecting the growth of chloroplasts and photosystem II, salt stress hinders transpiration and the photosynthetic machinery. Furthermore, it results in osmotic stress, lowers soil and leaf water potential, and decreases plant turgor pressure by altering water relation (Atieno et al. 2017). In order to control osmotic adjustment during salt stress, genes implicated in ion transport have been targeted using genome editing and genetic engineering techniques. Arabidopsis's ability to withstand salinity was enhanced by the overexpression of SOS1 (salt excessively sensitive 1) (Yue et al. 2012). Like SOS1, overexpression of Hordeum vulgare's HvHKT2;1 (subfamily II HKT transporter) increased Na^+ translocation, which improved barley's ability to withstand salt (Mian et al. 2011). In a different investigation, a significant level of salt tolerance was conferred in rice by the PARAQUAT TOLERANCE 3 mutants (OsPQT3) created using CRISPR/Cas9 (Alfatih et al. 2020). By using genome editing technologies to investigate the role of OsmiR535 in salt stress tolerance, it was proposed that rice's ability to withstand salinity may be enhanced by OsmiR535 deletion using CRISPR/Cas9. Furthermore, a homozygous five bp deletion in OsmiR535's coding sequence may be a viable target for enhancing rice's ability to withstand salinity (Yue et al. 2020). There is no denying that the CRISPR/Cas system is a promising tool for enhancing salt tolerance in many crops, despite the lack of report available on the possible implications of the CRISPR/Cas9 system toward the increase of salinity tolerance.

6.3 CRISPR/Cas9 for Mitigating the Impact of Heat Stress

The 10–15 °C rise in temperature over the ambient temperature which is necessary for normal growth and development is referred to as heat stress. Excessive heat stress has become a significant problem that causes significant yield losses and is predicted to get worse in the future (Yu et al. 2019). Plants exhibit bolting, wilting, fruit senescence, and leaf damage in response to heat stress. In addition to causing several physiological, biochemical, and molecular alterations that can negatively impact plant development and productivity, it can also create visual signs such as leaf browning and burn (Qiu et al. 2017). Thus, there is an urgent need for methods to lessen the devastating effects of heat stress. Rice with the heat-sensitive albino1 (OsHSA1) gene deleted showed a faster greening phenotype than the wild type, but it was also more susceptible to heat. Early chloroplast development has been shown to be significantly influenced by HSA1, and later, it has been shown to protect chloroplasts against heat stress (Yu et al. 2019). A fructokinase-like protein that is encoded by the OsHSA1 is involved in the development and protection of chloroplasts in rice at various stages of growth. Under heat stress, mutants produced by CRISPR/Cas9 in tomato Slcpk28 demonstrated a greater formation of ROS and protein oxidation as well as a decrease in the activity of antioxidant enzymes such as ascorbate peroxidase (Hu et al. 2021). By reducing ROS buildup and upregulating the expression of genes encoding heat shock proteins (HSPs) and heat stress transcription factors (HSFs), the CRISPR/Cas9-mediated deletion of SlMAPK3 demonstrated enhanced heat stress tolerance (Yu et al. 2019).

7 CRISPR in Plant-Microbe Interactions

Throughout the whole life cycle, a variety of microbial species interact with plants and animals in nature. Early on in their evolutionary history, humans have been exposed to a diverse range of microbes, which has increased their ability to adapt to a healthy lifestyle. Similarly, plants cohabit with microbes, including archaea, protists, bacteria, and fungi, together called microbiota (Van Der Ent et al. 2009). Our capacity to control plant-microbe interactions has been transformed by CRISPR technology, especially when it comes to strengthening symbiotic connections that enhance nutrient intake.

7.1 Mycorrhizal Fungi and Nutrient Uptake

By forming advantageous relationships with plant roots, mycorrhizal fungi improve the absorption of nutrients, particularly phosphorus. Genes involved in the signalling pathways between mycorrhizal fungus and plants can be edited using CRISPR (Mondal et al. 2022). Mycorrhizal connection formation can be improved by deleting genes involved in strigolactone production. Plant hormones called strigolactones increase nutrient absorption by encouraging mycorrhizal fungus to colonize roots. In Arabidopsis, mycorrhizal colonization was improved using CRISPR/Cas9-mediated gene deletion, which boosted phosphorus absorption (Verma and Srivastava 2021).

7.2 Rhizobium and Nitrogen Fixation

Legumes and Rhizobium bacteria develop symbiotic interactions that enable the legumes to fix nitrogen from the atmosphere. Nitrogen availability in soils can be greatly increased by altering these relationships. By eliminating negative nodulation regulators with CRISPR, legumes may produce more nodules, which improves nitrogen fixation (Gouda et al. 2017). Targeting the NIN gene can increase the effectiveness of nitrogen absorption because it is essential for nodulation signalling. Using CRISPR/Cas9 to alter the expression of certain receptor-like kinases in legumes led to increased nitrogen fixation and nodulation in a variety of environmental settings (Singh and Ramakrishna 2021).

7.3 Enhanced Stress Resistance and Engineered Microbial Partners

By encouraging healthy plant development, CRISPR may also be used to increase a plant's resistance to stress, which in turn can boost nutrient absorption. Enhancing the general health of the plant by gene editing linked to stress responses (such as drought or salt) may boost the plant's capacity to form symbiotic interactions with rhizobia and mycorrhizal fungi (Zheng et al. 2020a, b). Using CRISPR to target the DREB transcription factors increased wheat's tolerance to drought, which in turn boosted its mycorrhizal relationships and nutrient absorption. Additionally, CRISPR may be used to alter the microorganisms themselves, increasing their beneficial impact on plants (Van Der Heijden et al. 2015). Researchers can increase the effectiveness of rhizobial strains in symbiosis by genetically modifying them to exhibit particular features (such as improved nitrogen fixation or stress tolerance). Improving nitrogen availability for legumes may result from using CRISPR-based alterations to boost the effectiveness of nitrogenase enzymes in Rhizobium spp. By employing CRISPR to modify the nifA gene in a Rhizobium strain, nitrogenase activity was markedly elevated, improving symbiotic plant growth (Shendekar et al. 2024).

7.4 Plant Defence Mechanisms and Root Architecture and Exudation

Plant defence mechanisms can be altered by CRISPR to more effectively interact with advantageous microorganisms. The important thing is to preserve positive interactions while boosting the plant's immune response. More strong symbiotic partnerships can be achieved by knocking down certain genes that activate defence responses without inciting overwhelming plant defence systems that would prevent the colonization of beneficial microbes (Hammerbacher et al. 2019). By deleting defense-related genes in soybeans using CRISPR, more nodulation with Rhizobium occurs without endangering the health of the plant. Plant-microbe interactions can be improved by altering the genes that govern root architecture. Through their roots, plants emit a range of substances that may draw in helpful microorganisms (Van Der Ent et al. 2009). By changing the expression of genes related to root exudation, CRISPR can improve the recruitment of nitrogen-fixing bacteria or mycorrhizal fungus. Increased colonization by beneficial microorganisms and enhanced nutrient absorption in maize were the results of altering the expression of genes governing the composition of root exudate (Solanki et al. 2023).

8 Challenges and Future Prospects

8.1 Technical Challenges

Although CRISPR technology has transformed genetic engineering, there are still several technical obstacles to overcome, especially when dealing with intricate crop genomes. The following are some major areas of worry:

8.1.1 Off-target Effects

Off-target consequences happen when CRISPR/Cas9 accidentally modifies unwanted genomic locations, which might result in undesirable phenotypes

(Scheben et al. 2017). The very high frequency of off-target effects (OTEs), which have been found at a frequency of ≥50%, is a serious worry for using CRISPR/Cas9 for gene therapy. The target DNA sequence must be bound precisely by the guide RNA (gRNA) (Zhao et al. 2016). Off-target cleavage can result from even small mismatches. Detection: Extensive techniques such as targeted deep sequencing or whole-genome sequencing, which might be resource-intensive, are needed to identify off-target effects (Zhang et al. 2015).

8.1.2 Delivery Methods

To guarantee that CRISPR components (Cas9, gRNA) enter plant cells, efficient delivery methods are essential. Plant Cell Barriers: The inflexible cell wall and membrane provide major delivery obstacles (Zhao et al. 2016). Transformation Efficiency: The current techniques, such as biolistics and agrobacterium-mediated transformation, can be inefficient, particularly in some crops. A potential technique for the temporary production of CRISPR components is viral vectors (Lino et al. 2018).

8.1.3 Efficiency in Complex Crop Genomes

Many crops contain genomes made up of several sets of chromosomes, known as polyploids, which makes editing more difficult. Homologous Recombination: Because many similar gene copies exist in complicated genomes, making exact modifications might be difficult (Mushtaq et al. 2021). It may be difficult to navigate the regulatory environment around genetically modified organisms, which can slow down research and development. The effectiveness of regenerating whole plants from modified cells might differ greatly among species, which can impact overall success rates.

8.1.4 Horizontal Gene Transfer

HGT is the mechanism by which microbes exchange genetic material when they don't use conventional reproductive techniques. Transformation, transduction, or conjugation are possible ways for this to happen. In the context of microbiome editing, horizontal gene transfer (HGT) raises serious problems, especially regarding unforeseen ecological repercussions (Wu 2024). This is a summary, backed up by sources, of the implications of HGT in microbiome research. The transfer of genetic material between organisms by a process other than traditional reproduction is referred to as horizontal gene transfer, or HGT. HGT has a major impact on the makeup and function of microbial communities in microbiomes (Tao et al. 2022).

8.1.5 Unintended Ecological Consequences

- Disruption of Microbial Communities: Community dynamics may be unintentionally impacted when the genomes of certain bacteria within a microbiome are edited. Changes in microbial diversity and function might result, for instance, from the introduction of antibiotic resistance genes by horizontal gene transfer (HGT), which could enable modified organisms to outcompete natural species (Piergentili et al. 2021).

- Spread of Unintended Traits: Through the microbiome and perhaps the environment, edited microorganisms can spread characteristics like pathogenicity or resistance by transferring altered genes to wild-type strains. Managing microbial populations may become difficult as a result of this gene transfer (Caplan et al. 2015).
- Impact on Ecosystem Functions: HGT-induced alterations in the composition of microbial communities can modify vital ecological processes as carbon fixation, nutrient cycling, and pollutant degradation. Higher trophic levels and the general health of the ecosystem may be impacted in a cascade by these changes (Zhou et al. 2015).
- Emergence of Pathogenic Microbes: New infections may arise within the microbiome because of HGT's ability to help non-pathogenic microorganisms acquire virulence factors. When microbial populations are altered in clinical or agricultural contexts, this is especially troubling.

To mitigate potential ecological repercussions, it is essential to comprehend the ramifications of horizontal gene transfer in microbiome editing. HGT and its impacts on microbial communities and ecosystems will need to be carefully considered for responsible deployment as microbiome research and engineering progress. To evaluate the long-term effects of microbiome modification on environmental stability and health, more investigation and observation are required (Braddick and Ramarohetra 2020).

9 Regulatory and Ethical Considerations

9.1 Ethical Concerns and Public Acceptance

Many nations have regulated the production and commercialization of genetically modified plants due to serious public and scientific concerns about the hazards involved in cultivating genetically modified plants and using or consuming their products (Choudhary et al. 2014). The insertion of transgenes, or genes that would not naturally occur in these plants, is primarily responsible for these concerns (Piergentili et al. 2021). Proponents and developers of plants modified with CRISPR claim that normally, neither the plants nor the products derived from them contain any foreign DNA. Opponents of these methods argue that crop varieties created through them are genetically engineered and might be governed by GMO regulations (Turnbull et al. 2021). To allay public fears and establish appropriate biotechnological paths in agriculture, new plant breeding technologies (NPBTs) need to be regulated. But the possibility that CRISPR methods could alter a plant's genome in a way that is heritable raises concerns and need international consensus on regulations.

The possible threats to human health and the environment are the principal sources of public concern. Social influences, as well as ethical and religious convictions like the idea that genetic scientists are "playing God," can have an impact on how people perceive risk (Chatzopoulou et al. 2020). People believe that novel crops that emerge from NPBTs are not like traditional crops and that they pose unknown environmental dangers. Some of the reasons for these worries include a lack of public comprehension of science, mistrust of developers, insufficient laws, and a lack of clear communication about the advantages and risks (Chaturvedi 2004). Better public engagement would help close the gap between the opinions of the public and the biotechnology sector if the biotechnology, agricultural, and national governments believe genetic engineering can produce improved plant varieties that address concerns about food security (FAO 2020). Governments must, in any case, take public concerns regarding the advantages and disadvantages of novel crop kinds into account. Public acceptance of new crop varieties with high advantages and minimal dangers is higher. The public's acceptance of different genetically modified products and local knowledge of scientific concepts like genetic engineering, biotechnology, and cloning are influenced by the wide variations in educational attainment both within and between nations (Braddick and Ramarohetra 2020). More public education, sincere community consultation, open and honest regulatory procedures for evaluating new products, continuous research into possible threats to human health and the environment, and courteous communication between legislators, policymakers, and the public are all necessary for this (Ishii and Araki 2016).

9.2 Biosafety Regulations

Even though genetically modified crops offer many advantages, many people are worried about the possible risks connected to these plants. There may be health and environmental risks from gene transfer to non-target plants. Some GM plants have antibiotic-resistance genes added in addition to genes encoding desired new features (Chatzopoulou et al. 2020). These genes serve as markers and allow for the quick selection of plant cells that have successfully received gene transfers. Nonetheless, there is little possibility that microbes in the human gut could pick up these antibiotic-resistant genes. Concerns regarding the emergence of BT resistance in pests and insects that consume BT crops have been voiced by farmers (Turnbull et al. 2021). Many nations have tight laws controlling the sale and production of genetically modified organisms (GMOs) to allay concerns about possible hazards related to GM goods and procedures. Globally speaking, there are primarily two approaches to GMO regulatory frameworks: (i) product-based regulatory oversight and (ii) process-based regulations, which concentrate on the methods utilized to create GMOs with novel traits. One possible explanation for the disparities in GMO regulations is that certain nations haven't signed the Cartagena Protocol on Biosafety (CPB) (Chaturvedi 2004).

9.3 Global Regulations of GMOs and Genome-Edited Crops

Advanced molecular biology tools, such CRISPR/Cas9, have made precise and targeted alterations to plant genome loci popular globally. These modifications have various potential advantages over traditional plant-breeding methods for the generation of novel plant characteristics (Turnbull et al. 2021). Concerns over human health and environmental safety have limited the commercial use of genetically modified (GM) plants to a limited number of cultivated crops, even though GM plants have the potential to improve food security for the world's expanding population greatly. Table 13.3 summarizes the information on various crops regulated across the world by various countries along with different regulatory agencies that control the GMO generation, production, and cultivation. Concerns raised by the public and anticipated queries should be considered when using genome editing to improve crops (Lancet 2018). In response to public concerns and to regulate genetically modified organisms (GMOs), numerous nations have implemented diverse regulatory frameworks, given the over 100-fold growth in global GM crop production over the past 25 years. Global GMO rules fall into one of two categories: methods based on products or processes (Sánchez 2020). When it comes to GMO legislation, the USA, India, Japan, Canada, and Argentina all take a product-based approach, meaning that the framework evaluates the qualities of the final product regardless of how it is made. Process-based GMO laws, on the other hand, are predicated on the method of producing the organism. Currently, process-based laws govern GM crops in the EU and New Zealand. The regulation of gene-edited crops is not uniform and varies significantly across countries and regions (Hamza et al. 2021).

10 Future Directions

As a genome-editing technique, CRISPR (clustered regularly interspaced short palindromic repeat)-Cas9 has demonstrated significant promise in addressing new challenges in agriculture. This method can be used to change any DNA sequencing to give plants the desired traits (Wang et al. 2022). Additionally, this approach holds promise for creating non-transgenic gene-edited plants to meet the world's food needs and mitigate the negative consequences of climate change. Nearly all of the major crops, including barley, rice, and wheat, have benefited from CRISPR/Cas9 (Hamza et al. 2021). More importantly, a simple method for creating non-transgenic genome-edited plants to combat climate change and ensure future food security is CRISPR/Cas9. Researchers can create a robust and sustainable agricultural system using CRISPR/Cas9 to enhance plant growth and development and responsiveness to various environmental challenges. For improving agriculture, genes that adversely affect crop output can be eliminated using CRISPR/Cas9. Rodríguez-Leal and colleagues used CRISPR/Cas9 to target the tomato quantitative genes' promotor region. Their findings showed that modifying tomato plants' genes slightly enhanced the size, shape, and architecture of the fruit (Rodríguez-Leal et al. 2017). Li and colleagues used CRISPR/Cas9 to target the genes GS3, DEP1, and Gn1a in rice,

Table 13.3 Details of worldwide commercially approved genome-edited crops and legislation on the release of gene-edited plants

Country	Regulatory agencies	GMO commercial cultivation area (million hectares)	Approved genome edited crops	Approve year	Regulation governing the release of gene-edited crops	References
USA	USDA, APHIS, FDA, and EPA	75	Corn Tomato Soybean Mushroom Flax Non browning apple	2018 2018 2017 2016 2017	Coordinated framework for regulation of biotechnology, new SECURE rules (2020)	Herrera et al. (2017), Lombardo and Grando (2019) and Chaturvedi (2004)
Canada	Canadian Food Inspection Agency (CFIA)	11	Non browning potato Herbicide resistant canola	2016 2015	Directive 94–08 (Dir 94–08) assessment criteria for determining environmental safety of plants with novel traits	Lombardo and Grando (2019), Chaturvedi (2004), Smyth (2017) and Sánchez (2020)
European Union [only Spain and Portugal]		0.1	No approved crops		Directive 18/2001/EC (2001) after court decision in case C-528/16	Lombardo and Grando (2019) and Kolady and Herring (2014)
India	Indian Ministry of Science and Technology (2020), Genetic Engineering Appraisal Committee (GEAC)	11.4	Cotton	2002	Regulatory framework and guidelines for risk assessment (2020)	Lombardo and Grando (2019), Chaturvedi (2004) and Haque et al. (2018)
Japan	The Ministry of Agriculture, Forestry and Fisheries (MAFF)	No	Tomato	2021	GMO as defined under Cartagena Act (2019)	Chaturvedi (2004) and Dima and Inzé (2021)

(continued)

Table 13.3 (continued)

Country	Regulatory agencies	GMO commercial cultivation area (million hectares)	Approved genome edited crops	Approve year	Regulation governing the release of gene-edited crops	References
New Zealand	Food Standards Australia New Zealand (FSANZ)	No	No approved crops		Hazardous Substances and New Organisms Act (1998) after court decision NZHC 1067 (2014)	Chaturvedi (2004) and Bratlie et al. (2019)
Argentina	Argentine Biosafety Commission (CONABIA)	24.5	HB4 drought resistant wheat	2020	Resolution No. 173/15 (2015)	Lombardo and Grando (2019) and Chaturvedi (2004)

which have a deleterious effect on grain yield. Eliminating these genes considerably enhanced yield-related characteristics, such as grain size, number of grains, and thick erect panicles, according to the data (Li et al. 2016). In order to reduce grain weight, Xu and colleagues employed this method to eradicate the GW2, GW5, and TGW6 genes in rice. Their findings demonstrated that using this method for gene editing greatly increased grain weight (Xu et al. 2016). Additionally, the CRISPR/Cas9 system has been used to boost plant growth and development in order to increase agricultural productivity. One important enzyme that controls root growth is arginase, which inhibits nitric oxide synthase. Overexpression of arginase prevented the growth of lateral roots (Meng et al. 2015). Wang and colleagues used CRISPR/Cas9 to eradicate the ARG gene's activity in cotton in 2017. According to their findings, cotton lines with their genomes altered showed a considerable increase in lateral roots of 25% and total root surface area of 52% (Paul et al. 2019). Using this technology, Paul and colleagues also showed that the MADS-box transcription factor genes MADS78 and MADS79 are crucial regulators of endosperm cellularization and early rice seed development. Single knockout mutants of MADS78 or MADS79 showed early endosperm cellularization, however double mutants prevented seed development and were unable to produce viable rice seeds (Paul et al. 2019). In Arabidopsis, sar1b and sar1c genes were deleted using CRISPR/Cas9 to produce male sterility with changed pollen development (Liang et al. 2020a, b). Scientists may utilize it as their main tool in the future to alter crops to satisfy the world's food needs.

11 Conclusion

The CRISPR/Cas system has emerged as a transformative tool for genome editing, offering unparalleled precision and efficiency in addressing critical challenges in agriculture and food systems. In crops, CRISPR enables targeted modifications to enhance yield, improve resistance to biotic and abiotic stresses, and boost nutritional content, directly contributing to food security. It also accelerates the development of climate-resilient crops, essential in mitigating the effects of global climate change. Beyond plants, CRISPR/Cas technologies are unlocking opportunities in microbiome engineering, enhancing soil health and fostering beneficial plant-microbe interactions. This approach can promote sustainable agricultural practices by reducing dependence on chemical fertilizers and pesticides while increasing crop productivity. The integration of CRISPR/Cas genome editing with advanced breeding techniques and systems biology offers unprecedented potential to transform agriculture, enabling the creation of resilient, high-yielding, and nutritionally superior crops. However, these advancements must be accompanied by robust ethical considerations and balanced regulatory frameworks to ensure equitable access, societal acceptance, and environmental sustainability. By addressing these challenges, CRISPR/Cas technologies can play a pivotal role in achieving global food and nutrition security, empowering farmers, and fostering a resilient food system that meets the demands of a growing population while preserving natural resources for future generations.

References

Abdi G, Patil N, Mishra P, Tariq M, Dhariwal R, Parveen S, Mudgal G (2024) Genomic precision: unveiling the transformative role of genome editing in advancing genomics research and applications. In: Advances in genomics: methods and applications. Springer Nature Singapore, Singapore, pp 265–306. https://doi.org/10.1007/978-981-97-3169-5_13

Ajayi ND, Amoo NO, Obunukwu NGM (2024) Crispr and plant pathology: revolutionizing disease resistance in crops. World J Adv Res Rev 24(1):044–060. https://doi.org/10.30574/wjarr.2024.24.1.2996

Ajeng AA, Vejan P, Abdullah R (2024) The role of microbes in biofortification for a sustainable nutrient management and agricultural crop productivity: current trends and imminent challenges. In: Microbial technology for agro-ecosystems. Academic, pp 293–306. https://doi.org/10.1016/B978-0-443-18446-8.00009-7

Akhil S, Mohammad A, Rubby S, Sandip B (2024) Evaluating CRISPR/Cas9's progressive role in crop enhancement and sustainable agriculture. J Adv Biol Biotechnol. https://doi.org/10.9734/jabb/2024/v27i6949

Akköprü EP, Atlıhan R, Okut H, Chi H (2015) Demographic assessment of plant cultivar resistance to insect pests: a case study of the dusky-veined walnut aphid (Hemiptera: Callaphididae) on five walnut cultivars. J Econ Entomol 108(2):378–387. https://doi.org/10.1093/jee/tov011

Akram F, Sahreen S, Aamir F, Haq IU, Malik K, Imtiaz M, Naseem W, Nasir N, Waheed HM (2023) An insight into modern targeted genome-editing technologies with a special focus on CRISPR/Cas9 and its applications. Mol Biotechnol 65(2):227–242. https://doi.org/10.1007/s12033-022-00501-4

Alfatih A, Wu J, Jan SU, Zhang Z, Xia J, Xiang C (2020) Loss of rice PARAQUAT TOLERANCE 3 confers enhanced resistance to abiotic stresses and increases grain yield in field. Plant Cell Environ 43(11):2743–2754. https://doi.org/10.1111/pce.13856

Ali Z, Abulfaraj A, Idris A, Ali S, Tashkandi M, Mahfouz MM (2015) CRISPR/Cas9-mediated viral interference in plants. Genome Biol 16(1). https://doi.org/10.1186/s13059-015-0799-6

Atieno J, Li Y, Langridge P, Dowling K, Brien C, Berger B, Varshney RK, Sutton T (2017) Exploring genetic variation for salinity tolerance in chickpea using image-based phenotyping. Sci Rep 7(1). https://doi.org/10.1038/s41598-017-01211-7

Backer R, Rokem JS, Ilangumaran G, Lamont J, Praslickova D, Ricci E, Subramanian S, Smith DL (2018) Plant growth-promoting rhizobacteria: context, mechanisms of action, and roadmap to commercialization of biostimulants for sustainable agriculture. Front Plant Sci 9. https://doi.org/10.3389/fpls.2018.01473

Bai X, Correa VR, Toruño TY, Ammar E, Kamoun S, Hogenhout SA (2008) AY-WB phytoplasma secretes a protein that targets plant cell nuclei. Mol Plant-Microbe Interact 22(1):18–30. https://doi.org/10.1094/mpmi-22-1-0018

Baltes NJ, Gil-Humanes J, Cermak T, Atkins PA, Voytas DF (2014) DNA replicons for plant genome engineering. Plant Cell 26:151–163. https://doi.org/10.1105/tpc.113.119792

Bastet A, Zafirov D, Giovinazzo N, Guyon-Debast A, Nogué F, Robaglia C, Gallois J (2019) Mimicking natural polymorphism in eIF4E by CRISPR-Cas9 base editing is associated with resistance to potyviruses. Plant Biotechnol J 17(9):1736–1750. https://doi.org/10.1111/pbi.13096

Berg G, Grube M, Schloter M, Smalla K (2014) The plant microbiome and its importance for plant and human health. Front Microbiol 5. https://doi.org/10.3389/fmicb.2014.00491

Bhattacharjee S, Bhowmick R, Kant L, Paul K (2023) Strategic transgene-free approaches of CRISPR-based genome editing in plants. Mol Gen Genomics 298(3):507–520

Biswas S, Kundu DK, Mazumdar SP, Saha A, Majumdar B, Ghorai AK, Ghosh D, Yadav A, Saxena AK, Allied Fibres BI (2018) Study on the activity and diversity of bacteria in a New Gangetic alluvial soil (Eutrocrept) under rice-wheat jute cropping system. J Environ Biol 39:379–386. https://doi.org/10.22438/jeb/39/3/MRN-523

Bouis HE, Hotz C, McClafferty B, Meenakshi JV, Pfeiffer WH (2011) Biofortification: a new tool to reduce micronutrient malnutrition. Food Nutr Bull 32(1 Suppl):S31–S40. https://doi.org/10.1177/15648265110321S105

Braddick D, Ramarohetra RF (2020) Emergent challenges for CRISPR: biosafety, biosecurity, patenting, and regulatory issues. In: Genome engineering via CRISPR-Cas9 system. Academic, pp 281–307

Bratlie S, Halvorsen K, Myskja BK, Mellegård H, Bjorvatn C, Frost P, Heiene G, Hofmann B, Holst-Jensen A, Holsst-Larsen T, Malnes RS, Paus B, Sandvig B, Sjøli SI, Skarstein B, Thorseth MB, Vagstad N, Våge DI, Borge OJ (2019) A novel governance framework for GMO. EMBO Rep 20(5). https://doi.org/10.15252/embr.201947812

Brevik EC, Slaughter L, Singh BR, Steffan JJ, Collier D, Barnhart P, Pereira P (2020) Soil and human health: current status and future needs. Air Soil Water Res 13:117862212093444. https://doi.org/10.1177/1178622120934441

Bünemann EK, Bongiorno G, Bai Z, Creamer RE, De Deyn G, De Goede R, Fleskens L, Geissen V, Kuyper TW, Mäder P, Pulleman M, Sukkel W, Van Groenigen JW, Brussaard L (2018) Soil quality—a critical review. Soil Biol Biochem 120:105–125. https://doi.org/10.1016/j.soilbio.2018.01.030

Cao Y, Zhou H, Zhou X, Li F (2020) Control of plant viruses by CRISPR/Cas system-mediated adaptive immunity. Front Microbiol 11. https://doi.org/10.3389/fmicb.2020.593700

Caplan AL, Parent B, Shen M, Plunkett C (2015) No time to waste—the ethical challenges created by CRISPR: CRISPR/Cas, being an efficient, simple, and cheap technology to edit the genome of any organism, raises many ethical and regulatory issues beyond the use to manipulate human germ line cells. EMBO Rep 16(11):1421–1426

Carthey AJ, Gillings MR, Blumstein DT (2018) The extended genotype: microbially mediated olfactory communication. Trends Ecol Evol 33(11):885–894. https://doi.org/10.1016/j.tree.2018.08.010

Chandrasekaran J, Brumin M, Wolf D, Leibman D, Klap C, Pearlsman M et al (2016) Development of broad virus resistance in non-transgenic cucumber using CRISPR/Cas9 technology. Mol Plant Pathol 17:1140–1153. https://doi.org/10.1111/mpp.12375

Chaturvedi S (2004) Biosafety regulation: need for fine balancing. Econ Polit Wkly 39:3693–3697

Chatzopoulou S, Eriksson NL, Eriksson D (2020) Improving risk assessment in the European food safety authority: lessons from the European Medicines Agency. Front Plant Sci 11:349

Chen H, Min F, Hu X, Ma D, Huo Z (2023) Biochar assists phosphate solubilizing bacteria to resist combined Pb and Cd stress by promoting acid secretion and extracellular electron transfer. J Hazard Mater 452:131176. https://doi.org/10.1016/j.jhazmat.2023.131176

Chen F, Chen L, Yan Z, Xu J, Feng L, He N, Guo M, Zhao J, Chen Z, Chen H, Yao G, Liu C (2024) Recent advances of CRISPR-based genome editing for enhancing staple crops. Front Plant Sci 15. https://doi.org/10.3389/fpls.2024.1478398

Chialva M, Di Fossalunga AS, Daghino S, Ghignone S, Bagnaresi P, Chiapello M, Novero M, Spadaro D, Perotto S, Bonfante P (2018) Native soils with their microbiotas elicit a state of alert in tomato plants. New Phytol 220(4):1296–1308. https://doi.org/10.1111/nph.15014

Choudhary B, Gheysen G, Buysse J, van der Meer P, Burssens S (2014) Regulatory options for genetically modified crops in India. Plant Biotechnol J 12(2):135–146

Cordones MN, Mohamed S, Tanoi K, Natsuko Kobayashi NI, Takagi K, Vernet A et al (2017) Production of low-Cs rice plants by inactivation of the K transporter OsHAK1 with the CRISPR-Cas system. Plant J 92:43–56. https://doi.org/10.1111/tpj.13632

Cossart P (2020) The new microbiology: from microbiomes to CRISPR. Wiley

Dima O, Inzé D (2021) The role of scientists in policy making for more sustainable agriculture. Curr Biol 31(5):R218–R220. https://doi.org/10.1016/j.cub.2021.01.090

Doggalli G, Kavya N, Singh OB, Manojkumar HG, Srivastava M, Chaya GB, Karadagi AV, Vinodh Kumar PN (2024) Transformative gene editing methods: precision in genetically modified crops through trait modification. J Sci Res Rep 30(8):615–628. https://doi.org/10.9734/jsrr/2024/v30i82283

Dong OX, Yu S, Jain R, Zhang N, Duong PQ, Butler C, Li Y, Lipzen A, Martin JA, Barry KW, Schmutz J, Tian L, Ronald PC (2020) Marker-free carotenoid-enriched rice generated through targeted gene insertion using CRISPR-Cas9. Nat Commun 11(1):1178. https://doi.org/10.1038/s41467-020-14981-y

Erdogan İ, Cevher-Keskin B, Bilir Ö, Hong Y, Tör M (2023) Recent developments in CRISPR/Cas9 genome-editing technology related to plant disease resistance and abiotic stress tolerance. Biology 12(7):1037. https://doi.org/10.3390/biology12071037

Fan Y, Liu J, Lyu S, Wang Q, Yang S, Zhu H (2017) The soybean Rfg1 gene restricts nodulation by *Sinorhizobium fredii* USDA193. Front Plant Sci 8. https://doi.org/10.3389/fpls.2017.01548

Food and Agriculture Organization (FAO) (2020) The state of food security and nutrition in the world. Food and Agriculture Organization (FAO)

Gasiunas G, Barrangou R, Horvath P, Siksnys V (2012) Cas9–crRNA ribonucleoprotein complex mediates specific DNA cleavage for adaptive immunity in bacteria. Proc Natl Acad Sci 109(39). https://doi.org/10.1073/pnas.1208507109

Geisen S, Mitchell EAD, Adl S, Bonkowski M, Dunthorn M, Ekelund F, Fernández LD, Jousset A, Krashevska V, Singer D, Spiegel FW, Walochnik J, Lara E (2018) Soil protists: a fertile frontier in soil biology research. FEMS Microbiol Rev 42(3):293–323. https://doi.org/10.1093/femsre/fuy006

German-Retana S, Walter J, Doublet B, Roudet-Tavert G, Nicaise V, Lecampion C, Houvenaghel M, Robaglia C, Michon T, Gall OL (2008) Mutational analysis of plant cap-binding protein eIF4E reveals key amino acids involved in biochemical functions and potyvirus infection. J Virol 82(15):7601–7612. https://doi.org/10.1128/jvi.00209-08

Gill SS, Gill R, Trivedi DK, Anjum NA, Sharma KK, Ansari MW, Ansari AA, Johri AK, Prasad R, Pereira E, Varma A, Tuteja N (2016) *Piriformospora indica*: potential and significance in plant stress tolerance. Front Microbiol 7. https://doi.org/10.3389/fmicb.2016.00332

Gosavi G, Yan F, Ren B, Kuang Y, Yan D, Zhou X, Zhou H (2020) Applications of CRISPR technology in studying plant-pathogen interactions: overview and perspective. Phytopathol Res 2(1). https://doi.org/10.1186/s42483-020-00060-z

Gouda S, Kerry RG, Das G, Paramithiotis S, Shin H, Patra JK (2017) Revitalization of plant growth promoting rhizobacteria for sustainable development in agriculture. Microbiol Res 206:131–140. https://doi.org/10.1016/j.micres.2017.08.016

Hacquard S, Spaepen S, Garrido-Oter R, Schulze-Lefert P (2017) Interplay between innate immunity and the plant microbiota. Annu Rev Phytopathol 55(1):565–589. https://doi.org/10.1146/annurev-phyto-080516-035623

Hammerbacher A, Coutinho TA, Gershenzon J (2019) Roles of plant volatiles in defence against microbial pathogens and microbial exploitation of volatiles. Plant Cell Environ 42(10):2827–2843. https://doi.org/10.1111/pce.13602

Hamza A, Ahmad M, Tariq A, Din AU, Abbas Z, Bano M, Rafique RS (2021, December 8) CRISPR/CAS9: an approach to meet global food demand. https://ikprress.org/index.php/PCBMB/article/view/7221

Haque E, Taniguchi H, Hassan MM, Bhowmik P, Karim MR, Śmiech M, Zhao K, Rahman M, Islam T (2018) Application of CRISPR/Cas9 genome editing technology for the improvement of crops cultivated in tropical climates: recent progress, prospects, and challenges. Front Plant Sci 9. https://doi.org/10.3389/fpls.2018.00617

Herrera FF, Domené-Painenao O, Cruces JM (2017) The history of agroecology in Venezuela: a complex and multifocal process. Agroecol Sustain Food Syst 41(3–4):401–415. https://doi.org/10.1080/21683565.2017.1285842

Hoeksema JD, Chaudhary VB, Gehring CA, Johnson NC, Karst J, Koide RT, Pringle A, Zabinski C, Bever JD, Moore JC, Wilson GWT, Klironomos JN, Umbanhowar J (2010) A meta-analysis of context-dependency in plant response to inoculation with mycorrhizal fungi. Ecol Lett 13(3):394–407. https://doi.org/10.1111/j.1461-0248.2009.01430.x

Hu Z, Li J, Ding S, Cheng F, Li X, Jiang Y, Yu J, Foyer CH, Shi K (2021) The protein kinase CPK28 phosphorylates ascorbate peroxidase and enhances thermotolerance in tomato. Plant Physiol 186(2):1302–1317. https://doi.org/10.1093/plphys/kiab120

Huang W (2024) Nitrogen fixation in legumes: genetic mechanisms and agricultural applications. Field Crop 7(2):58–69. https://doi.org/10.5376/fc.2024.07.0007

Huang Y, Dong H, Shang M, Wang K (2021) CRISPR/Cas systems: the link between functional genes and genetic improvement. Crop J 9(3):678–687. https://doi.org/10.1016/j.cj.2021.03.004

Ibrahim S, Saleem B, Rehman N, Zafar SA, Naeem MK, Khan MR (2021) CRISPR/Cas9 mediated disruption of *Inositol Pentakisphosphate 2-Kinase 1* (*TaIPK1*) reduces phytic acid and improves iron and zinc accumulation in wheat grains. J Adv Res 37:33–41. https://doi.org/10.1016/j.jare.2021.07.006

Igalavithana A, Lee S, Niazi N, Lee Y, Kim K, Park J, Moon D, Ok Y (2017) Assessment of soil health in urban agriculture: soil enzymes and microbial properties. Sustain For 9(2):310. https://doi.org/10.3390/su9020310

Illouz-Eliaz N, Nissan I, Nir I, Ramon U, Shohat H, Weiss D (2020) Mutations in the tomato gibberellin receptors suppress xylem proliferation and reduce water loss under water-deficit conditions. J Exp Bot 71(12):3603–3612. https://doi.org/10.1093/jxb/eraa137

Ingle KP, Padole DA (2017) Phosphate solubilizing microbes: an overview. Int J Curr Microbiol App Sci 6(1):844–852. https://doi.org/10.20546/ijcmas.2017.601.099

Iqbal Z, Sattar MN, Shafiq M (2016) CRISPR/Cas9: a tool to circumscribe cotton leaf curl disease. Front Plant Sci 7:475. https://doi.org/10.3389/fpls.2016.00475

Ishii T, Araki M (2016) Consumer acceptance of food crops developed by genome editing. Plant Cell Rep 35:1507–1518

Jeon JE, Kim J, Fischer CR, Mehta N, Dufour-Schroif C, Wemmer K, Mudgett MB, Sattely E (2020) A pathogen-responsive gene cluster for highly modified fatty acids in tomato. Cell 180(1):176–187.e19. https://doi.org/10.1016/j.cell.2019.11.037

Ji J, Zhang C, Sun Z, Wang L, Duanmu D, Fan Q (2019) Genome editing in cowpea *Vigna unguiculata* using CRISPR-Cas9. Int J Mol Sci 20(10):2471. https://doi.org/10.3390/ijms20102471

Jiang WZ, Zhou HB, Bi HH, Fromm M, Yang B, Weeks DP (2013) Demonstration of CRISPR/Cas9/sgRNA-mediated targeted gene modification in Arabidopsis, Tobacco, Sorghum and rice. Nucleic Acids Res 41:e188. https://doi.org/10.1093/nar/gkt780

Karunarathna NL, Wang H, Harloff H, Jiang L, Jung C (2020) Elevating seed oil content in a polyploid crop by induced mutations in SEED FATTY ACID REDUCER genes. Plant Biotechnol J 18(11):2251–2266. https://doi.org/10.1111/pbi.13381

Kaur A (2020) Genetically modified crops in India: experiments with Bt Cotton to explore the road ahead. Open Agric 5(1):386–394

KhokharVoytas A, Shahbaz M, Maqsood MF, Zulfiqar U, Naz N, Iqbal UZ, Sara M, Aqeel M, Khalid N, Noman A, Zulfiqar F, Al Syaad KM, AlShaqhaa MA (2023) Genetic modification strategies for enhancing plant resilience to abiotic stresses in the context of climate change. Funct Integr Genomics 23(3):283. https://doi.org/10.1007/s10142-023-01202-0

Kim G, Son S, Yu H, Mun J (2019a) MtGA2ox10 encoding C20-GA2-oxidase regulates rhizobial infection and nodule development in *Medicago truncatula*. Sci Rep 9(1). https://doi.org/10.1038/s41598-019-42407-3

Kim Y, Moon H, Park C (2019b) CRISPR/Cas9-targeted mutagenesis of Os8N3 in rice to confer resistance to *Xanthomonas oryzae* pv. *oryzae*. Rice 12(1). https://doi.org/10.1186/s12284-019-0325-7

Kolady DE, Herring RJ (2014) Regulation of genetically engineered crops in India: implications of policy uncertainty for social welfare, competition, and innovation. Can J Agric Econ/Rev Can Agroecon 62(4):471–490. https://doi.org/10.1111/cjag.12055

Kwak M, Kong HG, Choi K, Kwon S, Song JY, Lee J, Lee PA, Choi SY, Seo M, Lee HJ, Jung EJ, Park H, Roy N, Kim H, Lee MM, Rubin EM, Lee S, Kim JF (2018) Rhizosphere microbiome structure alters to enable wilt resistance in tomato. Nat Biotechnol 36(11):1100–1109. https://doi.org/10.1038/nbt.4232

LeCun Y, Bengio Y, Hinton G (2015) Deep learning. Nature 521(7553):436–444. https://doi.org/10.1038/nature14539

Lekberg Y, Koide RT (2005) Is plant performance limited by abundance of arbuscular mycorrhizal fungi? A meta-analysis of studies published between 1988 and 2003. New Phytol 168(1):189–204. https://doi.org/10.1111/j.1469-8137.2005.01490.x

Li M, Li X, Zhou Z, Wu P, Fang M, Pan X, Lin Q, Luo W, Wu G, Li H (2016) Reassessment of the four yield-related genes Gn1a, DEP1, GS3, and IPA1 in rice using a CRISPR/Cas9 system. Front Plant Sci 7. https://doi.org/10.3389/fpls.2016.00377

Li S, Lin D, Zhang Y, Deng M, Chen Y, Lv B, Li B, Lei Y, Wang Y, Zhao L, Liang Y, Liu J, Chen K, Liu Z, Xiao J, Qiu J, Gao C (2022) Genome-edited powdery mildew resistance in wheat without growth penalties. Nature 602(7897):455–460. https://doi.org/10.1038/s41586-022-04395-9

Liang J, Liu J, Jia P, Yang T, Zeng Q, Zhang S, Liao B, Shu W, Li J (2020a) Novel phosphate-solubilizing bacteria enhance soil phosphorus cycling following ecological restoration of land degraded by mining. ISME J 14(6):1600–1613. https://doi.org/10.1038/s41396-020-0632-4

Liang X, Li S, Gong L, Li S, Zhang Y (2020b) COPII components Sar1b and Sar1c play distinct yet interchangeable roles in pollen development. Plant Physiol 183(3):974–985. https://doi.org/10.1104/pp.20.00159

Lino CA, Harper JC, Carney JP, Timlin JA (2018) Delivering CRISPR: a review of the challenges and approaches. Drug Deliv 25(1):1234–1257

Lio A (2023) Assessing the use of CRISPR-Cas9 in agriculture. Univ Toronto's J Sci Innov 2:45–56

Liu L, Zhang J, Xu J, Li Y, Guo L, Wang Z, Zhang X, Zhao B, Guo Y, Zhang N (2020) CRISPR/Cas9 targeted mutagenesis of SlLBD40, a lateral organ boundaries domain transcription factor, enhances drought tolerance in tomato. Plant Sci 301:110683. https://doi.org/10.1016/j.plantsci.2020.110683

Liu C, Li B, Dong Y, Lin H (2022) Endophyte colonization enhanced cadmium phytoremediation by improving endosphere and rhizosphere microecology characteristics. J Hazard Mater 434:128829. https://doi.org/10.1016/j.jhazmat.2022.128829

Lombardo L, Grando MS (2019) Genetically modified plants for nutritionally improved food: a promise kept? Food Rev Int 36(1):58–76. https://doi.org/10.1080/87559129.2019.1613664

Lopes MJDS, Dias-Filho MB, Gurgel ESC (2021) Successful plant growth-promoting microbes: inoculation methods and abiotic factors. Front Sustain Food Syst 5. https://doi.org/10.3389/fsufs.2021.606454

Lu H, Luo T, Fu H, Wang L, Tan Y, Huang J, Wang Q, Ye G, Gatehouse AMR, Lou Y, Shu Q (2018) Resistance of rice to insect pests mediated by suppression of serotonin biosynthesis. Nat Plants 4(6):338–344. https://doi.org/10.1038/s41477-018-0152-7

Lynch D, Vogel D (2001) The regulation of GMOs in Europe and the United States: a case-study of contemporary European regulatory politics. Council on Foreign Relations

Malnoy M, Viola R, Jung M, Koo O, Kim S, Kim J, Velasco R, Kanchiswamy CN (2016) DNA-free genetically edited grapevine and apple protoplast using CRISPR/Cas9 ribonucleoproteins. Front Plant Sci 7. https://doi.org/10.3389/fpls.2016.01904

Mao X, Zheng Y, Xiao K, Wei Y, Zhu Y, Cai Q et al (2018) OsPRX2 contributes to stomatal closure and improves potassium deficiency tolerance in rice. Biochem Biophys Res Commun 495:461–467. https://doi.org/10.1016/j.bbrc.2017.11.045

Martínez-Viveros O, Jorquera M, Crowley D, Gajardo G, Mora M (2010) Mechanisms and practical considerations involved in plant growth promotion by rhizobacteria. J Soil Sci Plant Nutr 10(3). https://doi.org/10.4067/s0718-95162010000100006

Meng Z, Meng Z, Zhang R, Liang C, Wan J, Wang Y, Zhai H, Guo S (2015) Expression of the rice arginase gene OsARG in cotton influences the morphology and nitrogen transition of seedlings. PLoS One 10(11):e0141530. https://doi.org/10.1371/journal.pone.0141530

Mian A, Oomen RJ, Isayenkov S, Sentenac H, Maathuis FJ, Véry A (2011) Over-expression of an Na+– and K+-permeable HKT transporter in barley improves salt tolerance. Plant J 68(3):468–479. https://doi.org/10.1111/j.1365-313x.2011.04701.x

Mishra J, Prakash J, Arora NK (2016) Role of beneficial soil microbes in sustainable agriculture and environmental management. Clim Change Environ Sustain 4(2):137. https://doi.org/10.5958/2320-642x.2016.00015.6

Mishra S, Nayak S, Tuteja N, Poosapati S, Swain DM, Sahoo RK (2024) CRISPR/Cas-mediated genome engineering in plants: application and prospectives. Plan Theory 13(14):1884. https://doi.org/10.3390/plants13141884

Mondal S, Halder SK, Mondal KC (2022) Tailoring in fungi for next generation cellulase production with special reference to CRISPR/CAS system. In: Systems microbiology and biomanufacturing. Springer, pp 1–17

Mushtaq M, Ahmad Dar A, Skalicky M, Tyagi A, Bhagat N, Basu U et al (2021) CRISPR-based genome editing tools: insights into technological breakthroughs and future challenges. Gene 12(6):797

Muzyukina P (2023). The role of the CRISPR-Cas system in the human pathogen Clostridioides difficile and its application for epidemiological monitoring and infection prevention. Molecular biology. Université Paris-Saclay. English. (NNT: 2023UPASL058)

Nanda M, Kumar V, Sharma D (2019) Multimetal tolerance mechanisms in bacteria: the resistance strategies acquired by bacteria that can be exploited to 'clean-up' heavy metal contaminants from water. Aquat Toxicol 212:1–10. https://doi.org/10.1016/j.aquatox.2019.04.011

Nekrasov V, Wang C, Win J, Lanz C, Weigel D, Kamoun S (2017) Rapid generation of a transgene-free powdery mildew resistant tomato by genome deletion. Sci Rep 7(1). https://doi.org/10.1038/s41598-017-00578-x

Ogata T, Ishizaki T, Fujita M, Fujita Y (2020) CRISPR/Cas9-targeted mutagenesis of OsERA1 confers enhanced responses to abscisic acid and drought stress and increased primary root growth under nonstressed conditions in rice. PLoS One 15(12):e0243376. https://doi.org/10.1371/journal.pone.0243376

Oliva R, Ji C, Atienza-Grande G, Huguet-Tapia JC, Perez-Quintero A, Li T, Eom J, Li C, Nguyen H, Liu B, Auguy F, Sciallano C, Luu VT, Dossa GS, Cunnac S, Schmidt SM, Slamet-Loedin IH, Cruz CV, Szurek B et al (2019) Broad-spectrum resistance to bacterial blight in rice using genome editing. Nat Biotechnol 37(11):1344–1350. https://doi.org/10.1038/s41587-019-0267-z

Osakabe Y, Watanabe T, Sugano SS, Ueta R, Ishihara R, Shinozaki K, Osakabe K (2016) Optimization of CRISPR/Cas9 genome editing to modify abiotic stress responses in plants. Sci Rep 6(1). https://doi.org/10.1038/srep26685

Park J, Dempewolf E, Zhang W, Wang Z (2017) RNA-guided transcriptional activation via CRISPR/dCas9 mimics overexpression phenotypes in Arabidopsis. PLoS One 12(6):e0179410. https://doi.org/10.1371/journal.pone.0179410

Paul P, Dhatt BK, Miller M, Folsom JJ, Wang Z, Krassovskaya I, Liu K, Sandhu J, Yu H, Zhang C, Obata T, Staswick P, Walia H (2019) MADS78 and MADS79 are essential regulators of early seed development in rice. Plant Physiol 182(2):933–948. https://doi.org/10.1104/pp.19.00917

Peng A, Chen S, Lei T, Xu L, He Y, Wu L, Yao L, Zou X (2017) Engineering canker-resistant plants through CRISPR/Cas9-targeted editing of the susceptibility gene CsLOB1 promoter in citrus. Plant Biotechnol J 15(12):1509–1519. https://doi.org/10.1111/pbi.12733

Philippot L, Raaijmakers JM, Lemanceau P, Van Der Putten WH (2013) Going back to the roots: the microbial ecology of the rhizosphere. Nat Rev Microbiol 11(11):789–799. https://doi.org/10.1038/nrmicro3109

Piergentili R, Del Rio A, Signore F, Umani Ronchi F, Marinelli E, Zaami S (2021) CRISPR-Cas and its wide-ranging applications: from human genome editing to environmental implications, technical limitations, hazards and bioethical issues. Cells 10(5):969

Prabhu N, Borkar S, Garg S (2019) Phosphate solubilization by microorganisms. In: Elsevier eBooks, pp 161–176. https://doi.org/10.1016/b978-0-12-817497-5.00011-2

Pyott DE, Sheehan E, Molnar A (2016) Engineering of CRISPR/Cas9 mediated potyvirus resistance in transgene-free Arabidopsis plants. Mol Plant Pathol 17:1276–1288. https://doi.org/10.1111/mpp.12417

Qiu Z, Kang S, He L, Zhao J, Zhang S, Hu J, Zeng D, Zhang G, Dong G, Gao Z, Ren D, Chen G, Guo L, Qian Q, Zhu L (2017) The newly identified heat-stress sensitive albino 1 gene affects chloroplast development in rice. Plant Sci 267:168–179. https://doi.org/10.1016/j.plantsci.2017.11.015

Raboy V (2001) Seeds for a better future: 'low phytate' grains help to overcome malnutrition and reduce pollution. Trends Plant Sci 6(10):458–462. https://doi.org/10.1016/s1360-1385(01)02104-5

Ren B, Wang X, Duan J, Ma J (2019) Rhizobial tRNA-derived small RNAs are signal molecules regulating plant nodulation. Science 365(6456):919–922. https://doi.org/10.1126/science.aav8907

Rodríguez-Leal D, Lemmon ZH, Man J, Bartlett ME, Lippman ZB (2017) Engineering quantitative trait variation for crop improvement by genome editing. Cell 171(2):470–480.e8. https://doi.org/10.1016/j.cell.2017.08.030

Samtiya M, Aluko RE, Dhewa T (2020) Plant food anti-nutritional factors and their reduction strategies: an overview. Food Prod Process Nutr 2:6. https://doi.org/10.1186/s43014-020-0020-5

Sánchez MA (2020) Chile as a key enabler country for global plant breeding, agricultural innovation, and biotechnology. GM Crops Food 11(3):130–139. https://doi.org/10.1080/21645698.2020.1761757

Sánchez-León S, Gil-Humanes J, Ozuna CV, Giménez MJ, Sousa C, Voytas DF, Barro F (2017) Low-gluten, nontransgenic wheat engineered with CRISPR/Cas9. Plant Biotechnol J 16(4):902–910. https://doi.org/10.1111/pbi.12837

Scheben A, Wolter F, Batley J, Puchta H, Edwards D (2017) Towards CRISPR/Cas crops—bringing together genomics and genome editing. New Phytol 216(3):682–698. https://doi.org/10.1111/nph.14702

Shan Q, Wang Y, Li J, Zhang Y, Chen K, Liang Z et al (2013) Targeted genome modification of crop plants using a CRISPR-Cas system. Nat Biotechnol 31:686–688. https://doi.org/10.1038/nbt.2650

Shen L, Wang C, Fu Y, Wang J, Liu Q, Zhang X et al (2018) QTL editing confers opposing yield performance in different rice varieties. J Integr Plant Biol 60:89–93. https://doi.org/10.1111/jipb.12501

Shendekar S, Mangla S, Gore V, Meshram M, Raut D (2024) The CRISPR revolution: editing plant genomes for abiotic stress resilience and a better tomorrow. In: Current research in biotechnology and biological sciences, vol 2. Bright Sky Publications, p 93

Shi J, Gao H, Wang H, Lafitte HR, Archibald RL, Yang M et al (2017) ARGOS8variants generated by CRISPR-Cas9 improve maize grain yield under field drought stress conditions. Plant Biotechnol J 15:207–216. https://doi.org/10.1111/pbi.12603

Shoresh M, Harman GE, Mastouri F (2010) Induced systemic resistance and plant responses to fungal biocontrol agents. Annu Rev Phytopathol 48(1):21–43. https://doi.org/10.1146/annurev-phyto-073009-114450

Shukla M, Al-Busaidi KT, Trivedi M, Tiwari RK (2018) Status of research, regulations and challenges for genetically modified crops in India. GM Crops Food 9(4):173–188

Singh S, Ramakrishna W (2021) Application of CRISPR–Cas9 in plant–plant growth-promoting rhizobacteria interactions for next green revolution. 3 Biotech 11(12):492

Smyth SJ (2017) Canadian regulatory perspectives on genome engineered crops. GM Crops Food 8(1):35–43

Solanki MK, Joshi NC, Singh PK, Singh SK, Santoyo G, de Azevedo LCB, Kumar A (2023) From concept to reality: transforming agriculture through innovative rhizosphere engineering for plant health and productivity. Microbiol Res 279:127553

Tang F, Yang S, Liu J, Zhu H (2015) Rj4, a gene controlling nodulation specificity in soybeans, encodes a thaumatin-like protein but not the one previously reported. Plant Physiol 170(1):26–32. https://doi.org/10.1104/pp.15.01661

Tao S, Chen H, Li N, Liang W (2022) The application of the CRISPR-Cas system in antibiotic resistance. Infect Drug Resist 15:4155–4168

The Lancet (2018) Global nutrition report. The Lancet

Tripathi S, Khatri P, Fatima Z, Pandey RP, Hameed S (2022) A landscape of CRISPR/Cas technique for emerging viral disease diagnostics and therapeutics: progress and prospects. Pathogens 12(1):56. https://doi.org/10.3390/pathogens12010056

Turnbull C, Lillemo M, Hvoslef-Eide TAK (2021) Global regulation of genetically modified crops amid the gene edited crop boom—a review. Front Plant Sci 12. https://doi.org/10.3389/fpls.2021.630396

United Nations Children's Fund (UNICEF) (2020) The state of the world's children. United Nations Children's Fund (UNICEF)

van Beljouw SPB, Sanders J, Rodríguez-Molina A, Brouns SJJ (2023) RNA-targeting CRISPR-Cas systems. Nat Rev Microbiol 21(1):21–34. https://doi.org/10.1038/s41579-022-00793-y

Van Der Ent S, Van Wees SC, Pieterse CM (2009) Jasmonate signaling in plant interactions with resistance-inducing beneficial microbes. Phytochemistry 70(13–14):1581–1588. https://doi.org/10.1016/j.phytochem.2009.06.009

Van Der Heijden MGA, Martin FM, Selosse M, Sanders IR (2015) Mycorrhizal ecology and evolution: the past, the present, and the future. New Phytol 205(4):1406–1423. https://doi.org/10.1111/nph.13288

Van Zeijl A, Wardhani TAK, Kalhor MS, Rutten L, Bu F, Hartog M, Linders S, Fedorova EE, Bisseling T, Kohlen W, Geurts R (2018) CRISPR/Cas9-mediated mutagenesis of four putative symbiosis genes of the tropical tree *Parasponia andersonii* reveals novel phenotypes. Front Plant Sci 9. https://doi.org/10.3389/fpls.2018.00284

Vaswani A, Shazeer N, Parmar N, Uszkoreit J, Jones L, Gomez AN, Kaiser L, Polosukhin I (2017) Attention is all you need. arXiv (Cornell University). https://doi.org/10.48550/arxiv.1706.03762

Verma S, Srivastava J (2021) Mycoremediation: a novel approach to rescue soil from heavy metal contamination. Mycoremediat Environ Sustain 3:117–142

Wang Y, Cheng X, Shan Q, Zhang Y, Liu J, Gao C, Qiu J (2014) Simultaneous editing of three homoeoalleles in hexaploid bread wheat confers heritable resistance to powdery mildew. Nat Biotechnol 32(9):947–951. https://doi.org/10.1038/nbt.2969

Wang F, Wang C, Liu P, Lei C, Hao W, Gao Y, Liu YG, Zhao K (2016a) Enhanced rice blast resistance by CRISPR/Cas9-targeted mutagenesis of the ERF transcription factor gene OsERF922. PLoS One 11(4):e0154027. https://doi.org/10.1371/journal.pone.0154027

Wang H, Wang H, Shao H, Tang X (2016b) Recent advances in utilizing transcription factors to improve plant abiotic stress tolerance by transgenic technology. Front Plant Sci 7. https://doi.org/10.3389/fpls.2016.00067

Wang L, Wang L, Tan Q, Fan Q, Zhu H, Hong Z, Zhang Z, Duanmu D (2016c) Efficient inactivation of symbiotic nitrogen fixation related genes in *Lotus japonicus* Using CRISPR-Cas9. Front Plant Sci 7. https://doi.org/10.3389/fpls.2016.01333

Wang L, Chen L, Li R, Zhao R, Yang M, Sheng J et al (2017) Reduced drought tolerance by CRISPR/Cas9-mediatedSlMAPK3 mutagenesis in tomato plants. J Agric Food Chem 65:8674–8682. https://doi.org/10.1021/acs.jafc.7b02745

Wang L, Zang X, Zhou J (2022) Synthetic biology: a powerful booster for future agriculture. Adv Agrochem 1(1):7–11. https://doi.org/10.1016/j.aac.2022.08.005

World Health Organization (2018) Nutrition and food systems. World Health Organization

World Health Organization (2020a) Malnutrition. World Health Organization

World Health Organization (2020b) Micronutrient deficiencies. World Health Organization

Wu Y (2024) The development of gene editing technology and controversial issues: a discussion. Highl Sci Eng Technol 91:123–130

Xie K, Yang Y (2013) RNA-guided genome editing in plants using a CRISPR-Cas system. Mol Plant 6:1975–1983. https://doi.org/10.1093/mp/sst119

Xu H, Xiao T, Chen CH, Li W, Meyer CA, Wu Q et al (2015) Sequence determinants of improved CRISPR sgRNA design. Genome Res 25:1147–1157. https://doi.org/10.1101/gr.191452.115

Xu R, Yang Y, Qin R, Li H, Qiu C, Li L, Wei P, Yang J (2016) Rapid improvement of grain weight via highly efficient CRISPR/Cas9-mediated multiplex genome editing in rice. J Genet Genomics 43(8):529–532. https://doi.org/10.1016/j.jgg.2016.07.003

Xu Z, Xu X, Gong Q, Li Z, Li Y, Wang S, Yang Y, Ma W, Liu L, Zhu B, Zou L, Chen G (2019) Engineering broad-spectrum bacterial blight resistance by simultaneously disrupting variable TALE-binding elements of multiple susceptibility genes in rice. Mol Plant 12(11):1434–1446. https://doi.org/10.1016/j.molp.2019.08.006

Yin X, Feng Q, Liu W, Zhu M, Zhang J, Li Y, Yang L, Zhang C, Cui M, Zheng X, Li Y (2023) Assessment and mechanism analysis of plant salt tolerance regulates soil moisture dynamics and controls root zone salinity and sodicity in seasonally irrigated agroecosystems. J Hydrol 617:129138. https://doi.org/10.1016/j.jhydrol.2023.129138

Yu W, Wang L, Zhao R, Sheng J, Zhang S, Li R, Shen L (2019) Knockout of SlMAPK3 enhances tolerance to heat stress involving ROS homeostasis in tomato plants. BMC Plant Biol 19(1). https://doi.org/10.1186/s12870-019-1939-z

Yuan J, Zhao J, Wen T, Zhao M, Li R, Goossens P, Huang Q, Bai Y, Vivanco JM, Kowalchuk GA, Berendsen RL, Shen Q (2018) Root exudates drive the soil-borne legacy of aboveground pathogen infection. Microbiome 6(1). https://doi.org/10.1186/s40168-018-0537-x

Yue Y, Zhang M, Zhang J, Duan L, Li Z (2012) SOS1 gene overexpression increased salt tolerance in transgenic tobacco by maintaining a higher K+/Na+ ratio. J Plant Physiol 169(3):255–261. https://doi.org/10.1016/j.jplph.2011.10.007

Yue E, Cao H, Liu B (2020) OsmiR535, a potential genetic editing target for drought and salinity stress tolerance in *Oryza sativa*. Plan Theory 9(10):1337. https://doi.org/10.3390/plants9101337

Zaidi SS, Mukhtar MS, Mansoor S (2018) Genome editing: targeting susceptibility genes for plant disease resistance. Trends Biotechnol 36(9):898–906. https://doi.org/10.1016/j.tibtech.2018.04.005

Zhai Y, Yu K, Cai S, Hu L, Amoo O, Xu L, Yang Y, Ma B, Jiao Y, Zhang C, Khan MHU, Khan SU, Fan C, Zhou Y (2019) Targeted mutagenesis of BnTT8 homologs controls yellow seed coat development for effective oil production in *Brassica napus* L. Plant Biotechnol J 18(5):1153–1168. https://doi.org/10.1111/pbi.13281

Zhang H, Zhang J, Wei P, Zhang B, Gou F, Feng Z, Mao Y, Yang L, Zhang H, Xu N, Zhu J (2014) The CRISPR/Cas9 system produces specific and homozygous targeted gene editing in rice in one generation. Plant Biotechnol J 12(6):797–807. https://doi.org/10.1111/pbi.12200

Zhang XH, Tee LY, Wang XG, Huang QS, Yang SH (2015) Off-target effects in CRISPR/Cas9-mediated genome engineering. Mol Ther Nucleic Acids 4:e264

Zhang T, Zheng Q, Yi X, An H, Zhao Y, Ma S, Zhou G (2018) Establishing RNA virus resistance in plants by harnessing CRISPR immune system. Plant Biotechnol J 16(8):1415–1423. https://doi.org/10.1111/pbi.12881

Zhang T, Zhao Y, Ye J, Cao X, Xu C, Chen B, An H, Jiao Y, Zhang F, Yang X, Zhou G (2019) Establishing CRISPR/Cas13a immune system conferring RNA virus resistance in both dicot and monocot plants. Plant Biotechnol J 17(7):1185–1187. https://doi.org/10.1111/pbi.13095

Zhao Y, Zhang C, Liu W, Gao W, Liu C, Song G, Li W, Mao L, Chen B, Xu Y, Li X, Xie C (2016) An alternative strategy for targeted gene replacement in plants using a dual-sgRNA/Cas9 design. Sci Rep 6(1). https://doi.org/10.1038/srep23890

Zheng Y, Li J, Wang B, Han J, Hao Y, Wang S, Ma X, Yang S, Ma L, Yi L, Peng W (2020a) Endogenous type I CRISPR-Cas: from foreign DNA defense to prokaryotic engineering. Front Bioeng Biotechnol 8:62. https://doi.org/10.3389/fbioe.2020.00062

Zheng Z, Zhang Y, Liu Z, Dong Z, Xie C, Bravo A, Peng D (2020b) The CRISPR-Cas systems were selectively inactivated during evolution of *Bacillus cereus* group for adaptation to diverse environments. ISME J 14(6):1479–1493

Zhou J, Peng Z, Long J, Sosso D, Liu B, Eom J, Huang S, Liu S, Cruz CV, Frommer WB, White FF, Yang B (2015) Gene targeting by the TAL effector PthXo2 reveals cryptic resistance gene for bacterial blight of rice. Plant J 82(4):632–643. https://doi.org/10.1111/tpj.12838

The manufacturer's authorised representative in the EU is Springer Nature Customer Service Centre GmbH, Europaplatz 3, 69115 Heidelberg, Germany. If you have any concerns regarding our products, please contact ProductSafety@springernature.com

Printed and bound by CPI Group (UK) Ltd, Croydon, CR0 4YY
01/10/2026
02225928-0009